SPACE TELESCOPE SCIENCE INSTITUTE

SYMPOSIUM SERIES: 8
Series Editor S. Michael Fall, Space Telescope Science Institute

THE ANALYSIS OF EMISSION LINES

SPACE
TELESCOPE
SCIENCE
INSTITUTE

THE ANALYSIS OF EMISSION LINES

A Meeting in Honour of the
70th Birthdays of D. E. Osterbrock & M. J. Seaton

Proceedings of the Space Telescope Science Institute Symposium,
held in Baltimore, Maryland
May 16–18, 1994

Edited by

ROBERT WILLIAMS
Space Telescope Science Institute, Baltimore, MD

MARIO LIVIO
Space Telescope Science Institute, Baltimore, MD

Published for the
Space Telescope Science Institute

CAMBRIDGE
UNIVERSITY PRESS

CAMBRIDGE UNIVERSITY PRESS
Cambridge, New York, Melbourne, Madrid, Cape Town, Singapore,
São Paulo, Delhi, Dubai, Tokyo, Mexico City

Cambridge University Press
The Edinburgh Building, Cambridge CB2 8RU, UK

Published in the United States of America by Cambridge University Press, New York

www.cambridge.org
Information on this title: www.cambridge.org/9780521480819

© Cambridge University Press 1995

First published 1995

A catalogue record for this publication is available from the British Library

ISBN 978-0-521-48081-9 Hardback
ISBN 978-0-521-67560-4 Paperback

Contents

Participants

Peter Albrecht	Universitäts-Sternwarte
Nahum Arav	University of Colorado
Bruce Balick	University of Washington
Stephi Baum	Space Telescope Science Institute
Manuel Bautista	Ohio State University
Luc Binette	European Southern Observatory
William Blair	Johns Hopkins University
Bill Blass	NASA/Goddard Space Flight Center
Robert Blum	Ohio State University
Peng Bo	Dwingeloo Observatory, NFRA
Kazimierz Borkowski	University of Maryland
James Braatz	University of Maryland
Tomas Brage	NASA/Goddard Space Flight Center
Michael Brotherton	University of Texas at Austin
Monsignori-Fossi Brunella	Osservatorio Astrofisico di Arcetri
Eugene Capriotti	Michigan State University
Wesley Chen	Johns Hopkins Univerity
Ross Cohen	University of California, CASS
Edward Colbert	University of Maryland
Marcella Contini	Tel Aviv University
Nancy Cox	San Francisco
Alexander Dalgarno	Harvard College Observatory
Darren DePoy	Ohio State University
Ken Dere	Naval Research Laboratory
Ralph Dettmar	Space Telescope Science Institute
Eric Deutsch	University of Washington
Sperello di Serego Alighieri	Osservatorio Astrofisico di Arcetri
Matthias Dietrich	Universitäts-Sternwarte
Harriet Dinerstein	University of Texas at Austin
Megan Donahue	Space Telescope Science Institute
Michael Dopita	Mt. Stromlo Siding Spring Observatory
Stephen Drake	NASA/Goddard Space Flight Center
Janet Drew	Oxford University
Hilmar Duerbeck	Universitaet Muenster
Reginald Dufour	Rice University
Florence Durret	Institut d'Astrophysique
Brian Espey	Johns Hopkins University
Ian Evans	Space Telescope Science Institute
Mike Fanelli	NASA/Goddard Space Flight Center
James Felten	NASA/Goddard Space Flight Center
Gary Ferland	University of Kentucky
Holland Ford	Space Telescope Science Institute
Jeff Gardner	University of Washington
Donald Garnett	University of Minnesota
Martin Gaskell	University of Nebraska
Richard Gelderman	NASA/Goddard Space Flight Center
Fred Hamann	University of California, CASS
Brian Handy	Montana State University

Patrick Harrington	University of Maryland
Tim Heckman	Johns Hopkins University
Ronald Hes	Kapteyn Institute
Jeff Hester	Arizona State University
Luis Ho	University of California, Berkeley
Keith Horne	University of Utrecht
Ivan Hubeny	NASA/Goddard Space Flight Center
Christian Hummel	US Naval Observatory
David Hummer	Universitäts-Sternwarte Munchen
Una Hwang	Center for Space Research, MIT
Garik Israelian	Sterrenkundig Institute
Sverneric Johansson	University of Lund
Brian Judd	Johns Hopkins University
Philip Judge	High Altitude Observatory, NCAR
Mary Beth Kaiser	Johns Hopkins University
Tim Kallman	NASA/Goddard Space Flight Center
Michael Kaufman	Johns Hopkins University
Randy Kimble	NASA/Goddard Space Flight Center
Anne Kinney	Space Telescope Science Institute
Yuan-Kuen Ko	NASA/Goddard Space Flight Center
Stefanie Komossa	Ruhr University
Anuradha Koratkar	Space Telescope Science Institute
Kirk Korista	Space Telescope Science Institute
Gerard Kriss	Johns Hopkins University
Julian Krolik	Johns Hopkins University
Nancy Jo Lame	Ohio State University
Martin Laming	NRL/SFA, Inc.
Wayne Landsman	NASA/Goddard Space Flight Center
Barry Lasker	Space Telescope Science Institute
Xiaowei Liu	University College London
Mario Livio	Space Telescope Science Institute
Knox Long	Space Telescope Science Institute
Gordon MacAlpine	University of Michigan
Stephen Mahan	University of Tennessee
Alessandro Marconi	Universitá di Firenze
William Martin	National Institute Standards & Technology
Smita Mathur	Center for Astrophysics
Stephan McCandliss	Johns Hopkins University
Peter Meikle	Imperial College
Luis Mendoza	Space Telescope Science Institute
Joseph Miller	Lick Observatory
Peter Milne	Clemson University
Ulisse Munari	Osservatorio Astronomico di Padova
C. Muqoz-Tuqon	Instituto di Astrofisica de Canarias
Gerard Muratorio	Observatoire de Marseille
Richard Mushotzky	NASA/Goddard Space Flight Center
Hagai Netzer	Tel Aviv University
David Neufeld	Johns Hopkins University
Colin Norman	Johns Hopkins University
Dara Norman	University of Washington

C. R. O'Dell	Rice University
Susanne Och	European Southern Observatory
Ernesto Oliva	Osservatorio Astrofisico di Arcetri
Knut Olsen	University of Washington
Marina Orio	University of Wisconsin
Donald Osterbrock	Lick Observatory
Nino Panagia	Space Telescope Science Institute
Robert Parker	NASA-HQ
William Parkinson	Center for Astrophysics
Miriani Pastoriza	Instituto di Física, UFRGS
Brooke Patterson	University of Washington
Manuel Peimbert	Instituto de Astronomia
Jianfang Peng	Ohio State University
Daniel Pequignot	Observatoire de Meudon
Lars Peterson	Aarhus University
Philip Pinto	University of Arizona
Richard Pogge	Ohio State University
Anil Pradhan	Ohio State University
Reuven Ramaty	NASA/Goddard Space Flight Center
David Reiss	University of Washington
Matthew Richter	University of California, Berkeley
Dorothea Rosa	Heidelberg, Germany
Michael Rosa	European Southern Observatory
William Rose	University of Maryland
Jean-René Roy	Universite Laval
Robert Rubin	NASA/Ames Research Center
Vera Rubin	Department of Terrestrial Magnetism
Eric Schlegel	NASA/Goddard Space Flight Center
Hartmut Schulz	Ruhr University
Francois Schweizer	Carnegie Institution of Washington
Paul Scowen	Arizona State University
Michael Seaton	University College London
Waltraut Seitter	Universitaet Muenster
Pierluigi Selvelli	Osservatorio Astronomico di Trieste
Richard Shaw	Space Telescope Science Institute
Joseph Shields	University of Arizona
George Sonneborn	NASA/Goddard Space Flight Center
Peter Stockman	Space Telescope Science Institute
Thaisa Storchi Bergmann	Instituto de Fisica, UFRGS
Ralph Sutherland	University of Colorado, JILA
Ewa Szuszkiewicz	International School for Advanced Studies
Antonio Talavera	ESA—IUE Observatory
Guillermo Tenorio-Tagle	Instituto di Astrofisica di Canarias
Silvia Torres-Peimbert	Instituto de Astronomia
Virginia Trimble	University of California
Todd Tripp	University of Wisconsin
Zlatan Tsvetanov	Johns Hopkins University
Peter van Hoof	Kapteyn Institute
Emanuel Vassiliadis	Space Telescope Science Institute
Sylvain Veilleux	KPNO/NOAO

Marianne Vestergaard	Center for Astrophysics
Sueli Viegas	Royal Greenwich Observatory
Montserrat Villar	European Southern Observatory
Roberto Viotti	Instituto di Astrofisica Spaziale
Saeqa Vrtilek	University of Maryland
Donald Walter	Rice University
Zheng Wen	Rice University
Ray Weymann	Carnegie Observatories
Belinda Wilkes	Center for Astrophysics
Bob Williams	Space Telescope Science Institute
Beverley Wills	University of Texas at Austin
Lodewijk Woltjer	Observatoire de Haute-Provence
Bruce Woodgate	NASA/Goddard Space Flight Center
Junhan You	Shanghai Jiao Tong University
Wei Zheng	Johns Hopkins University
Esther Zirbel	Space Telescope Science Institute
Dan Zucker	University of Washington

Preface

Apart from stars and those objects which radiate reflected starlight, most of the objects in the Universe radiate an emission spectrum. It was the astronomers interest in analyzing the spectrum of the sun and other stars in the last century that motivated the development of radiative transfer, and with the newly formulated macroscopic relations of LTE early in this century, that led to our understanding of absorption spectra. The original observational stimulus for this activity had been Fraunhofer's study of the solar spectrum almost a century before.

Interest in emission-line spectra came later, when spectrographs coupled to telescopes enabled the spectra of fainter gaseous emission regions to be observed. They revealed a totally different type of spectrum than that which had been observed from stars. The fact that local thermodynamic equilibrium does not hold for emission regions has complicated the interpretation of their spectra. Huggins' initial discovery of 'nebulium' in gaseous nebulae and its subsequent identification with ionized oxygen by Bowen had demonstrated that rarefied conditions must pertain in nebulae. Stromgren's subsequent 1939 paper in the Astrophysical Journal was a landmark in demonstrating how far-UV continuum radiation from a hot star was absorbed by surrounding gas and converted into visible Balmer line radiation.

In the decades that followed, the realization that many interesting objects such as supernova remnants, active galactic nuclei, and quasars radiated an emission-line spectrum, motivated the analysis of emission regions. For the past 35 years two names have stood out amidst those who have been involved in the efforts to determine physical conditions in objects that emit emission lines. Don Osterbrock and Mike Seaton have both made important contributions to our understanding of emission-line analysis, not the least of which is the fact that they have inspired a generation of investigators who in turn have focussed on this topic. Observations have been one of the more important parts of Don Osterbrock's work, whereas Mike Seaton has concentrated on the atomic data and processes. However, each of them has drawn upon both the observations and data to interpret spectra so that important information could be deduced from the objects under study. The 1960 article "Planetary Nebulae" from Reports on Progress in Physics, and the book "The Astrophysics of Gaseous Nebulae" have been constant companions in the attache cases of many of us over the years.

Much of what we know about some of the more interesting objects in the sky come from analyses of their emission spectra. The reviews which constitute the current symposium on emission-line analysis are a testimony to the vitality and diversity of this subject, and they serve as an appropriate tribute to the dedicated efforts of the two scientists whom we honor. In appreciation of the fundamental contributions they have made to this field, we are pleased to dedicate this symposium and these proceedings to Profs. Donald E. Osterbrock and Michael J. Seaton in celebration of their 70th birthdays.

Bob Williams, Mario Livio
Space Telescope Science Institute
Baltimore, Maryland
May, 1995

Emission Lines: Past and Future

By L. WOLTJER

Observatoire de Haute Provence, F-04870 Saint Michel l'Observatoire, France

In the first half of this century many emission lines were or had been identified. Noteworthy moments were the identification of the Nebulium lines (λ 4959/5007) as forbidden lines of O^{++} (Bowen 1927) and of the strong solar green coronal line λ 5303 as due to Fe^{13+} (Edlen 1942). In addition, a first quantitative understanding of some aspects of nebular spectra was obtained: the Balmer decrement was calculated by Menzel and associates (1937), the temperatures of the central stars of planetary nebulae were inferred by Zanstra (1927), and the first information on elemental abundances in nebulae was gained.

In the second half of this century a much more detailed understanding of emission spectra was acquired. Emission lines assumed a fundamental role for the diagnostics of conditions in nebulae. As a result electron densities N_e and temperatures T_e as well as elemental abundances became known in many objects. Excitation and ionization conditions in nebulae were found to be frequently radiative (photoionization), but shocks and perhaps fast particles were also found to play a role. Non-equilibrium conditions were seen to be important especially in the hot, tenuous plasmas revealed by X-ray observations: the ionization state was often different from that expected from the temperature, and even T_e and the temperature of the proton gas could be different.

Chemistry was found to play a role in many emission nebulae. Numerous new molecules were observed, especially by radio observations in cool, dense media. Dust was found to strongly affect the appearance of spectra, not only by absorption and scattering but also by its effects on abundances. And finally observations of emission lines gave much information on velocity fields in nebulae and also on magnetic and electric fields.

These developments resulted from improvements in atomic and molecular parameters, in observations and in modelling. The calculation and sometimes measurement of transition probabilities, photoionization cross-sections, collisional cross-sections and recombination coefficients were an obvious prerequisite for the quantitative analysis of emission line spectra. Observations improved in precision, and perhaps almost more important in dynamic range, by the introduction of new detectors, in particular CCD's. And the initially limited wavelength range over which optical emission lines had been observed was extended to include lines in the MHz region of the radio spectrum to the gamma-ray lines associated with nuclear rather than atomic processes.

The interpretation of these data was much strengthened by the introduction of sophisticated modelling techniques: Self-consistent photoionization models play an increasing role, radiative transfer calculations have been substantially improved, hydrodynamics has become a new ingredient, reaction networks of chemical or nuclear nature have been included as having non-equilibrium processes. As a result, a deeper insight has been obtained in the origin and evolution of emission nebulae.

It is particularly appropriate that this symposium be held in honor of Mike Seaton and Don Osterbrock, who have contributed decisively to give the subject its present more quantitative shape.

Seaton's first paper was, I believe, written in 1949 (with Bates) on the continuous absorption of O, N, and C. In 1953 and 1954 followed two papers which were fundamental for the understanding of the spectra of planetary nebulae, HII regions and supernova remnants. The first paper contained the calculation of the electron collision cross-sections of the ground states of N^+, O°, O^+, O^{++}, Ne^{++}, S^+, and in the second paper the blue [OII] and the [SII] doublet ratios as a function of N_e were explicitly given. With the data of these two papers it became possible to quantitatively understand the stronger forbidden lines in the nebular spectra.

Osterbrock started out in 1951 with a paper on transition probabilities of [Ca II], [Fe XV], etc. Four years later he measured the [OII] doublet ratios in the Orion Nebula and confirmed Seaton's calculations. In a subsequent joint paper (Seaton & Osterbrock 1957) they studied the red [OII] doublet, which again gives valuable information on N_e. The same year Osterbrock determined N_e in the filaments of the Crab Nebula from the λ 3727 doublet ratio.

I came into the subject myself in 1957 with the analysis of Mayall's spectra of the Crab Nebula and was lucky enough to find that all the atomic data needed to determine the conditions in the Nebula had just been calculated. This led to the conclusion that abundances in the Crab Nebula were rather normal (except for helium) and that the ionization could be understood as being due to the ultraviolet extension of the synchrotron radiation spectrum of the Nebula.

This point was taken up again by Williams (1967) who constructed a photoionization model for the "Ionization and thermal equilibrium of a gas excited by uv synchrotron radiation" from which line intensities were predicted. This was the first of the power-law type photoionization models which has blossomed into a whole industry for Active Galactic Nuclei. In the meantime similar approaches had been followed by Hjellming (1966) for HII regions excited by hot stars and for Planetary Nebulae by Goodson (1967). The main additions to these models made since then are the charge exchange reactions and the di-electronic recombination.

While as a result of these developments no major problems remained in understanding the intensities of the lines of the more common elements, a curious anomaly was noted by Dennefeld and Péquignot (1983) who measured the [Ni II] lines λ 7378/7411 in the Crab Nebula and found an overabundance of Ni with respect to iron of a factor of around 60. It then was noticed that similar but less strong anomalies existed in other supernova remnants and even in the Orion Nebula (see also Osterbrock *et al.* 1992 for the latter object). The matter was discussed by Henry (1984) who attempted to see if there could be problems with the ionization equilibrium of Ni, with the atomic parameters or with differential condensation into grains of Fe and Ni. None of these approaches was found to be very promising. Also in Seyfert galaxies (Halpern and Oke 1986) and in Herbig-Haro objects the anomaly was found. Stahl and Wolf (1986) found the same in the Luminous Blue Variables in the Large Magellanic Cloud. The most extreme result was obtained by Johnson *et al.* (1992) who found in the P Cygni Nebula an overabundance of Ni by a factor of 2000!

As noted by Lucy (1994), the latter result makes it impossible to believe that real overabundances are the cause of the observed line strengths. Instead he considers the radiative excitation (of 284 levels) by continuum radiation in the 900–2200 Å range and calculates their effect on the level populations. A good fit is obtained for the P Cygni Nebula with a predicted intensity ratio λ 7378/λ 7411 of 3.5, close to the observed value (Barlow *et al.* 1994), but far from the factor of 10–11 for pure collisional excitation. However, Lucy also finds that radiative excitation is inadequate in the case of the highly diffuse radiation field in the Crab Nebula. This would then indicate a real overabundance which could be produced during the supernova outburst by nuclear reactions in an unusually neutron-rich environment. It is, of course, the case that in all supernova explosions nickel is produced in the form of ^{56}Ni, but this rapidly decays into Co and then

Fe. However, before this conclusion can be entirely convincing, it would be necessary to understand the curious geometry of the [Ni II] emission; according to results reported at this meeting by Mac Alpine the [Ni II] emission is strong on the side of the filaments which faces the pulsar. Also the results of Hudgins *et al.* (1990) from the [Ni II] line at 1.19 μm indicate a smaller apparent overabundance (factor of 6) of Ni. While on the subject of the Crab Nebula, I also note that Murdin (1994) appears to have detected the long sought for halo. Though extremely faint, it could contain several solar masses of hydrogen and represent a relic of the stellar wind in an earlier evolutionary phase (Chevalier 1977). The spectroscopic detection of the narrow Hβ emission and the absence of the braod [O III] lines make it rather clear that scattered light from the Nebula is not a problem.

Returning to the photoionization models, these have been particularly elaborated in the context of quasars and Seyfert galaxies. Various complications have been added to the models, including optical depth effects, clouds of high N_e, scattering of electrons and/or dust, and an intercloud medium. Perry and Dyson (1985) have added shocks in the Seyfert winds, while Osterbrock and Parker (1965) tried to add ionization by fast protons.

Perhaps the most enduring problem for the pure photoionization models has been to explain the very strong Fe II emission seen in Seyferts and quasars. Already in 1979 Collin-Souffrin *et al.* concluded that "the Fe II region should be thermally heated, excited and ionized," essentially because with photoionization most Fe would be Fe^{++}. While in the meantime the models have been stretched to explain some of the spectra, cases like PHL 1092 with its almost pure Fe II spectrum (Bergeron & Kunth 1980) remain far beyond the possibilities. Very strong Fe II emission has also been observed in some luminous IRAS galaxies and interpreted as due to shocks in a starburst environment (Lipari *et al.* 1993). The foregoing discussion shows the difficulties of dealing with elements like Fe and Ni with their complex atomic structures which result in thousands of lines.

As suggested by Antonucci & Miller (1985), scattering may help to explain the Seyfert 2 galaxies as heavily obscured Sy 1's. Lower ionization than in the Seyferts is found in the LINERS identified as a group by Heckman (1980). At first, these were believed to be shock excited objects, but Ferland & Netzer (1983) showed that photoionization with a softer radiation field could also explain the spectra.

Shocks produce a hot gas which tends to have a rather narrow range of states of ionization, unless very different shock velocities occur in the medium. Photoionization by power-law type radiation fields tends to give a broader range of stages of ionization, the energy of the highest ionization stages corresponding to the hardest photons in the ionizing spectrum. As a result, coronal lines may appear. Fe 6^+ was detected in the spectrum of NGC 1068 by Seyfert (1943), following its earlier identification in the planetary NGC 7027 by Bowen and Wyse (1939). In some Seyferts also Fe^{9+} and Fe^{10+} are seen in the optical. Many such lines are also found in the IR. Osterbrock *et al.* (1990) found [S VIII] at 0.99 μm, and Oliva *et al.* (1994) observed lines of Si^{5+}, Si^{6+}, Si^{8+}, S^{8+} and Ca^{7+} in the 1–4 μm region. Observations of such lines are particularly useful to ascertain the shape of the ionizing radiation field.

Variability in the recombination lines of a Seyfert galaxy was first found in NGC 3516 (Andrillat & Souffrin 1968), where the Hβ line had much weakened relative to [O III] since the first observations by Seyfert (1943). Many variable Seyferts have been found since. From IUE observations the variations of the ionizing radiation have been inferred. The variations in the emission lines show a time lag and this allows a determination of the characteristic size of the Broad Line Region which is found to be of the order of a pc for luminous Seyferts or weak quasars. Precise observations over long enough periods allow mapping of the BLR to be done. If moreover accurate line profiles are measured, also information about the kinematics in different parts of the BLR may be obtained. These variations provide a unique tool for resolving the BLR. Unfortunately they take much telescope time and are difficult to schedule at the larger telescope facilities. Recent results include the multi-author studies on NGC 3516 (Wanders *et al.* 1993), NGC 3783 (Reichert *et al.* 1994) and NGC 5548 (Peterson *et al.* 1994), as well as the line profile studies of Rosenblatt *et al.* (1994).

Absorption by dust makes the interpretation of the optical spectra more difficult and may render invisible part of the nuclear region of an AGN. In this respect observations in the IR have a major advantage. For example in Cygnus A an absorption of $A_V = 20$–80 magnitudes has been inferred (Djorgovski *et al.* 1991) for the nucleus which begins to be visible only for $\lambda > 2.4$ μm. IR observations are also important for abundance observations. Some stages of ionization of common elements have no accessible lines in the optical part of the spectrum (Ne^+, etc.). Also the broader wavelength range obtained by including the IR sometimes gives very sensitive diagnostics. For example the intensity ratio of the [Ne III] lines at 15.5 μm and at 3869 Å changes by a factor of 100 if T_e varies from 5000 to 15000 K, essentially independently of N_e (Pottasch *et al.* 1984), while the ratio of the [Ne V] lines at 14.3 μm and at 3425 Å changes by a factor of 25 if N_e changes from 3×10^3 to 5×10^5 cm^{-3}, with only a relatively weak dependence on T_e (Pottasch *et al.* 1986).

Photoionized plasmas tend to have temperatures around 10^4 K. Much hotter conditions are found in supernova remnants, shock heated to 10^6–10^8 K or more, and in the chromospheres and coronae of stars (10^4–10^7 K). In both cases collisional ionization and excitation predominate, and accurate collisional cross-sections are essential.

Still much atomic physics is needed to understand the uv and X-ray spectra. For example, Arnaud & Raymond (1992) recently recalculated the ionization equilibrium of iron at 4×10^6 K. The abundance of Fe^{18+} turns out to be six times larger and that of Fe^{15+} six times smaller than in previous calculations, mainly because of improved dielectronic recombination coefficients. Such results are of importance in the calculation of abundances which in the solar corona sometimes show unexpected behavior. For instance it is found that the ratio of the Ne and Mg abundances in the solar wind is ten times smaller than in the photosphere (Widing & Feldman 1992).

Other effects that may be of importance include non equilibrium ionization and unequal temperatures for different plasma constituents. Delayed ionization in shocks was first noted by Itoh (1977) and turns out to be of essential importance in the interpretation of the X-ray spectra of supernova remnants. Inferred abundances of elements like Ca or Si could be in error by a factor of 100 or more if these effects are not properly taken into account. Similar effects may occur in the solar corona due to microflares (see the review by Mason and Monsignori Fossi 1994).

In young, rapidly expanding supernova remnants even the temperatures of the neutral H and of the proton gas may be very different. Chevalier and Raymond (1978) showed that charge exchange between the hot protons behind the shock separating the remnant and the interstellar medium and neutral H swept up by the shock may produce broad wings in the Balmer lines. These have actually been observed in Tycho's SNR and in SNR 1006 as well as in some remnants in the Large Magellanic Cloud. The narrow components in the Balmer lines would be due to collisional excitation of the neutral H before its ionization. However, the width of these narrow components (40 km s^{-1}) is much larger than would be expected for interstellar H atoms and some kind of a precursor would be needed to stir up the gas; the nature of this precursor is not understood (Smith *et al.* 1994).

Emission lines from cool gas (T < 100 K) are also of much importance in the interstellar medium and in particular in dense molecular clouds. Lines of H_2, CO and a plethora of molecules have been discovered at mm and submm wavelengths. Again, laboratory measurements are needed for line identifications, collisional cross-sections and transition probabilities. The large number of chemical reactions between complex molecules and the importance of the surface effects on dust grains make this a complex subject outside the range of this review. Of particular interest are the possibilities for measuring isotope ratios, but also here chemistry plays a confusing role.

At the other end of the spectrum are the gamma-ray lines resulting from nuclear decays. Observations are difficult because of the low photon fluxes and high instrumental backgrounds. The main observational results to-date include the lines from the ^{56}Ni $\rightarrow$ ^{56}Co $\rightarrow$ ^{56}Fe decay in SN 1987A which lead to quantitative results concerning the iron production in supernovae (Teegarden 1994), the line of ^{26}Al in the interstellar medium resulting from (super)novae, W.R. stars, etc. (Diehl *et al.* 1994), the e^+e^- annihilation lines occurring in some ill understood objects (Churazov *et al.* 1994), and the decay lines of ^{44}Ti. The latter decay ^{44}Ti $\rightarrow$ ^{44}Sc $\rightarrow$ ^{44}Ca has been observed in Cas A (Iyudin *et al.* 1994). Two different measured values for the lifetime of ^{44}Ti of 78 and 96 years lead to values of 3.2 respectively 1.4×10^{-4} $M_\odot$ for the production of Ti in the supernova. Again, the essential importance of accurate values for the nuclear constants is much in evidence.

The future

Order of magnitude improvements in the observation of emission lines may be expected in the near future, largely as a result of a number of forthcoming space missions. In the submm and IR part of the spectrum the Infrared Space Observatory—a 60 cm cryogenically cooled telescope—is scheduled to be launched late in 1995 by the European Space Agency. Among the instruments are spectrographs covering the wavelength range of 2.5–197 μm with spectral resolutions mainly between 100 and 20000. ISO should take observations for about 18 months. While much of the program will be concerned with cooler gas, observations of the emission line spectra of supernova remnants and of (dusty) AGN should also take place. ISO possibly will be followed by SOFIA (airplane) and SIRTF, two infrared facilities under consideration at NASA. FIRST—a 3-m telescope for submm observations (85–900 μm)—should be launched by ESA in 2006.

HST, IUE and the Extreme Ultraviolet Explorer should continue to produce spectral results, HST with high spatial resolution and IUE with easier availability for long time series. The spectroscopic capabilities of HST should become very substantially increased when STIS (the imaging spectrograph) is installed by NASA in 1997. STIS should give spectral resolutions from around 10^3 to 10^4 in the optical and 10^5 in the uv in combination with spatial resolutions of $0.''12$–$0.''06$. IUE has now been functioning for nearly two decades and there is some fear it may not live forever; its loss would be a severe blow to the mapping of the BLR in Seyferts.

The X-ray domain will be well served: Astro E (Japan), XMM (ESA) and AXAF (NASA) are all three scheduled for launch around 1999. The spectral resolution in these missions would be much better than available today (in the range of 100–1000), and as a result individual lines should be resolved in AGN, SNR (also with improved spatial resolution) and stars. More gamma-ray line observations should come from continued use of the GRO (NASA) and from INTEGRAL to be launched in 2002 by ESA.

Finally detailed spectra of the solar chromosphere and corona in the wavelength range 160–1600 Å should be obtained by SOHO, an ESA/NASA cooperative project to be launched in 1995. Spectral resolutions range up to 40,000 with angular resolution in the range of 2–3 arcsec.

As noted earlier, the IR is a particularly propitious wavelength region for the observation of emission lines in Seyferts, quasars and other objects, in part because it is still very much underexplored and in part because of the much reduced interstellar absorption. Space missions by necessity have relatively small telescopes and therefore poor angular resolution in the IR. It is in this area that the new large ground-based telescopes can make a major contribution. Also their high photon collecting power will allow the full dynamic range of the new optical detectors to be exploited and thereby fainter lines to be observed. As reported at this meeting by Péquignot, the first lines of elements beyond the iron group have now been observed in planetary nebulae and this may become a subject of importance for nucleogenesis.

Seeing the continuing stream of high quality data and new results on AGN obtained by Osterbrock and his many students, we can still look forward to much in the future. But all the observations of emission lines will be of no avail without reliable atomic data. We can only wish that Seaton and his many followers will continue the good work they have been doing during the past half century for the benefit of all of us.

Acknowledgements

I am indebted to Dr. L. Lucy for a discussion on the nickel problem. Part of this paper was written during a visit to the Arcetri Observatory in Florence through the hospitality of Prof. F. Pacini.

REFERENCES

ANDRILLAT, Y., & SOUFFRIN, S. 1968 *Ap. Lett.* **1**, 111

ANTONUCCI, R. R., & MILLER, J. S. 1985 *ApJ* **297**, 621

ARNAUD, M., & RAYMOND, J. C. 1992 *ApJ* **398**, 39

BARLOW, M. J., DREW, J. E., MEABURN, J., & MASSEY, R. M. 1994 *MNRAS* **268**, L29

BATES, D. R., & SEATON, M. J. 1949 *MNRAS* **109**, 698

BERGERON, J., & KUNTH, D. 1980 *AA* **85**, L11

BOWEN, I. 1927 *PASP* **39**, 295

BOWEN, I. S., & WYSE, A. B. 1939 *Lick Obs. Bull.* **19**, 1

CHEVALIER, R. 1977 *Supernovae* (ed. D. Schramm) Reidel, Dordrecht

CHEVALIER, R., & RAYMOND, J. C. 1978 *ApJ* **225**, L27

CHURAZOV, E., & 15 CO-AUTHORS 1994 *ApJS* **92**, 381

COLLIN-SOUFFRIN, S., JOLY, M., HEIDMANN, N., & DUMONT, S. 1979 *AA* **72**, 293

DENNEFELD, M. & PÉQUIGNOT, D. 1983 *AA* **127**, 42

DIEHL, R. & 15 CO-AUTHORS 1994 *ApJS* **92**, 429

DJORGOVSKI, S., WEIR, N., MATTHEWS, K., & GRAHAM, J. R. 1991 *ApJ* **372**, L67

EDLEN, B. 1942 *Zs. f. Astrophysik* **22**, 30

FERLAND, G. J. & NETZER, H. 1983 *ApJ* **264**, 105

GOODSON, W. L. 1967 *Zs. f. Astrophysik* **66**, 118

HALPERN, J. P. & OKE, J. B. 1986 *ApJ* **301**, 753

HECKMAN, T. M. 1980 *AA* **87**, 152

HENRY, R. B. C. 1984 *ApJ* **281**, 644

HJELLMING, R. M. 1966 *ApJ* **143**, 420

HUDGINS, D., HERTER, T. & JOYCE, R. J. 1990 *ApJ* **354**, L57

ITOH, H. 1977 *PASJ* **29**, 813

IYUDIN, A. F., DIEHL, R., BLOEMEN, H., HERMSEN, W., LICHTI, G. G., MORRIS, D., RYAN, J., SCHÖNFELDER, V., STEINLE, H., VARENDORFF, M., DE VRIES, C., & WINKLER, C. 1994 *AA* **284**, L1

JOHNSON, D. R. H., BARLOW, M. J., DREW, J. E., & BRINKS, E. 1992 *MNRAS* **255**, 261

LIPARI, S., TERLEVICH, R., & MACCHETTO, F. 1993 *ApJ* **406**, 451

LUCY, L. 1994 *AA* to be published

MASON, H. & MONSIGNORI FOSSI, B. C. 1994 *AA Review* 6

MENZEL, D. H. 1937 *ApJ* **85**, 330 (also BAKER, J. G. & MENZEL, D. H. 1938 *ApJ* **88**, 52)

MURDIN, P. 1994 *MNRAS* **269**, 89

OLIVA, E., SALVATI, M., MOORWOOD, A. F. M., & MARCONI, A. 1994 *AA* in press

OSTERBROCK, D. E. 1951 *ApJ* **114**, 469

OSTERBROCK, D. E. 1957 *PASP* **69**, 227

OSTERBROCK, D. E. & PARKER, R. A. R. 1965 *ApJ* **141**, 892

OSTERBROCK, D. E., SHAW, R. A. & VEILLEUX, S. 1990 *ApJ* **352**, 561

OSTERBROCK, D. E., TRAN, H. D., & VEILLEUX, S. 1992 *ApJ* **389**, 305

PERRY, J. J. & DYSON, J. E. 1985 *MNRAS* **213**, 665

PETERSON, B. M. & 42 CO-AUTHORS 1992 *ApJ* **425**, 622

POTTASCH, S. R., BEINTEMA, D. A., RAIMOND, E., BAUD, B., VAN DUINEN, R., HABING, H. J., HOUCK, J. R., DE JONGH, T., JENNINGS, R. E., OLNON, F. M., & WESSELIUS, P. R. 1984 *ApJ* **278**, L33

POTTASCH, S. R., PREITE-MARTINEZ, A., OLNON, F. M., JING-ER, M., & KINGMA, S. 1986 *AA* **161**, 363

REICHERT, G. A. & 64 CO-AUTHORS 1994 *ApJ* **425**, 582

ROSENBLATT, E. I., MALKAN, M. A., SARGENT, W. L. W., & READHEAD, A. C. S. 1994 *ApJS* **93**, 73

SEATON, M. J. 1953 *Proc. Roy. Soc. A* **218**, 400

SEATON, M. J. 1954 *Ann. d'Astrophysique* **17**, 74

SEATON, M. J. & OSTERBROCK, D. E. 1957 *ApJ* **125**, 66

SEYFERT, C. K. 1943 *ApJ* **97**, 28

SMITH, R. C., RAYMOND, J. C., & LAMING, J. M. 1994 *ApJ* **420**, 286

STAHL, O. & WOLF, B. 1986 *AA* **158**, 371

TEEGARDEN, B. J. 1994 *ApJS* **92**, 363

WANDERS, I. & 31 CO-AUTHORS 1993 *AA* **39**, 53

WIDING, K. G. & FELDMAN, U. 1992 *Proc. Solar Wind Seven Conference* (eds. E. Marsch & R. Schwenn) 405

WILLIAMS, R. E. 1967 *ApJ* **147**, 556

ZANSTRA, H. 1927 *ApJ* **65**, 50

Atomic Data for the Analysis of Emission Lines

By ANIL K. PRADHAN AND JIANFANG PENG

Dept. of Astronomy, Ohio State University, Columbus, Ohio 43210.

Within the last decade or so new atomic data has become available for most atomic systems of interest in astrophysics. Recent progress in atomic processes relevant to spectral formation is reviewed and the data sources are listed. Recommended transition probabilities and effective collision strengths are presented for a number of nebular emission lines.

1. Introduction

In a pioneering study on the electron impact excitation of atomic oxygen, Seaton (1953) formulated the now well known close-coupling approximation of atomic collision theory, which he termed the "continuum state Hartree-Fock method", reflecting the physical picture that the new method was an extension of the bound state method to the continuum region that encompassed electron-ion scattering and photoionization phenomena. For nearly three decades, the close coupling approximation has been widely employed to calculate the most accurate low-energy cross sections for excitation and photoionization, and radiative transition probabilities. Large computational packages were developed, mainly at University College London and the Queen's University of Belfast, to carry out the enormous task of fulfilling the needs of astrophysicists and plasma physicists. In particular, the R-matrix method developed by Burke and associates (Burke *et al.* 1971) has proved to be computationally very efficient for large-scale calculations.

A huge amount of radiative atomic data was produced, during last 10 years or so, under the auspices of an international collaboration of atomic physicists and astrophysicists, called the Opacity Project, led by Seaton (Seaton *et al.* 1994). Photoionization cross sections and oscillator strengths were calculated using the R-matrix method (Seaton 1987; Berrington *et al.* 1987) for almost all astrophysically abundant elements in various ionization stages. The calculations were carried out for LS multiplets; fine structure was not considered as it is relatively less important for the calculation of stellar opacities, which was the express aim of the Project. Recently, employing the basic tecniques of the Opacity Project and new developments incorporating relativistic effects into the R-matrix method, a new project called the Iron Project has been initiated (Hummer *et al.* 1993) that aims to compute accurate cross sections for electron impact excitation of most astrophysical ions including fine structure. The main aim of the Iron Project is the calculation of precise atomic data for the Iron group elements that are astrophysically very important but for which little reliable data is presently available. Thus the Opacity Project and the Iron Project are now providing much of the atomic data needed by astronomers.

This report consists of a review of data sources for:
• Electron impact excitaton • Photoionization • Radiative transition probabilities • Electron-ion recombination

For completeness, data sources are also given for line broadening, recombination lines, electron impact ionization, charge exchange, proton impact excitation, isotopic and hyperfine structure, and energy levels and wavelengths. With the exception of energy levels,

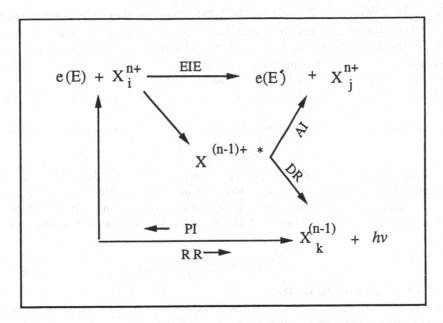

FIGURE 1. Inter-related radiative and collisional atomic processes: EIE - electron impact excitation, AI - autoionization, DR - dielectronic recombination, PI - photoionization, RR - radiative recombination

very little of the data is experimental; we confine ourselves mainly to a discussion of the theoretical data. A general bibliography for most of these atomic processes has been presented by Butler (1993).

The prominent spectral lines in nebular astrophysics are mostly due to forbidden transition among low-lying levels of atomic ions. In an attempt to update and extend the extremely useful compilation by Mendoza (1983) more than a decade ago, recommended data for electron impact excitation and transition probabilities for emission lines in AGN's, nebulae and other sources are tabulated.

2. Theoretical considerations

Electron-ion interactions may be depicted by the following diagram:

Quantum mechanically *all* of the atomic processes shown may be treated by a wavefunction expansion that represents the total e+ion system in terms of the coupled eigenfunctions of the "target" or the "core" states of the ion, i.e.

$$\Psi(E) = A \sum_i \chi_i \theta_i + \sum_j c_j \Phi_j, \tag{2.1}$$

where χ_i is the target ion wave function in a specific state $S_i\ L_i$ and θ_i is the wave function for the free electron in a channel labeled as $S_i L_i k_i^2 \ell_i (SL\pi)$; k_i^2 being its incident kinetic energy. While the close coupling approximation (e.g. the R-matrix method) includes coupling between the target states of the ion, simpler approximations such as the distorted wave or the central field approximations neglect the coupling effects which may be important at low energies.

One particularly important coupling effect manifests itself as autoionizing resonances, from doubly-excited states of the e+ion system (the center of the Fig. 1), that can substantially enhance the cross sections and rates for excitation, photoionization and recombination. The near threshold region of the cross sections, that dominates the rate of electron excitation and line formation for forbidden and intercombination transitions, may be particularly affected by broad and extensive resonance structures. Dipole allowed transitions are less affected. In general therefore it is necessary to employ the accurate close coupling approximation for the forbidden and the intercombination lines, but for the dipole allowed transitions the distorted wave approximation often suffices to obtain rates to 10–30% uncertainty.

A brief discussion of a few points related to the atomic processes is given below.

2.1. *Electron impact excitation*

The quantity usually computed for electron impact excitation is denoted as $\Omega(i, j)$, which was introduced by Hebb & Menzel (1940) and later termed the "collision strength" by Seaton, in analogy with the line strength for radiative transitions. The Ω is dimensionless and symmetric with respect to initial and final states. It is related to the cross section as

$$Q(i, j; k_i^2) = \frac{\Omega(i, j)}{k_i^2 g_i} (\pi a_o^2),$$ (2.2)

in units of the area of the H atom.

The usually tabulated quantity is the maxwellian averaged collision strength, also called the effective collision strength,

$$\Upsilon(i, j) = \int_0^\infty \Omega(i, j) exp \left(\frac{-\epsilon}{kT} \right) d \left(\frac{\epsilon}{kT} \right).$$ (2.3)

The excitation rate coefficient, in $cm^3 sec^{-1}$, is defined as

$$q(i, j) = \frac{8.63 \times 10^{-6}}{g_i T^{1/2}} \Upsilon(i, j) exp \left(\frac{-\Delta E_{ij}}{kT} \right),$$ (2.4)

related to the de-excitation rate coefficient $(E_i < E_j)$ as

$$q(j, i) = \frac{g_j}{g_i} q(i, j).$$ (2.5)

The influence of autoionizing resonances may be seen in Fig. 2.

A detailed discussion on the analysis of collision strengths and rate coefficients is given by Burgess & Tully (1992), who describe analytic fitting procedures to $\Omega(E)$ and $\Upsilon(T)$ for interpolation and extrapolation in a compact form for the different types of transitions.

2.2. *Photoionization cross sections*

The availability of a very large number of photoionization cross sections from the Opacity Project is likely to influence considerably the modeling of radiative and collisional plasmas. There are two main features of these data: (i) inclusion of autoionizing resonances in the near-threshold region, (ii) cross sections for many excited states, typically several hundred bound states for each atom or ion. Prior to the Opacity Project, such data was available mainly for the ground states of relatively few ions, and the excited states were treated in either the hydrogenic approximation or with some Coulomb screening (such as in the quantum defect method).

Fig. 3 shows the photoionization cross section of a complex system, Fe II, illustrating the effect of resonances and the covergence of the close coupling wavefunction expansion. A total of 83 states of the core ion Fe III were included in the R-matrix (Nahar & Pradhan 1994a). In comparison, the central field calculations of Reilman & Manson (1979) underestimate the cross sections by several factors in the near threshold region, although at higher energy the discrepancy is relatively small. Comparison is also made

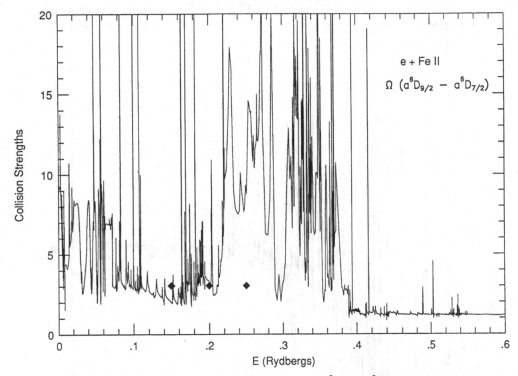

FIGURE 2. Collision strength for the forbidden transition $^6D_{9/2} - {}^6D_{7/2}$ in Fe II (Pradhan & Zhang 1993). The $\Upsilon(T)$, at T = 10,000 K, using the close coupling collision strengths shown in Fig. 2 is approximately a factor of three higher than the one calculated using the distorted wave values (diamonds) calculated by Nussbaumer & Storey (1980).

with previous R-matrix calculations of Sawey & Berrington (1992) whose wavefunction expansion had not converged as it did not include the channels that contribute to the phtoionization of the dominant 3d shell; consequently their values are up to two orders of magnitude lower.

Another important effect in photoionization, first described by Yu & Seaton (1987), is the formation of large resonances due to photoexcitation-of-core (PEC). The PEC resonances appear most strongly in the photoionization of excited bound states along a Rydberg series, characterized by a core ion state and a Rydberg electron, i.e. $(S_cL_c)n\ell$. At incident photon frequencies corresponding to dipole core transitions a doubly excited autoionizing state is formed, resulting in a large resonance. Fig. 4 illustrates the magnitude of PEC resonances in the photoionization of two excited Rydberg states of Fe III (Nahar, private communication).

We have

$$h\nu + Fe\ III \longrightarrow e^- + Fe\ IV$$
$$h\nu + 3d^5(^6S)ns[^7S] \longrightarrow 3d^44p(^6P^o)ns[^7P^o]$$

The dipole transition $^6S \rightarrow {}^6P^o$ between the Fe IV core states is responsible for the PEC resonance, while the Rydberg electron essentially remains a "spectator" owing to the weak interaction with the core states (the process is basically the same as in di-electronic recombination). The PEC resonances attenuate the photoionization cross

Fe II, $3d^6 4s\ ^6D$

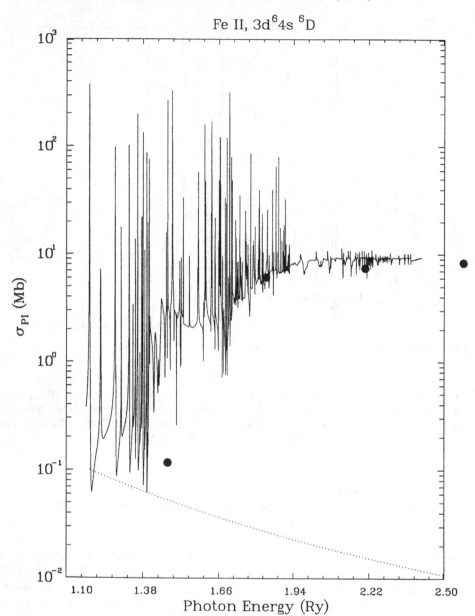

FIGURE 3. Photoionization cross section of the ground state $3d^6 4s(^6D)$ of Fe II (Nahar & Pradhan 1994a); filled circles—Reilman & Manson (1979); dashed line—Sawey & Berrington (1992).

sections by orders of magnitude, and the cross sections for apparently hydrogenic excited states deviate substantially from what may otherwise by a hydrogenic background.

2.3. *Electron-ion recombination*

Recombination rates are usually calculated for two processes, radiative recombination (RR) and di-electronic recombination (DR), and summed together to obtain the total. It is assumed that the RR values are derived from photoionization cross sections for the "background" cross sections (i.e. without autoionizing resonances), and the DR

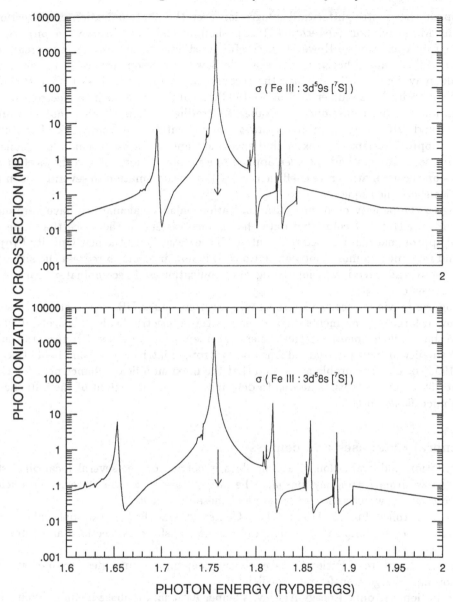

FIGURE 4. Large resonances in photoionization cross sections due to photoexcitation-of-core (PEC). The arrow denotes the peak position of the PEC resonance, at the energy of the dipole transition $3d^5(^6S) \rightarrow 3d^44p(^6P^o)$ in the core ion S IV.

values correspond to the contribution from resonances as they converge to excited states of the recombining ion, at electron temperatures sufficiently high to affect excitation of the ion states. The rates for RR and DR are calculated separately, the former using "background," resonance-free photoionization cross sections obtained via approximations such as the Central Field (e.g. Reilman & Manson 1979), and the latter using the Burgess (1965) General Formula, or more recent works by Jacobs *et al.* (1977 onwards) and others (references are given later).

However the two processes, RR and DR, are in principle unified. With the availability of the Opacity Project photoionization cross sections, including resonances, it might at first appear that a straightforward integration over the photoionization cross sections, using the Milne relation (Osterbrock 1989), should yield the effective (e+ion) recombination rate coefficients. However, a careful consideration shows that the treatment is rather difficult and laborious. For example, low-lying resonances close to ionization threshold may significantly enhance the recombination rate, i.e. low-temperature DR first discussed by Nussbaumer & Storey (1983). A unified approach to electron-ion recombination has been developed by Nahar & Pradhan (1992, 1994b), that subsumes both RR and DR processes in an *ab initio* manner within the framework of the close coupling approximation, thus extending excitation and photoionization calculations to recombination. The method has been applied to a number of ions. Fig. 5 shows the total electron-ion recombination rate coefficient $\alpha_R(T)$ for recombination to several ions in the carbon isoelectronic sequence.

Owing to the paucity of atomic data, ionization balance calculations have heretofore employed radiative and collisional data that is inconsistent in the sense that different physical approximations are used to calculate all the data. With the new Opacity Project data, and ongoing recombination calculations, it is now possible to redress the situation in the case of radiative ionization, using photoionization and recombination rates that are self-consistent.

During the previous decade, experimental measurements of DR cross sections for some ions showed large enhancements due to weak, external electric fields. Theoretical works indicate that while for neutral atoms the enhancement could be several factors, the effect is much smaller for mutiply charged ions due to stronger intrinsic Coulomb field. Badnell *et al.* (1993) discuss the problem and find that the maximum field enhancement for C IV is about 40%. Further work is needed to determine the precise extent of the influence of plasma microfields on DR.

3. General databases and centers

Compilation and evaluation of atomic data is carried out at several locations listed below. Apart from astronomy, perhaps the largest user of atomic data is the nuclear fusion community which is served by some of these data centers.

• The Controlled Fusion Atomic Data Center at Oak Ridge National Laboratory; data include electron impact ionization and charge transfer cross sections and rates and analytic fits thereof.

• National Institute of Fusion Science, Nagoya, Japan; data includes fits to excitation, ionization and charge transfer rate coefficients.

• International Atomic Energy Agency, Vienna, Austria; database includes excitation, ionization and charge exchange data. Janev *et al.* (1988, 1989) and Janev (1992) describe the atomic and molecular database relevant to fusion, which is obtainable in the ALADDIN database system (Hulse 1990). Limited data is available for electron and neutral H collisions with H and H, He^{q+}, Li^{q+}, Be^{q+}, B^{q+}, C^{q+}, O^{q+}, Si^{q+}, Ti^{q+}, Fe^{q+}, Cr^{q+}, Ni^{q+}, Mo^{q+}, W^{q+} ions, and with H_2, H_2^+, CH_4, CH_4^+, C_mH_n, $C_mH_n^+$, O_2, O_2^+, CO, CO^+, H_2O, H_2O^+ molecules (Phaneuf 1993).

• The wavelengths and energy levels continue to be provided by the National Institute for Standards and Technology, Gaithersburg, Maryland. An electronic database is in preparation and can be accessed on Internet (IP: atm.phy.nist.gov; username: asd).

• TOPBASE: This is the most extensive electronic database containing radiative data, oscillator strengths and photoionization cross sections, from the Opacity Project (Seaton *et al.* 1994; Cunto *et al.* 1993).

The latest version 0.7 of TOPBASE may be accessed from two nodes:

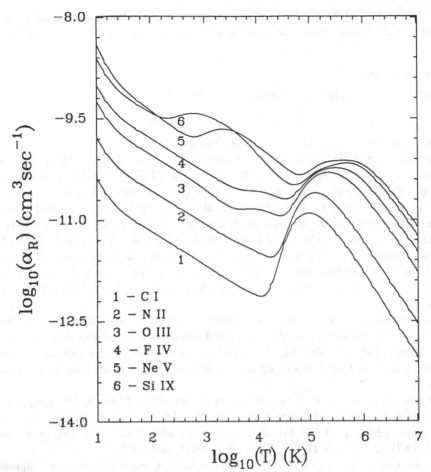

FIGURE 5. Total electron-ion recombination rate coefficients for C-like ions (Nahar & Pradhan 1994b).

▷ Centre de Donnees Astronomique de Strasbourg (CDS);
IP: 130.79.128.5, login: topbase, pw: Seaton+
▷ NASA-NSSDC, Goddard Space Flight Center;
IP: 128.183.101.54, login: topbase, pw: Seaton+
(Note the uppercase 'S' in the password). An on-line manual is available.

Some of the TOPBASE data is being upgraded through new calculations by the Iron Project, in particular for the important low ionization stages of Iron (Fe I–V). The photoionization cross sections obtained using the low energy close coupling approximation with the R-matrix method, and incorporated in TOPBASE, are of high accuracy in the important near-threshold region containing the extensive autoionizing resonance structures. However, in the high energy region the cross sections are often inaccurate owing to the presence of pseudo resonances. Thus the TOPBASE photoionization cross sections were fitted to a simple ν^{-3} "tail" above all thresholds of ionization included in the close coupling calculations; these are not reliable for applications that need such data (e.g. X-ray opacities that involve inner-shell ionizations). A new effort is under way to improve the high energy "tails" of the photoionization cross sections and will be incorporated into TOPBASE in due course.

Further extensions of TOPBASE are also planned to include collisional data from the Iron Project (TIPBASE), as well as data for electron-ion recombination rate coefficients.

4. Data sources

This section gives published references with some discussion.

4.1. *Electron impact excitation*

There are three recent compilations of data sources:

1. A special issue of *Atomic Data and Nuclear Data Tables* entitled "Electron Excitation Data for Analysis of Spectral Line Radiation from Infrared to X-ray Wavelengths: Reviews and Recommendations" (**57**, No. 1/2, May/July 1994; Ed: James Lang), contains assessments of data, up to March 1992, along several isoelectronic and isonuclear sequences. The atomic systems (and reviewers) are: H and H-like (Callaway), He and He-like (Dubau), Li and Li-like (McWhirter), Be and Be-like (Berrington), B and B-like (Sampson, Zhang & Fontes), C and C-like (Monsignori Fossi & M. Landini), N and N-like (Kato), O-like (Lang & Summers), F-like (Bhatia), Si II–Si IV and S II–S IV (Dufton & Kingston), Fe II–Fe VIII (Pradhan), Fe IX–Fe XIV (Mason), Fe XV–Fe XVII (Badnell & Moores).

2. While reference 1. concentrates on recently computed data, nearly all of the previously available data sources in literature has been evaluated by Pradhan & Gallagher (At. Data. Nucl. Data. Tables, **52**, 227, 1992). A discussion of the various methods of atomic collision theory, and important atomic effects, is also given (copies are available from AKP).

3. The most recent review of excitation cross sections of Carbon, Oxygen, Iron and rare gas ions is by Tayal, Pradhan & Pindzola (in *Atomic and Molecular Processes in Magnetic Fusion Edge Plasmas*, Ed: R. K. Janev, Plenum 1995). The ions reviewed are: C II–C VI, O II–O VIII, Fe II–Fe VIII, Ar VII–VIII, and Kr VII.

Large-scale calculations, primarily for electron impact excitation, are in progress under the Iron Project. The results are being published in *Astronomy and Astrophysics, Supplement Series* (some in the main A&A journal). An up-to-date list of publications follows.

Atomic data from the IRON Project:

I. Goals and methods. D. G. Hummer, K. A. Berrington, W. Eissner, Anil K. Pradhan, H. E. Saraph, J. A. Tully, A&A 279, 298–309, 1993.

II. Effective collision strengths for infrared transitions in carbon-like ions. D. J. Lennon, V. M. Burke, A&A Suppl. Ser. 103, 273–277, 1994.

III. Rate coefficients for electron impact excitation of Boron-like ions: Ne VI, Mg VIII, Al IX, Si X, S X, Ar XIV, Ca XVI and Fe XXII, Hong Lin Zhang, Mark Graziani & Anil K. Pradhan, A&A 283, 319–330, 1994.

IV. Electron excitation of the $^2P^o_{3/2} - {}^2P^o_{1/2}$ fine structure transition in fluorine-like ions. H. E. Saraph, J. A. Tully, A&A Suppl. Ser. (in press).

V. Effective collision strengths for transitions in the ground configuration of oxygen-like ions. K. Butler, C. J. Zeippen, A&A Suppl. Ser. (in press).

VI. Collision strengths and rate coefficients for Fe II. Hong Lin Zhang & Anil K. Pradhan, A&A (in press).

VII. Radiative transition probabilities for Fe II. Sultana N. Nahar, A&A (in press).

VIII. Electron excitation of the $3d^4({}^5D_j)$ ground state fine structure transition in Ti-like ions V II, Cr III, Mn IV, Fe V, Co VI and Ni VII. K. A. Berrington, A&A Suppl. Ser. (submitted).

IX. Electron excitation of the $^2P^o_{3/2-1/2}$ fine structure transitions in chlorine-like ions from Ar II to Ni XII. J. C. Pelan & K. A. Berrington, A&A Suppl. Ser. (submitted).

X. Effective collision strengths for infrared transitions in silicon- and sulphur-like ions. M. E. Galavis, C. Mendoza and C. J. Zeippen, A&A Suppl. Ser. (submitted).

XI. The $^2P^o_{1/2-3/2}$ fine-structure lines of Ar VI, K VII and Ca VIII. H. E. Saraph & P. J. Storey, A&A Suppl. Ser. (submitted).

In addition to the Iron Project work there are some other recent calculations, also using the R-matrix method, for important astrophysical ions. Two of these are:

O II: McLaughlin & Bell (ApJS 1994, in press) have presented analytical fits to their earlier data for $\Upsilon(T)$ for the transitions between the 11 states $2p^3(^4S^o, ^2D^o, ^2P^o)$, $2s2p^4(^4P, ^2D, ^2S, ^2P)$, $2p^34s(^4P, ^2D, ^2S, ^2P)$. Work on fine structure transitions is in progress.

S II: Cai & Pradhan (ApJS, **88**, 329, 1993) have presented $\Upsilon(T)$ for all 378 transitions between 28 fine structure levels of the 12 lowest LS terms $3p^3(^4S^o, ^2D^o, ^2P^o)$, $3s3p^2(^4P, ^2D, ^2S)$, $3p^23d(^2P, ^4F, ^4D, ^2F0, 3p^24s(^4P, ^2P)$.

These works update previous data for the optical and the UV lines of O II and S II, with some significant changes. For example the rate coefficient for the S II $\lambda1256$ UV multiplet $3p^3(^4S^o)_{3/2} \rightarrow 3s3p^4(^4P_{1/2,3/2,5/2})$, observed by the Hopkins Ultraviolet Telescope in the Io torus, is calculated by Cai and Pradhan to be nearly a factor of two lower than the ealier results of Ho & Henry (1990), bringing the observed intensity and abundance in better agreement with observations.

4.2. Radiative transition probabilities

A general bibliography of forbidden transitions was prepared by Biemont & Zeippen, in proceedings of the Meudon Atomic Data Workshop (1989) published in Journal de Physique, Vol. 1, Coll. 1, Suppl. II, No. 3, 1991 (Ed: C. J. Zeippen & M. LeDourneuf). This volume (hereafter referred to as ZLD) also contains a review by Wiese.

Several evaluated compilations, including the data, are:

1. "Forbidden lines in ns^2np^k ground configuration and nsnp excited configurations of Beryllium through Molybdenum atoms and ions," Victor Kaufman & Jack Sugar, *J. Phys. Chem. Ref. Data*, **15**, 321 (1986). This work is complemented by additional compilations in *J. Phys. Chem. Ref. Data* **16** Suppl. 3, and **17** Suppl. 4, by J. R. Fuhr, W. C. Martin & W. L. Wiese.

2. "Atomic transition probabilities," J. Fuhr & W. Wiese in the *CRC Handbook of Chemistry and Physics, 71st Edition*, 1990. This is a critical compilation including both forbidden and allowed lines.

3. "Atomic data for resonance absorption lines I. wavelengths longward of the Lyman limit," D. C. Morton, ApJSupp, **77**, 119 (1991).

4. TOPBASE contains a large number of oscillator strengths (A-values may also be obtained) for *dipole* transitions between LS multiplets for all isoelectronic sequences up to Si-like ions, with $Z \leq 26$. The TOPBASE data is with calculated (not observed) wavelengths, and does not include fine structure components within the multiplets. Future revisions of TOPBASE intend to include such data.

Some of the Opacity Project data has been extended to incorporate fine structure through algebraic transformation, from LS multiplet to LSJ components, and using experimentally observed energy levels: Si I, S III, Ar V and Ca VII (Nahar 1993) and Fe II (Nahar 1994); the Fe II data includes 21,587 fine structure transitions. Another calculation using the R-matrix method is by Bell *et al.* (1994, ApJS in press) for 26 dipole fine structure transitions in O II.

An updated version of the old NBS tables is in preparation at the National Institute for Standards and Technology (NIST). This is an evaluated compilation with some data from experiments and other sources, but primarily incorporating the Opacity Project data for some of the light elements (with fine structure included as in Nahar's work). The preliminary reference is: "Atomic Transition Probabilities of Carbon, Nitrogen, and Oxygen—A Critical Data Compilation," W. L. Wiese, J. R. Fuhr, & T. M. Deters, Monograph Series of *J. Phys. Chem. Ref. Data*, 1995.

Experimental measurements are relatively few. Among the recent works are the intersystem A-values measured by the Harvard-Nevada groups for several ions (Smith & Parkinson, paper presented at this meeting); dipole transitions in Si I and Fe II using the laser resonance-fluorescence method (e.g. O'Brian & Lawler 1991), and for dipole allowed transitions in O II, O III and other ions using Beam-Foil spectroscopy (Engström 1993). The experimental techniques employed by these groups are highly accurate, with uncertainties within a few percent for lifetimes and A-values.

4.3. *Photoionization cross sections*

A comprehensive bibliography has been compiled by LeDourneuf (ZLD, 1991). General data sources are as follows.

1. TOPBASE provides detailed photoionization cross sections for $Z = 1\text{--}14$ isoelectronic sequences up to Iron. As mentioned above, these cross sections may not be accurate in the high energy, non-resonant region where they have been fitted to a simple analytic form that does not correspond to actual photoionization calculations. The photoionization data for some of the heavier elements in low ionization stages (e.g. Fe I–V) is being updated, as the older data is not accurate (users are advised to check the NEWS file).

2. "Subshell photoionization cross sections and ionization energies of atoms and ions from He to Zn," have been computed by Verner *et al.* (1993) using the Hartee-Dirac-Slater method, analogous to the central field calculations but including some relativistic effects. Ground state photoionization cross sections are tabulated in analytic form up to high energies, $E \leq 50$ kev, including inner-shell ionization edges.

3. "Analytic fits for partial photoionization cross sections," by Verner & Yakovlev (1994, A&A in press), gives partial cross sections for the ground state shells of all atoms from H to Zn up to 100 kev.

4. "Photoabsorption cross sections for positive atomic ions with $Z \leq 30$," by Reilman & Manson (1979) are obtained through Hartree-Slater central-field calculations. These include only the ground state data and do not include autoionizing resonances, or coupling effects, and consequently are not generally accurate in the low energy region of photoionization. However, at high energies and for multiply charged ions these data reproduce well the background cross sections, compared to TOPBASE or other R-matrix calculations.

5. "X-ray interactions: photoionization, scattering, transmission, and reflection at $E = 50\text{--}30{,}000$ eV, $Z = 1\text{--}92$," by Henke *et al.* (1993) describes the absorption and scattering of X-rays in terms of atomic scattering factors calculated from photoabsorption cross sections (both of which are tabulated). This work should be useful in applications involving high energy photoabsorption.

For astrophysical applications, a combination of 1. and 2. is recommended, for low and high energies respectively, until more accurate calculations are completed in the high energy range and incorporated into TOPBASE.

6. H-ions: A computer program for analytic calculations for H-like ions has been written by Storey & Hummer (1991).

7. Photoionization of heavy atomic systems is discussed by Kelly (1990).

4.4. *Electron-ion recombination*

General reviews are:

1. "Electron-ion recombination," Flannery (1993).

2. "Di-electronic recombination and related resonance processes," Hahn & Lagattuta (1988).

Scaling laws for DR are discussed by McLauglin & Hahn (1991). References to works on RR and DR are given in the following recent papers.

3. Analytic fits to RR rates: Pequignot *et al.* (1991).

4. Ionization balance calculations with references to previous collisional ionization and RR and DR rates: Arnaud & Rothenflug (1985) and Arnaud & Raymond (1992).

5. More recent works on DR rates are: C IV (Badnell *et al.* 1993), C V (Kilgus *et al.* 1993), O II (Badnell 1992), $S^{q+}(q = 1 - 5)$ (Badnell 1991). The influence of external electric fields on DR is discussed in the work on C IV.

6. Unified, total rate coefficients are being calculated by Nahar & Pradhan, and are available for carbon-sequence ions: C I, N II, O III, F IV, Ne V, Mg VII, Si IX, S XI, and for C II, S I, S II, S III and Si I, SI II, SI III. These recombination rate coefficients incorporate both the RR and the DR processes in a unified manner and are consistent with the Opacity Project photoionization cross sections.

4.5. *Recombination lines*

The Hydrogen recombination lines are covered thoroughly by Hummer & Storey (1987) and Storey & Hummer (1988), including dust opacity (Hummer & Storey 1992).

The Helium lines have recently been computed by Smits (1991), with references to previous work by Brocklehurst and others (see also, Osterbrock 1989).

Recent work on recombination lines of O II by Storey (1994) and Liu *et al.* (1994) extends the treatment to complex ions, based on new Opacity Project data.

4.6. *Electron impact ionization*

The availabilility of ion accelerators and storage rings have vastly increased the ability to measure ionization cross sections (e.g. Mueller *et al.* 1985; Müller *et al.* 1991). But although there have been a much larger number of experiments for ionization than for electron impact excitation, most of the atomic systems, and their ions, have not yet been measured. Much of the experimental work has been carried out at Oak Ridge National Laboratory and the measured cross sections have been compiled in a recent work "Electron-Impact Ionization of Multicharged Ions at ORNL: 1985-1992" by D. C. Gregory & M. E. Bannister (1994, ORNL/TM-12729). The experimental data have clearly established the importance of excitation-autoionization (E-A) resonances that can enhance the cross section by orders of magnitude above the background. Theoretical calculations are now carried out for both the direct ionization and the E-A processes. Recent work includes the important iron isonuclear sequence (Pindzola *et al.* 1987), used by Arnaud & Raymond (1992) in their ionization balance calculations for Iron. Following are general references to available data (see also references in Arnaud & Rothenflug 1985).

1. Critical reviews are by Lennon *et al.* (1988), Itikawa (1992), Tawara & Kato (1987).

2. Limited data for excited state ionization is given by Golden & Sampson (1980).

3. Hydrogen and Helium ionization is considered by Jones *et al.* (1991).

4.7. *Line broadening*

The most recent compilation is the "Bibliography on Atomic Line Shapes and Shifts" by J. R. Fuhr & A. Lesage (NIST Special Publication 366, Supplement 4, 1993), including extensive literature references.

As part of the Opacity Project, Seaton (1987, 1988, 1989, 1990) has developed widely applicable approximations to hydrogenic and non-hydrogenic systems, the former including Stark plus electron impact broadening, and the latter with electon impact.

A general disucssion is given by Dimitrijević & Konjević (1987).

4.8. *Charge exchange*

1. A recent, exhaustive compilation, "Bibliography on electron transfer processes in Ion-Ion/Atom-Molecule collisions" by H. Tawara (1993), is available from the Research Information Center, National Institute for Fusion Science, Nagoya 464-01, Japan.

While the above reference includes charge exchange data sources for a variety of atomic and molecular species as reactants, most of the following references are for charge exchange between H and positive ions.

2. General references are: Dalgarno (1985), Butler & Dalgarno (1980).

3. A computer program LZRATE has been published by Bienstock (1983).

4. Butler, Heil & Dalgarno (1980) have presented data for C II, C III, C IV, N III, N IV, O III, O IV, Ne III, Ne IV.

5. H I and He II data is given by Jackson *et al.* (1992), H I data by Jouin & Harel (1991), Errea *et al.* (1992), and He I data by Slim *et al.* (1991).

6. Other works are: O VI (Andersson *et al.* 1991), Al III (Gargaud *et al.* 1990), Si II (Gargaud *et al.* 1982), Si III (McCaroll *et al.* 1976), Si V (Opradolce *et al.* 1985), Ar VII (Opradolce *et al.* 1983).

Detailed quantum mechanical calculations, as in these papers, are difficult to carry out. Approximate Landau-Zener rate coefficients may be obtained relatively simply and are given for some Al and Ca ions by Kingdon *et al.* (1995).

4.9. *Proton Impact Excitation*

In another pioneering paper, Seaton (1955) pointed out the importance of proton impact excitation for transitions at temperarures where $\Delta E/kT << 1$, e.g. the 2s-2p transition in H (and other Rydberg transitions) at nebular temperatures. Even for highly charged ions, fine structure transitions may be driven to a considerable extent by proton scattering at high temperatures, such as the excitation of coronal lines $Fe^{+13}3p_{1/2-3/2}$ (Seaton 1964). A number of studies have since been carried out and it has been demonstrated by semi-classical and quantum mechanical calculations that proton impact excitation may contribute significantly to the formation of spectral lines due to transitions between closely spaced levels (Dalgarno 1983). A review by Walling and Weisheit (1988) given many references to the relevant data on ion-ion excitation.

4.10. *Hyperfine structure and isotopic data*

Kurucz (1993) has reviewed isotopic data for nearly all naturally occuring elements. Asymmetries in emission line profiles due to hyperfine components and consequent errors in the determination of velocity fields and abundances are discussed.

Other effects, such as Doppler shifts, and scaling laws are discussed by Hühnermann (1993).

5. Effective collision strengths and A-values

In addition to the up-to-date data sources described above, it is of some interest to have the basic collisional and radiative data available for a limited number of atomic transitions that give rise to nebular lines formed at low temperatures in optically thin plasmas. A widely used compilation of such data was provided by Mendoza (1983) over a decade ago. New data are now available for most of the transitions. Table 1 presents an updated and extended dataset of maxwellian averaged collision strengths, $\Upsilon(T)$, the Einstein A-values, and experimental wavelengths $\lambda(\text{Å})$. The Υ are tabulated at four temperatures: 5000, 10000, 15000 and 20000 K. In some instances the temperature dependent values are unavailable. The LS collision strengths may be subdivided into fine structure components in a straightforward manner according to the statistical weights if either $S = 0$ or $L = 0$ for one of the terms. This procedure is employed for available data where applicable. Original references are marked as superscripts and should be quoted by users in literature.

The available atomic data for these and other ions far exceeds the data presented herein. Users should consult these when additional data is required. For instance, the $\Upsilon(T)$ for Fe II given in Table 1 is a very small subset of the data for 10,012 transitions for the IR, O, and UV transitions calculated by Zhang & Pradhan (1994). An even larger dataset is available for Fe III. We have concentrated only on low temperatures but the original references often contain additional data at higher temperatures and at a finer temperature mesh.

Acknowledgements

It is a pleasure to acknowledge with gratitude the guidance that AKP has received from Prof. Mike Seaton for over two decades, and the continuing inspiration from Prof. Don Osterbrock. This work was supported in part by a grant from the National Science Foundation (PHY-9115057) and the NASA LTSA program (NAGW-3315).

REFERENCES

ANDERSSON, L. R., GARGAUD, M., & MCCARROLL, R. 1991 *J. Phys. B* **24**, 2073

ARNAUD, M., & RAYMOND, J. 1992 *ApJ* **398**, 394

ARNAUD, M., & ROTHENFLUG. R. 1985 *A&AS* **60**, 425

BADNELL, N. R. 1991 *ApJ* **379**, 356

BADNELL, N. R. 1992 *Phys. Rev A* **46**, 660

BADNELL, N. R., PINDZOLA, M. S., DICKSON, W. J., SUMMERS, H. P., GRIFFIN, D. C., & LANG, J. 1993 *ApJL* **407**, L91

BURGESS, A. 1965 *ApJ* **141**, 1588

BERRINGON, K. A., BURKE, P. G., BUTLER, K., SEATON, M. J., STOREY, P. J., TAYLOR, K. T., YU, Y. 1987 *J. Phys. B* **20**, 6379

BIENSTOCK, S. 1983 *Comput. Phys. Commun.* **29**, 333

BURGESS, A. & TULLY, J. A. 1992 *A&A* **254**, 436

BURKE, P. G., HIBBERT, A., & ROBB, D. 1971 *J. Phys. B* **4**, 153

BUTLER, K. 1993 in *Planetary Nebulae*, IAU Symposium No. 155 (eds. R. Weinberger & A. Acker), Kluwer, p. 73

BUTLER, S. E, & DALGARNO, A. 1980 *ApJ* **241**, 838

BUTLER, S. E., HEIL, T. G., & DALGARNO, A. 1980 *ApJ* **241**, 442

CUNTO, W., MENDOZA, C., OCHSENBEIN, F., & ZEIPPEN, C. 1993 *A&A* **275**, L5

DALGARNO, A. 1983 in *Atoms In Astrophysics*, (eds. P. G. Burke, W. Eissner, D. G. Hummer & I. C. Percival) Plenum Publishing Corporation

DALGARNO, A. 1985 *Nucl. Instrum. & Methods. Phys. Res. Sect. B* **B9**, 662

ENGSTRÖM, L. 1993 *Physica Scripta* **T47**, 49

ERREA, L. F., LÓPEZ, A., MÉNDEZ, L., & RIERA, S. 1992 *J. Phys. B* **25**, 811

FLANNERY, M. R. 1993 in *Adv. At. Molec. Phys.* **32**, 117

GARGAUD, M., MCCARROLL, R., LENNON, M. A., WILSON, S. M., MCCULLOUGH, R. W., & GILBODY, H. B. 1990 *J. Phys. B* **23**, 505

GOLDEN, L. B., & SAMPSON, D. H. 1980 *J. Phys. B* **13**, 2645

HAHN, Y., & LAGATTUTA, K. J. 1988 *Phys. Repts.* **166**, 196

HEBB, M. H. & MENZEL, D. A. 1940 *ApJ* **92**, 408

HENKE, B. L., GULLIKSON, E. M., & DAVIS, J. C. 1993 *At. Data. Nucl. Data. Tables* **54**, 181

HÜHNERMANN, H. 1993 *Physica Scripta* **T47**, 70

HULSE, R. A. 1990 in *Atomic Processes in Plasmas, Gaithersburg, MD, 1989* (eds. Y.-K. Kim & R. C. Elton) AIP Conf. Proc., 206

HUMMER, D. G., & STOREY, P. J. 1987 *MNRAS* **224**, 801

HUMMER, D. G., & STOREY, P. J. 1992 *MNRAS* **254**, 277

ITIKAWA, Y. 1992 *At. Data. Nucl. Data. Tables* **49**, 209

JACOBS, V. L., DAVIS, J., KEPPLE, P. C., & BLAHA, M. 1977 *ApJ* **211**, 605

JANEV, R. K. (ed.) 1992 *Atomic and Plasma-Material Interaction Data for Fusion, Vol. 3* (Suppl. to Nuclear Fusion, November, 1992)

JANEV, R. K., PHANEUF, R. A., & HUNTER, H. 1988 *At. Data. Nucl. Data. Tables* **40**, 249

JANEV, R. K., HARRISON, M. F. A., & DRAWIN, H. W. 1989 *Nucl. Fusion* **29**, 109

JACKSON, D., SLIM, H. A., BRANSDEN, B. H., & FLOWER, D. R. 1992 *J. Phys. B* **25**, L127

JONES, S., MADISON, D. H., & SRIVASTAVA, M. K. 1991 *J. Phys. B* **24**, 1899

JOUIN, H., & HAREL, C. 1991 *J. Phys. B* **24**, 3219

KILGUS, G., HABS, D., SCHWALM, D., WOLF, A., SCHUCH, R., & BADNELL, N. R. 1993 *Phys. Rev. A* **47**, 4859

KINGDON, J., FERLAND, G. J., & FEIBELMAN, W. 1995 *ApJ* (submitted)

KURUCZ, R. L. 1993 *Physica Scripta* **T47**, 110

LENNON, M. A., BELL, K. L., GILBODY, H. B., HUGHES, J. G., KINGSTON, A. E., MURRAY, M. J., SMITH, F. J. 1988 *Jour. Phys. Chem. Ref. Data* **17**, 1285

LIU, X. W., STOREY, P. J., BARLOW, M. J. & CLEGG, R. E. S. 1994 *A&A* in press

MCLAUGLIN, D. J. & HAHN, Y. 1991 *Phys. Rev. A* **43**, 1313

MCCARROLL, R., & VALIRON, P. 1976 *A&A* **53**, 83

MENDOZA, C. 1983 in *Planetary Nebulae* (ed. D. R. Flower), Dordrecht: Reidel

MÜLLER, A., SCHENNACH, S., WAGNER, M., HASELBAUER, J., UWIRA, O., SPIES, W., JENNEWEIN, E., BECKER, R., KLEINOD, M., PRÖBSTEL, U., ANGERT, N., KLABUNDE, J., MOKLER, P. H., SPÄDTE, P. & WOLF, B. 1991 *Physica Scripta* **T37**, 62

MÜLLER, D. W., MORGAN, T. J., DUNN, G. H., GREGORY, D. C. & CRANDALL, D. H. 1985 *Phys. Rev. A* **31**, 2095

NAHAR, S. N. 1993 *Physica Scripta* **48**, 297

NAHAR, S. N. 1994 *A&A* in press

NAHAR, S. N. & PRADHAN, A. K. 1992 *Phys. Rev. Lett.* **68**, 1488

NAHAR, S. N. & PRADHAN, A. K. 1994a *J. Phys. B* **27**, 429

NAHAR, S. N. & PRADHAN, A. K. 1994b *Phys. Rev. A* **49**, 1816

NUSSBAUMER, H. & STOREY, P. J. 1980 *A&A* **89**, 308

NUSSBAUMER, H. & STOREY, P. J. 1983 *A&A* **126**, 75

O'BRIAN, T. R. & LAWLER, J. E. 1991 *Phys. Rev. A* **44**, 7134

OPRADOLCE, L., MCCARROLL, R., & VALIRON, P. 1985 *A&A* **148**, 229

OPRADOLCE, L., VALIRON, P., & MCCARROLL, R. 1983 *J. Phys. B* **16**, 2017

OSTERBROCK, D. E. 1989 *Astrophysics of Gaseous Nebulae and Active Galactic Nuclei*, University Science Books, Mill Valley CA

PEQUIGNOT, D., PETITJEAN, P., & BOISSON, C. 1991 *A&A* **251**, 280

PHANEUF, R. A. 1993 *Physica Scripta* **T47**, 124

PINDZOLA, M. S., BADNELL, N. R., & GRIFFIN, D. C. 1990 *Phys. Rev. A* **42**, 282

REILMAN, R. F. & MANSON, S. T. 1979 *ApJS* **40**, 815

SAWEY, P. M. J. & BERRINGTON, K. A. 1992 *J. Phys. B* **25**, 1451

SEATON, M. J. 1955 *Proc. Phys. Soc.* **68**, 457

SEATON, M. J. 1964 *MNRAS* **127**, 191

SEATON, M. J. 1987 *J. Phys. B* **20**, 6431

SEATON, M. J. 1988 *J. Phys. B* **21**, 3033

SEATON, M. J. 1989 *J. Phys. B* **22**, 3603

SEATON, M. J. 1990 *J. Phys. B* **23**, 3255

SEATON, M. J., Y. YU, MIHALAS, D., & PRADHAN, A. K. 1994 *MNRAS* **266**, 805

SEATON, M. J. 1987 *J. Phys. B* **20**, 6363

SLIM, H. A., HECK, E. L., BRANSDEN, B. H., & FLOWER, D. R. 1991 *J. Phys. B* **24**, 1683

SMITS, D. P. 1991 *MNRAS* **248**, 193

STOREY, P. J. 1994 *A&A* in press

STOREY, P. J. & HUMMER, D. G. 1988 *MNRAS* **231**, 1139

STOREY, P. J. & HUMMER, D. G. 1991 *Comput. Phys. Commun.* **66**, 129

TAWARA, H., & KATO, T. 1987 *At. Data. Nucl. Data. Tables* **36**, 167

WALLING, R. S. & WEISHEIT, J. C. 1988 *Phys. Repts.* **162**, 1

YU, Y. & SEATON, M. J. 1987 *J. Phys. B* **20**, 6409

ZHANG, H. L. & PRADHAN, A. K. 1993 *ApJL* **409**, L77

Ion	Transition	λ (Å)	A(sec^{-1})	$\Upsilon(T \times 10^4$K$)$			
				$T = 0.5$	1.0	1.5	2.0
H I	$1s - 2s$	1215.67[bb]	8.23 + 0[bb]	2.55 − 1	2.74 − 1	2.81 − 1	2.84 − 1[aa]
	$1s - 2p$	1215.66[bb]	6.265 + 8[bb]	4.16 − 1	4.72 − 1	5.28 − 1	5.85 − 1[aa]
He I	$1^1S - 2^3S$	625.48[ab]	1.13 − 4[bc]	6.50 − 2	6.87 − 2	6.81 − 2	6.72 − 2[ab]
	$1^1S - 2^1S$	601.30[ab]	5.13 + 1[bf]	3.11 − 2	3.61 − 2	3.84 − 2	4.01 − 2[ab]
	$1^1S - 2^3P^o$	591.29[ab]	1.76 + 2[bd]	1.60 − 2	2.27 − 2	2.71 − 2	3.07 − 2[ab]
	$1^1S - 2^1P^o$	584.21[ab]	1.80 + 9[bb]	9.92 − 3	1.54 − 2	1.98 − 2	2.40 − 2[ab]
	$2^3S - 2^1S$	15553.7[ab]	1.51 − 7[bf]	2.24 + 0	2.40 + 0	2.32 + 0	2.20 + 0[ab]
	$2^3S - 2^3P^o$	10817.0[ab]	1.02 + 7[bb]	1.50 + 1	2.69 + 1	3.74 + 1	4.66 + 1[ab]
	$2^3S - 2^1P^o$	8854.5[ab]	1.29 + 0[bf]	7.70 − 1	9.75 − 1	1.05 + 0	1.08 + 0[ab]
	$2^1S - 2^3P^o$	35519.5[ab]	2.70 − 2[bd]	1.50 + 0	1.70 + 0	1.74 + 0	1.72 + 0[ab]
	$2^1S - 2^1P^o$	20557.7[ab]	1.98 + 6[bb]	9.73 + 0	1.86 + 1	2.58 + 1	3.32 + 1[ab]
	$2^3P^o - 2^1P^o$	48804.3[ab]	−−	1.45 + 0	2.07 + 0	2.40 + 0	2.60 + 0[ab]
He II	$1S - 2S$	303[af]	5.66 + 2[ag]	1.60 − 1	1.59 − 1	1.57 − 1	1.56 − 1[ad]
	$1S - 2P$	303[af]	1.0 + 10	3.40 − 1	3.53 − 1	3.63 − 1	3.73 − 1[ad]
Li II	$1^1S - 2^3S$	210.11[ah]	2.039 − 2[ag]	5.54 − 2	5.49 − 2	5.43 − 2	5.38 − 2[ah]
	$1^1S - 2^1S$	−−	1.95 + 3[ag]	3.81 − 2	3.83 − 2	3.85 − 2	3.86 − 2[ah]
	$1^1S - 2^3P^o$	202.55[ah]	3.289 − 7[bd]	9.07 − 2	9.17 − 2	9.26 − 2	9.34 − 2[ah]
	$1^1S - 2^1P^o$	199.30[ah]	2.56 + 2[bb]	3.82 − 2	4.05 − 2	4.28 − 2	4.50 − 2[ah]
C I	$^1D_2 - ^3P_0$	9811.03[ac]	7.77 − 8[bh]	6.03 − 1	1.14 + 0	1.60 + 0	1.96 + 0[bi]
	$^1D_2 - ^3P_1$	9824.12[ee]	7.79 − 5[bh]				
	$^1D_2 - ^3P_2$	9850.28[ee]	2.30 − 4[ee]				
	$^1S_0 - ^3P_1$	4621.57[ee]	2.60 − 3[ee]	1.49 − 1	2.52 − 1	3.20 − 1	3.65 − 1[bi]
	$^1S_0 - ^3P_2$	4628.64[ac]	2.00 − 5[bh]				
	$^1S_0 - ^1D_2$	8727.18[ee]	5.01 − 1[ee]	1.96 − 1	2.77 − 1	3.40 − 1	3.92 − 1[bi]
	$^3P_1 - ^3P_0$	6.094 + 6[ee]	7.95 − 8[ee]	2.43 − 1	3.71 − 1[al]	−−	−−
	$^3P_2 - ^3P_0$	2304147[ac]	1.71 − 14[bh]	1.82 − 1	2.46 − 1[al]	−−	−−
	$^3P_2 - ^3P_1$	3704140[ee]	2.65 − 7[ee]	7.14 − 1	1.02 + 0[al]	−−	−−
	$^5S^o_2 - ^3P_1$	2965.70[ac]	6.94 + 0[bk]	4.75 − 1	6.71 − 1	8.22 − 1	9.50 − 1[bj]
	$^5S^o_2 - ^3P_2$	2968.08[ac]	1.56 + 1[bk]				
C II	$^2P^o_{3/2} - ^2P^o_{1/2}$	1.5774 + 5[ee]	2.29 − 6[ee]	1.89 + 0	2.15 + 0	2.26 + 0	2.28 + 0[db]
	$^4P_{1/2} - ^2P^o_{1/2}$	2325[cc]	7.0 + 1[cb]	2.43 − 1	2.42 − 1	2.46 − 1	2.48 − 1[db]
	$^4P_{1/2} - ^2P^o_{3/2}$	2329[cc]	6.3 + 1[cb]	1.74 − 1	1.77 − 1	1.82 − 1	1.84 − 1[db]
	$^4P_{3/2} - ^2P^o_{1/2}$	2324[cc]	1.4 + 0[cb]	3.61 − 1	3.62 − 1	3.68 − 1	3.70 − 1[db]
	$^4P_{3/2} - ^2P^o_{3/2}$	2328[cc]	9.4 + 0[cb]	4.72 − 1	4.77 − 1	4.88 − 1	4.93 − 1[db]
	$^4P_{3/2} - ^4P_{1/2}$	4.55 + 6[bk]	2.39 − 7[bk]	6.60 − 1	8.24 − 1	9.64 − 1	1.06 + 0[db]
	$^4P_{5/2} - ^2P^o_{1/2}$	2323	−−	2.29 − 1	2.34 − 1	2.42 − 1	2.45 − 1[db]
	$^4P_{5/2} - ^2P^o_{3/2}$	2326[cc]	5.1 + 1[ca]	1.02 + 0	1.02 + 0	1.04 + 0	1.05 + 0[db]
	$^4P_{5/2} - ^4P_{1/2}$	1.99 + 6[bk]	3.49 − 14[bk]	7.30 − 1	8.53 − 1	9.32 − 1	9.71 − 1[db]
	$^4P_{5/2} - ^4P_{3/2}$	3.53 + 6[bk]	3.67 − 7[bk]	1.65 + 0	1.98 + 0	2.23 + 0	2.39 + 0[db]
C III	$^3P^o_2 - ^1S_0$	1907[bl]	5.19 − 3[bl]	1.12 + 0	1.01 + 0	9.90 − 1	9.96 − 1[bm]
	$^3P^o_1 - ^1S_0$	1909[cc]	1.21 + 2[cd]				
	$^3P^o_0 - ^1S_0$	1909.6[bl]	−−				
	$^1P^o_1 - ^1S_0$	977.02[bl]	1.79 + 9[bl]	3.85 + 0	4.34 + 0	4.56 + 0	4.69 + 0[bm]
	$^3P^o_1 - ^3P^o_0$	4.22 + 6[ee]	3.00 − 7[ee]	8.48 − 1	9.11 − 1	9.75 − 1	1.03 + 0[bm]
	$^3P^o_2 - ^3P^o_0$	1.25 + 6[bl]	−−	5.79 − 1	6.77 − 1	7.76 − 1	8.67 − 1[bm]
	$^3P^o_2 - ^3P^o_1$	1.774 + 6[ee]	2.10 − 6[bl]	2.36 + 0	2.66 + 0	2.97 + 0	3.23 + 0[bm]
C IV	$^2P^o_{3/2} - ^2S_{1/2}$	1548.2[bb]	2.65 + 8[bb]	−−	8.88 + 0	−−	8.95 + 0[ba]
	$^2P^o_{1/2} - ^2S_{1/2}$	1550.8[bb]					

TABLE 1. Effective Collision strengths and A-values

Ion	Transition	$\lambda(\text{Å})$	$A(\text{sec}^{-1})$	$\Upsilon(0.5)$	$\Upsilon(1.0)$	$\Upsilon(1.5)$	$\Upsilon(2.0)$
N I	$^2D^o_{5/2} - {}^4S^o_{3/2}$	5200.4^{ee}	$6.92 - 6^{ee}$	$1.55 - 1$	$2.90 - 1$	$--$	$4.76 - 1^{bo}$
	$^2D^o_{3/2} - {}^4S^o_{3/2}$	5197.9^{ee}	$1.62 - 5^{ee}$	$1.03 - 1$	$1.94 - 1$	$--$	$3.18 - 1^{bo}$
	$^2P^o_{3/2} - {}^4S^o_{3/2}$	3466.5^{ee}	$6.18 - 3^{ee}$	$5.97 - 2$	$1.13 - 1$	$--$	$1.89 - 1^{bo}$
	$^2P^o_{1/2} - {}^4S^o_{3/2}$	3466.5^{ee}	$2.46 - 3^{ee}$	$2.98 - 2$	$5.67 - 2$	$--$	$9.47 - 2^{bo}$
	$^2D^o_{5/2} - {}^2D^o_{3/2}$	$1.148 + 7^{ee}$	$1.07 - 8^{ee}$	$1.28 - 1$	$2.69 - 1$	$--$	$4.65 - 1^{bo}$
	$^2P^o_{3/2} - {}^2P^o_{1/2}$	$2.59 + 8^{ee}$	$5.17 - 13$	$3.29 - 2$	$7.10 - 2$	$--$	$1.53 - 1^{bo}$
	$^2P^o_{3/2} - {}^2D^o_{5/2}$	10397.7^{ee}	$5.48 - 2^{ee}$	$1.62 - 1$	$2.66 - 1$	$--$	$4.38 - 1^{bp}$
	$^2P^o_{3/2} - {}^2D^o_{3/2}$	10407.2^{ee}	$2.47 - 2^{ee}$	$8.56 - 2$	$1.47 - 1$	$--$	$2.52 - 1^{bp}$
	$^2P^o_{1/2} - {}^2D^o_{5/2}$	1040.1^{bn}	$3.45 - 2^{bp}$	$6.26 - 2$	$1.09 - 1$	$--$	$1.90 - 1^{bp}$
	$^2P^o_{1/2} - {}^2D^o_{3/2}$	10407.6^{ee}	$4.71 - 2^{ee}$	$6.01 - 2$	$9.70 - 2$	$--$	$1.57 - 1^{bp}$
N II	$^1D_2 - {}^3P_0$	6529.0^{ac}	$5.35 - 7^{bh}$	$2.57 + 0$	$2.64 + 0$	$2.70 + 0$	$2.73 + 0^{dc}$
	$^1D_2 - {}^3P_1$	6548.1^{ee}	$1.04 - 3^{ee}$				
	$^1D_2 - {}^3P_2$	6583.4^{ee}	$3.02 - 3^{ee}$				
	$^1S_0 - {}^3P_1$	3062.9^{ee}	$3.40 - 2^{ee}$	$2.87 - 1$	$2.93 - 1$	$3.00 - 1$	$3.05 - 1^{dc}$
	$^1S_0 - {}^3P_2$	3071.4^{ac}	$1.51 - 4^{bh}$				
	$^1S_0 - {}^1D_2$	5754.6^{ee}	$1.08 + 0^{bh}$	$9.59 - 1$	$8.34 - 1$	$7.61 - 1$	$7.34 - 1^{dc}$
	$^3P_1 - {}^3P_0$	$2.055 + 6^{ee}$	$2.07 - 6^{ee}$	$3.71 - 1$	$4.08 - 1$	$4.29 - 1$	$4.43 - 1^{dc}$
	$^3P_2 - {}^3P_0$	$7.65 + 5^{ac}$	$1.16 - 12^{bh}$	$2.43 - 1$	$2.72 - 1$	$3.01 - 1$	$3.16 - 1^{dc}$
	$^3P_2 - {}^3P_1$	$1.22 + 6^{ee}$	$7.47 - 6^{ee}$	$1.01 + 0$	$1.12 + 0$	$1.21 + 0$	$1.26 + 0^{dc}$
	$^5S^o_2 - {}^3P_1$	2144^{cc}	$4.20 + 1^{ce}$	$1.19 + 0$	$1.19 + 0$	$1.21 + 0$	$1.21 + 0^{dc}$
	$^5S^o_2 - {}^3P_2$	2140^{cc}	$4.20 + 1^{ce}$				
N III	$^2P^o_{3/2} - {}^2P^o_{1/2}$	$5.73 + 5^{ee}$	$4.77 - 5^{ee}$	$1.32 + 0$	$1.45 + 0$	$1.55 + 0$	$1.64 + 0^{db}$
	$^4P_{1/2} - {}^2P^o_{1/2}$	1748^{cc}	$3.39 + 2^{bt}$	$1.89 - 1$	$1.98 - 1$	$2.04 - 1$	$2.07 - 1^{db}$
	$^4P_{1/2} - {}^2P^o_{3/2}$	1754^{cc}	$3.64 + 2^{bt}$	$1.35 - 1$	$1.51 - 1$	$1.62 - 1$	$1.68 - 1^{db}$
	$^4P_{3/2} - {}^2P^o_{1/2}$	1747^{cc}	$8.95 + 2^{bt}$	$2.81 - 1$	$2.98 - 1$	$3.09 - 1$	$3.16 - 1^{db}$
	$^4P_{3/2} - {}^2P^o_{3/2}$	1752^{cc}	$5.90 + 1^{bt}$	$3.67 - 1$	$3.99 - 1$	$4.23 - 1$	$4.35 - 1^{db}$
	$^4P_{3/2} - {}^4P_{1/2}$	$1.68 + 6^{ac}$	$--$	$1.01 + 0$	$1.10 + 0$	$1.14 + 0$	$1.16 + 0^{db}$
	$^4P_{5/2} - {}^2P^o_{1/2}$	1744.4^{ac}	$--$	$1.78 - 1$	$2.01 - 1$	$2.19 - 1$	$2.29 - 1^{db}$
	$^4P_{5/2} - {}^2P^o_{3/2}$	1747^{cc}	$3.08 + 2^{cf}$	$7.93 - 1$	$8.44 - 1$	$8.80 - 1$	$8.98 - 1^{db}$
	$^4P_{5/2} - {}^4P_{1/2}$	$7.10 + 5^{ac}$	$--$	$6.12 - 1$	$6.67 - 1$	$6.95 - 1$	$7.11 - 1^{db}$
	$^4P_{5/2} - {}^4P_{3/2}$	$1.23 + 6^{ac}$	$--$	$1.88 + 0$	$2.04 + 0$	$2.12 + 0$	$2.16 + 0^{db}$
N IV	$^3P^o_2 - {}^1S_0$	1483.3^{ac}	$1.15 - 2^{bu}$	$9.37 - 1$	$9.05 - 1$	$8.79 - 1$	$8.58 - 1^{bv}$
	$^3P^o_1 - {}^1S_0$	1486.4^{ac}	$5.77 + 2^{bu}$				
	$^3P^o_0 - {}^1S_0$	1487.9^{ac}	$--$				
	$^1P^o_1 - {}^1S_0$	765.15^{dd}	$2.40 + 9^{dd}$	$3.84 + 0$	$3.53 + 0$	$3.41 + 0$	$3.36 + 0^{bv}$
	$^3P^o_1 - {}^3P^o_0$	$1.585 + 6^{ee}$	$6.00 - 6^{ee}$	$--$	$--$	$--$	$--$
	$^3P^o_2 - {}^3P^o_0$	$4.83 + 5^{ac}$	$--$	$--$	$--$	$--$	$--$
	$^3P^o_2 - {}^3P^o_1$	$6.94 + 5^{ee}$	$3.63 - 5^{ee}$	$--$	$--$	$--$	$--$
N V	$^2P^o_{3/2} - {}^2S_{1/2}$	1238.8^{dd}	$3.41 + 8^{dd}$	$6.61 + 0$	$6.65 + 0$	$6.69 + 0$	$6.72 + 0^{bw}$
	$^2P^o_{1/2} - {}^2S_{1/2}$	1242.8^{dd}	$3.38 + 8^{dd}$				
O I	$^1D_2 - {}^3P_0$	6393.5^{ac}	$7.23 - 7^{ac}$	$1.24 - 1$	$2.66 - 1$	$--$	$5.01 - 1^{bo}$
	$^1D_2 - {}^3P_1$	6363.8^{ee}	$1.65 - 3^{ee}$				
	$^1D_2 - {}^3P_2$	6300.3^{ee}	$5.11 - 3^{ee}$				
	$^1S_0 - {}^3P_1$	2972.3^{ee}	$6.68 - 2^{ee}$	$1.53 - 2$	$3.24 - 2$	$--$	$6.07 - 2^{bo}$
	$^1S_0 - {}^3P_2$	2959.2^{ac}	$2.88 - 4^{ac}$				
	$^1S_0 - {}^1D_2$	5577.3^{ee}	$1.34 + 0^{ac}$	$7.32 - 2$	$1.05 - 1$	$--$	$1.48 - 1^{bo}$
	$^3P_0 - {}^3P_1$	$1.46 + 6^{ee}$	$1.75 - 5^{ee}$	$1.12 - 2$	$2.65 - 2$	$--$	$6.93 - 2^{bg}$
	$^3P_0 - {}^3P_2$	$4.41 + 5^{ac}$	$1.00 - 10^{ac}$	$1.48 - 2$	$2.92 - 2$	$--$	$5.36 - 2^{bg}$
	$^3P_1 - {}^3P_2$	$6.32 + 5^{ee}$	$8.91 - 5^{ee}$	$4.74 - 2$	$9.87 - 2$	$--$	$2.07 - 1^{bg}$

TABLE 1. (*Continued.*)

Ion	Transition	$\lambda(\text{Å})$	$A(\text{sec}^{-1})$	$\Upsilon(0.5)$	$\Upsilon(1.0)$	$\Upsilon(1.5)$	$\Upsilon(2.0)$
O II	$^2D^o_{5/2} - {}^4S^o_{3/2}$	3728.8^{ee}	$3.82-5^{ac}$	$7.95-1$	$8.01-1$	$8.10-1$	$8.18-1^{be}$
	$^2D^o_{3/2} - {}^4S^o_{3/2}$	3726.0^{ee}	$1.69-4^{ee}$	$5.30-1$	$5.34-1$	$5.41-1$	$5.45-1^{be}$
	$^2P^o_{3/2} - {}^4S^o_{3/2}$	2470.3^{ee}	$5.95-2^{ee}$	$2.65-1$	$2.70-1$	$2.75-1$	$2.80-1^{be}$
	$^2P^o_{1/2} - {}^4S^o_{3/2}$	2470.2^{ee}	$2.38-2^{ee}$	$1.33-1$	$1.35-1$	$1.37-1$	$1.40-1^{be}$
	$^2D^o_{5/2} - {}^2D^o_{3/2}$	$4.97+6^{ee}$	$1.25-7^{ee}$	$1.22+0$	$1.17+0$	$1.14+0$	$1.11+0^{be}$
	$^2P^o_{3/2} - {}^2P^o_{1/2}$	$5.00+7^{ee}$	$4.39-12^{ee}$	$2.80-1$	$2.87-1$	$2.93-1$	$3.00-1^{be}$
	$^2P^o_{3/2} - {}^2D^o_{5/2}$	7319.9^{ee}	$1.15-1^{ee}$	$7.18-1$	$7.30-1$	$7.41-1$	$7.55-1^{be}$
	$^2P^o_{3/2} - {}^2D^o_{3/2}$	7330.7^{ee}	$6.14-2^{ee}$	$4.01-1$	$4.08-1$	$4.14-1$	$4.22-1^{be}$
	$^2P^o_{1/2} - {}^2D^o_{5/2}$	7321.8^{bn}	$6.15-2^{bn}$	$2.90-1$	$2.95-1$	$3.00-1$	$3.05-1^{be}$
	$^2P^o_{1/2} - {}^2D^o_{3/2}$	7329.6^{ee}	$1.01-1^{ee}$	$2.70-1$	$2.75-1$	$2.81-1$	$2.84-1^{be}$
O III	$^1D_2 - {}^3P_0$	4932.6^{de}	$2.74-6^{df}$	$2.13+0$	$2.29+0$	$2.45+0$	$2.52+0^{dc}$
	$^1D_2 - {}^3P_1$	4958.9^{ee}	$6.74-3^{ac}$				
	$^1D_2 - {}^3P_2$	5006.7^{ee}	$1.96-2^{ac}$				
	$^1S_0 - {}^3P_1$	2321.0^{ee}	$3.27-1^{ee}$	$2.72-1$	$2.93-1$	$3.17-1$	$3.29-1^{dc}$
	$^1S_0 - {}^3P_2$	2332.1^{de}	$7.85-4^{df}$				
	$^1S_0 - {}^1D_2$	4363.2^{ee}	$2.65+0^{ee}$	$4.94-1$	$5.82-1$	$6.10-1$	$6.10-1^{dc}$
	$^3P_1 - {}^3P_0$	883562^{ee}	$2.61-5^{ee}$	$5.24-1$	$5.45-1$	$5.59-1$	$5.63-1^{dc}$
	$^3P_2 - {}^3P_0$	326611^{de}	$3.02-11^{df}$	$2.58-1$	$2.71-1$	$2.83-1$	$2.89-1^{dc}$
	$^3P_2 - {}^3P_1$	518145^{ee}	$9.69-5^{ee}$	$1.23+0$	$1.29+0$	$1.34+0$	$1.35+0^{dc}$
	$^5S^o_2 - {}^3P_1$	1660.8^{de}	$2.12+2^{df}$	$1.07+0$	$1.21+0$	$1.25+0$	$1.26+0^{dc}$
	$^5S^o_2 - {}^3P_2$	1666.1^{de}	$5.22+2^{df}$				
O IV	$^2P^o_{3/2} - {}^2P^o_{1/2}$	$2.587+5^{ee}$	$5.18-4^{bb}$	$2.02+0$	$2.40+0$	$2.53+0$	$2.57+0^{db}$
	$^4P_{1/2} - {}^2P^o_{1/2}$	1426.46^{di}	$1.81+3^{di}$	$1.21-1$	$1.33-1$	$1.42-1$	$1.48-1^{db}$
	$^4P_{1/2} - {}^2P^o_{3/2}$	1434.07^{di}	$1.77+3^{di}$	$8.67-2$	$1.02-1$	$1.15-1$	$1.24-1^{db}$
	$^4P_{3/2} - {}^2P^o_{1/2}$	1423.84^{di}	$2.28+1^{di}$	$1.80-1$	$2.00-1$	$2.16-1$	$2.28-1^{db}$
	$^4P_{3/2} - {}^2P^o_{3/2}$	1431.42^{di}	$3.28+2^{di}$	$2.36-1$	$2.68-1$	$2.98-1$	$3.18-1^{db}$
	$^4P_{3/2} - {}^4P_{1/2}$	$1.68+6^{ac}$	$--$	$1.04+0$	$1.09+0$	$1.13+0$	$1.16+0^{db}$
	$^4P_{5/2} - {}^2P^o_{1/2}$	1420.19^{di}	$--$	$1.15-1$	$1.36-1$	$1.55-1$	$1.69-1^{db}$
	$^4P_{5/2} - {}^2P^o_{3/2}$	1427.78^{di}	$1.04+3^{di}$	$5.08-1$	$5.67-1$	$6.15-1$	$6.48-1^{db}$
	$^4P_{5/2} - {}^4P_{1/2}$	$3.26+5^{di}$	$--$	$7.14-1$	$6.88-1$	$7.06-1$	$7.36-1^{db}$
	$^4P_{5/2} - {}^4P_{3/2}$	$5.62+5^{di}$	$1.02-4^{dh}$	$2.04+0$	$2.05+0$	$2.12+0$	$2.20+0^{db}$
O V	$^3P^o_2 - {}^1S_0$	1213.8^{bu}	$2.16-2^{bu}$	$7.33-1$	$7.21-1$	$6.74-1$	$6.39-1^{bm}$
	$^3P^o_1 - {}^1S_0$	1218.3^{bu}	$2.25+3^{bu}$				
	$^3P^o_0 - {}^1S_0$	1220.4^{bu}	$--$				
	$^1P^o_1 - {}^1S_0$	629.7^{dd}	$2.80+9^{dd}$	$2.66+0$	$2.76+0$	$2.82+0$	$2.85+0^{bm}$
	$^3P^o_1 - {}^3P_0$	$7.35+5^{ee}$	$5.81-5^{ee}$	$7.26-1$	$8.39-1$	$8.65-1$	$8.66-1^{eg}$
	$^3P^o_2 - {}^3P_0$	$2.26+5^{bu}$	$--$	$2.74-1$	$6.02-1$	$7.51-1$	$8.16-1^{eg}$
	$^3P^o_2 - {}^3P_1$	$3.26+5^{ee}$	$3.55-4^{ee}$	$3.19+0$	$2.86+0$	$2.80+0$	$2.77+0^{eg}$
O VI	$^2P^o_{3/2} - {}^2S_{1/2}$	1031.9^{dd}	$4.15+8^{dd}$	$4.98+0$	$5.00+0$	$5.03+0$	$5.05+0^{bw}$
	$^2P^o_{1/2} - {}^2S_{1/2}$	1037.6^{dd}	$4.08+8^{dd}$				
Ne II	$^2P^o_{1/2} - {}^2P^o_{3/2}$	$1.28+5^{ee}$	$8.55-3^{ee}$	$2.96-1$	$3.03-1$	$3.10-1$	$3.17-1^{ae}$
Ne III	$^1D_2 - {}^3P_0$	4012.8^{ac}	$8.51-6^{ac}$	$1.63+0$	$1.65+0$	$1.65+0$	$1.64+0^{ap}$
	$^1D_2 - {}^3P_1$	3967.5^{ee}	$5.95-2^{ac}$				
	$^1D_2 - {}^3P_2$	3868.8^{ee}	$1.39-1^{ee}$				
	$^1S_0 - {}^3P_1$	1814.6^{ee}	$2.76+0^{ee}$	$1.51-1$	$1.69-1$	$1.75-1$	$1.79-1^{ap}$
	$^1S_0 - {}^3P_2$	1793.7^{ac}	$3.94-3^{ac}$				
	$^1S_0 - {}^1D_2$	3342.5^{ee}	$4.28+0^{ee}$	$2.00-1$	$2.26-1$	$2.43-1$	$2.60-1^{ap}$
	$^3P_0 - {}^3P_1$	$3.60+5^{ee}$	$1.15-3^{ee}$	$3.31-1$	$3.50-1$	$3.51-1$	$3.50-1^{ap}$
	$^3P_0 - {}^3P_2$	$1.07+5^{ac}$	$2.18-8^{ac}$	$3.00-1$	$3.07-1$	$3.03-1$	$2.98-1^{ap}$
	$^3P_1 - {}^3P_2$	$1.56+5^{ee}$	$5.97-3^{ee}$	$1.09+0$	$1.65+0$	$1.65+0$	$1.64+0^{ap}$

TABLE 1. (*Continued.*)

Ion	Transition	$\lambda(\text{Å})$	$A(\sec^{-1})$	$\Upsilon(0.5)$	$\Upsilon(1.0)$	$\Upsilon(1.5)$	$\Upsilon(2.0)$
Ne IV	$^2D^o_{5/2} - ^4S^o_{3/2}$	2420.9^{ee}	$6.03 - 4^{ee}$	$8.45 - 1$	$8.43 - 1$	$8.32 - 1$	$8.24 - 1^{ao}$
	$^2D^o_{3/2} - ^4S^o_{3/2}$	2418.2^{ee}	$2.65 - 3^{ee}$	$5.63 - 1$	$5.59 - 1$	$5.55 - 1$	$5.50 - 1^{ao}$
	$^2P^o_{3/2} - ^4S^o_{3/2}$	1601.5^{ee}	$1.41 + 0^{ee}$	$3.07 - 1$	$3.13 - 1$	$3.12 - 1$	$3.09 - 1^{ao}$
	$^2P^o_{1/2} - ^4S^o_{3/2}$	1601.7^{ee}	$5.90 - 1^{ee}$	$1.53 - 1$	$1.56 - 1$	$1.56 - 1$	$1.55 - 1^{ao}$
	$^2D^o_{5/2} - ^2D^o_{3/2}$	$2.237 + 6^{ee}$	$1.44 - 6^{ee}$	$1.37 + 0$	$1.36 + 0$	$1.35 + 0$	$1.33 + 0^{ao}$
	$^2P^o_{3/2} - ^2P^o_{1/2}$	$1.56 + 7^{ee}$	$2.36 - 9^{bn}$	$3.17 - 1$	$3.43 - 1$	$3.58 - 1$	$3.70 - 1^{ao}$
	$^2P^o_{3/2} - ^2D^o_{5/2}$	4714.3^{ee}	$6.19 - 1^{ee}$	$8.56 - 1$	$9.00 - 1$	$9.08 - 1$	$9.09 - 1^{ao}$
	$^2P^o_{3/2} - ^2D^o_{3/2}$	4724.2^{ee}	$6.41 - 1^{bn}$	$4.73 - 1$	$5.09 - 1$	$5.15 - 1$	$5.16 - 1^{ao}$
	$^2P^o_{1/2} - ^2D^o_{5/2}$	4717.0^{bn}	$1.15 - 2^{bn}$	$3.40 - 1$	$3.68 - 1$	$3.73 - 1$	$3.74 - 1^{ao}$
	$^2P^o_{1/2} - ^2D^o_{3/2}$	4725.6^{ee}	$5.92 - 1^{ee}$	$3.24 - 1$	$3.36 - 1$	$3.39 - 1$	$3.39 - 1^{ao}$
Ne V	$^1D_2 - ^3P_0$	3301.3^{bh}	$2.37 - 5^{bh}$	$2.13 + 0$	$2.09 + 0$	$2.11 + 0$	$2.14 + 0^{dc}$
	$^1D_2 - ^3P_1$	3345.8^{ee}	$1.24 - 1^{ee}$				
	$^1D_2 - ^3P_2$	3425.9^{ee}	$4.36 - 1^{ee}$				
	$^1S_0 - ^3P_1$	1574.8^{ee}	$5.50 + 0^{ee}$	$2.54 - 1$	$2.46 - 1$	$2.49 - 1$	$2.51 - 1^{dc}$
	$^1S_0 - ^3P_2$	1592.3^{bh}	$6.69 - 3^{bh}$				
	$^1S_0 - ^1D_2$	2972.8^{ee}	$4.39 + 0^{ee}$	$6.63 - 1$	$5.77 - 1$	$6.10 - 1$	$6.49 - 1^{dc}$
	$^3P_1 - ^3P_0$	$2.428 + 5^{ee}$	$1.27 - 3^{ee}$	$1.68 + 0$	$1.41 + 0$	$1.19 + 0$	$1.10 + 0^{dc}$
	$^3P_2 - ^3P_0$	90082^{bh}	$5.08 - 9^{bh}$	$2.44 + 0$	$1.81 + 0$	$1.42 + 0$	$1.26 + 0^{dc}$
	$^3P_2 - ^3P_1$	$1.432 + 5^{ee}$	$4.59 - 3^{ee}$	$7.59 + 0$	$5.82 + 0$	$4.68 + 0$	$4.20 + 0^{dc}$
	$^5S^o_2 - ^3P_1$	1137.0^{ac}	$2.37 + 3^{ac}$	$1.11 + 0$	$1.43 + 0$	$1.39 + 0$	$1.34 + 0^{dc}$
	$^5S^o_2 - ^3P_2$	1146.1^{ac}	$6.06 + 3^{ac}$				
Ne VI	$^2P^o_{3/2} - ^2P^o_{1/2}$	$7.642 + 4^{ee}$	$2.02 - 2^{bb}$	$3.22 + 0$	$2.72 + 0$	$2.37 + 0$	$2.15 + 0^{aj}$
	$^4P_{1/2} - ^2P^o_{1/2}$	1003.6^{di}	$1.59 + 4^{di}$	$1.54 - 1$	$1.37 - 1$	$1.26 - 1$	$1.18 - 1^{aj}$
	$^4P_{1/2} - ^2P^o_{3/2}$	1016.6^{di}	$1.43 + 4^{di}$	$1.85 - 1$	$1.53 - 1$	$1.36 - 1$	$1.27 - 1^{aj}$
	$^4P_{3/2} - ^2P^o_{1/2}$	999.13^{di}	$3.20 + 2^{di}$	$2.69 - 1$	$2.32 - 1$	$2.10 - 1$	$1.96 - 1^{aj}$
	$^4P_{3/2} - ^2P^o_{3/2}$	1012.0^{di}	$3.33 + 3^{di}$	$4.51 - 1$	$3.73 - 1$	$3.32 - 1$	$3.08 - 1^{aj}$
	$^4P_{3/2} - ^4P_{1/2}$	$2.24 + 5^{di}$	$--$	$5.34 - 1$	$5.73 - 1$	$5.95 - 1$	$6.22 - 1^{aj}$
	$^4P_{5/2} - ^2P^o_{1/2}$	992.76^{di}	$--$	$2.56 - 1$	$2.11 - 1$	$1.89 - 1$	$1.76 - 1^{aj}$
	$^4P_{5/2} - ^2P^o_{3/2}$	1005.5^{di}	$1.14 + 4^{di}$	$7.88 + 0$	$6.75 - 1$	$6.09 - 1$	$5.68 - 1^{aj}$
	$^4P_{5/2} - ^4P_{1/2}$	92166^{di}	$--$	$4.12 - 1$	$4.23 - 1$	$4.36 - 2$	$4.56 - 1^{aj}$
	$^4P_{5/2} - ^4P_{3/2}$	$1.56 + 5^{di}$	$--$	$1.10 + 0$	$1.16 + 0$	$1.20 + 0$	$1.26 + 0^{aj}$
Ne VII	$^3P^o_2 - ^1S_0$	887.22^{bu}	$5.78 - 2^{bu}$	$1.29 - 1$	$1.72 - 1$	$2.05 - 1$	$2.28 - 1^{dj}$
	$^3P^o_1 - ^1S_0$	895.12^{bu}	$1.98 + 4^{bu}$				
	$^3P^o_0 - ^1S_0$	898.76^{bu}	$--$				
	$^1P^o_1 - ^1S_0$	465.22^{bu}	$4.09 + 9^{bu}$	$1.39 + 0$	$1.56 + 0$	$1.63 + 0$	$1.66 + 0^{dj}$
	$^3P^o_1 - ^3P^o_0$	$2.20 + 5^{bu}$	$1.99 - 3^{ee}$	$--$	$--$	$--$	$--$
	$^3P^o_2 - ^3P^o_0$	69127.6^{bu}	$--$	$--$	$--$	$--$	$--$
	$^3P^o_2 - ^3P^o_1$	$1.01 + 5^{ee}$	$1.25 - 2^{ee}$	$--$	$--$	$--$	$--$
Na III	$^2P^o_{1/2} - ^2P^o_{3/2}$	$7.319 + 4^{ee}$	$4.59 - 2^{ee}$		$3.00 - 1^{br}$		
Na IV	$^1D_2 - ^3P_0$	3417.2^{ac}	$2.24 - 5^{ac}$		$1.17 + 0^{br}$		
	$^1D_2 - ^3P_1$	3362.2^{ee}	$2.03 - 1^{ee}$				
	$^1D_2 - ^3P_2$	3241.7^{ee}	$5.75 - 1^{ee}$				
	$^1S_0 - ^3P_1$	1529.3^{ee}	$9.48 + 0^{ee}$		$1.63 - 1^{br}$		
	$^1S_0 - ^3P_2$	1503.8^{ac}	$1.05 - 2^{ac}$				
	$^1S_0 - ^1D_2$	2803.7^{ee}	$5.43 + 0^{ee}$		$1.57 - 1^{br}$		
	$^3P_0 - ^3P_1$	$2.129 + 5^{ee}$	$5.58 - 3^{ee}$		$1.77 - 1^{br}$		
	$^3P_0 - ^3P_2$	62467.9^{ac}	$1.67 - 7^{ac}$		$1.11 - 1^{br}$		
	$^3P_1 - ^3P_2$	90391.4^{ee}	$3.04 - 2^{ee}$		$4.71 - 1^{br}$		
Na V	$^2D^o_{5/2} - ^4S^o_{3/2}$	2068.4^{ee}	$1.73 - 3^{ee}$		$5.51 - 1^{br}$		
	$^2D^o_{3/2} - ^4S^o_{3/2}$	2066.9^{ee}	$1.78 - 2^{ee}$		$3.68 - 1^{br}$		

TABLE 1. (Continued.)

Ion	Transition	$\lambda(\text{Å})$	$A(\sec^{-1})$	$\Upsilon(0.5)$	$\Upsilon(1.0)$	$\Upsilon(1.5)$	$\Upsilon(2.0)$
	$^2P^o_{3/2} - {}^4S^o_{3/2}$	1365.1^{ee}	$4.74+0^{ee}$		$2.39-1^{br}$		
	$^2P^o_{1/2} - {}^4S^o_{3/2}$	1365.8^{ee}	$1.96+0^{ee}$		$1.20-1^{br}$		
	$^2D^o_{5/2} - {}^2D^o_{3/2}$	$2.78+6^{ee}$	$7.50-7^{ee}$		$6.96-1^{br}$		
	$^2P^o_{3/2} - {}^2P^o_{1/2}$	$2.70+6^{ee}$	$4.55-7^{ee}$		$4.38-1^{br}$		
	$^2P^o_{3/2} - {}^2D^o_{5/2}$	4010.9^{ee}	$1.40+0^{ee}$		$5.02-1^{br}$		
	$^2P^o_{3/2} - {}^2D^o_{3/2}$	4016.7^{ee}	$1.91+0^{ee}$		$2.79-1^{br}$		
	$^2P^o_{1/2} - {}^2D^o_{5/2}$	4017.9^{ac}	$1.41-1^{bn}$		$2.01-1^{br}$		
	$^2P^o_{1/2} - {}^2D^o_{3/2}$	4022.7^{ee}	$1.43+0^{ee}$		$1.90-1^{br}$		
Na VI	$^1D_2 - {}^3P_0$	2816.1^{ee}	--	$1.55+0$	$1.45+0$	$1.39+0$	$1.38+0^{dc}$
	$^1D_2 - {}^3P_1$	2872.7^{ee}	$4.06-1^{ee}$				
	$^1D_2 - {}^3P_2$	2971.9^{ee}	$1.27+0^{ee}$				
	$^1S_0 - {}^3P_1$	1356.6^{ee}	$1.69+1^{ee}$	$1.73-1$	$1.72-1$	$1.72-1$	$1.73-1^{dc}$
	$^1S_0 - {}^3P_2$	1343.9^{ee}	--				
	$^1S_0 - {}^1D_2$	2568.9^{ee}	$5.27+0^{ee}$	$1.07-1$	$1.16-1$	$1.28-1$	$1.39-1^{dc}$
	$^3P_1 - {}^3P_0$	$1.43+5^{ee}$	$6.14-3^{ee}$	$7.24-1$	$7.70-1$	$7.73-1$	$7.58-1^{dc}$
	$^3P_2 - {}^3P_0$	$5.37+4^{ee}$	--	$5.02-1$	$5.21-1$	$5.08-1$	$4.94-1^{dc}$
	$^3P_2 - {}^3P_1$	$8.61+4^{ee}$	$2.11-2^{ee}$	$2.03+0$	$2.13+0$	$2.10+0$	$2.05+0^{dc}$
Mg I	$^3P^o_2 - {}^1S_0$	4563.9^{ac}	$4.13-4^{dl}$				
	$^3P^o_1 - {}^1S_0$	4572.4^{ac}	$1.80+2^{dl}$				
	$^3P^o_0 - {}^1S_0$	4576.6^{ac}	--				
	$^1P^o_1 - {}^1S_0$	2852.1^{dd}	$4.95+8^{dd}$				
	$^3P^o_1 - {}^3P^o_0$	$4.986+6^{ee}$	$1.00-7^{ee}$				
	$^3P^o_2 - {}^3P^o_0$	$1.64+6^{ac}$	$4.08-12^{dl}$				
	$^3P^o_2 - {}^3P^o_1$	$2.456+6^{ee}$	$9.00-7^{ee}$				
Mg II	$^2P^o_{3/2} - {}^2S_{1/2}$	2795.5^{dd}	$2.6+8^{dd}$	$1.59+1$	$1.69+1$	$1.78+1$	$1.86+1^{ch}$
	$^2P^o_{1/2} - {}^2S_{1/2}$	2802.7^{dd}	$2.6+8^{dd}$				
Mg IV	$^2P^o_{1/2} - {}^2P^o_{3/2}$	$4.487+4^{ee}$	$1.99-1^{ee}$	$3.44-1$	$3.46-1$	$3.49-1$	$3.51-1^{ae}$
Mg V	$^1D_2 - {}^3P_0$	2993.1^{ac}	$5.20-5^{ac}$	$1.31+0$	$1.33+0$	$1.32+0$	$1.30+0^{an}$
	$^1D_2 - {}^3P_1$	2928.0^{ee}	$5.85-1^{ee}$				
	$^1D_2 - {}^3P_2$	2782.7^{ee}	$1.86+0^{ee}$				
	$^1S_0 - {}^3P_1$	1324.4^{ee}	$2.79+1^{ee}$	$1.42-1$	$1.48-1$	$1.46-1$	$1.44-1^{an}$
	$^1S_0 - {}^3P_2$	1293.9^{ac}	$2.45-2^{ac}$				
	$^1S_0 - {}^1D_2$	2417.5^{ee}	$6.59+0^{ee}$	$1.91-1$	$1.97-1$	$2.02-1$	$2.08-1^{an}$
	$^3P_0 - {}^3P_1$	$1.354+5^{ee}$	$2.17-2^{ee}$	$2.48-1$	$3.00-1$	$3.18-1$	$3.18-1^{an}$
	$^3P_0 - {}^3P_2$	39654.2^{ac}	$1.01-6^{ac}$	$2.31-1$	$2.92-1$	$3.04-1$	$2.99-1^{an}$
	$^3P_1 - {}^3P_2$	$5.608+4^{ee}$	$1.27-1^{ee}$	$8.30-1$	$1.03+0$	$1.08+0$	$1.07+0^{an}$
Mg VII	$^1D_2 - {}^3P_0$	2441.4^{ee}	--	$7.96-1$	$8.57-1$	$9.11-1$	$9.42-1^{dc}$
	$^1D_2 - {}^3P_1$	2509.2^{ee}	$1.17+0^{ee}$				
	$^1D_2 - {}^3P_2$	2629.1^{ee}	$3.36+0^{ee}$				
	$^1S_0 - {}^3P_1$	1189.8^{ee}	$4.58+1^{ee}$	$2.08-1$	$1.85-1$	$1.75-1$	$1.73-1^{dc}$
	$^1S_0 - {}^3P_2$	1174.3^{ee}	--				
	$^1S_0 - {}^1D_2$	2261.5^{ee}	$6.16+0^{ee}$	$5.25-1$	$4.46-1$	$3.90-1$	$3.82-1^{dc}$
	$^3P_1 - {}^3P_0$	$9.03+4^{ee}$	$2.44-2^{ee}$	$2.75-1$	$3.37-1$	$3.95-1$	$4.14-1^{dc}$
	$^3P_2 - {}^3P_0$	$3.42+4^{ee}$	--	$1.90-1$	$3.01-1$	$3.88-1$	$4.09-1^{dc}$
	$^3P_2 - {}^3P_1$	$5.50+4^{ee}$	$8.09-2^{ee}$	$7.69+0$	$1.08+0$	$1.32+0$	$1.39+0^{dc}$
Al II	$^3P^o_2 - {}^1S_0$	2661.1^{dw}	--	$3.062+0$	$3.564+0$	$3.612+0$	$3.54+0^{dk}$
	$^3P^o_1 - {}^1S_0$	2669.9^{dw}	--				
	$^3P^o_0 - {}^1S_0$	2674.3^{dw}	--				
	$^1P^o_1 - {}^1S_0$	1670.8^{dw}	$1.46+9^{dd}$	$2.045+0$	$3.251+0$	$4.096+0$	$4.717+0^{ea}$
	$^3P^o_1 - {}^3P^o_0$	$1.6426+6^{dw}$	$4.10-6^{ee}$	--	--	--	--
	$^3P^o_2 - {}^3P^o_0$	$5.4124+5^{dw}$	- - -	--	--	--	--

TABLE 1. (*Continued.*)

Ion	Transition	$\lambda(\text{Å})$	$A(\sec^{-1})$	$\Upsilon(0.5)$	$\Upsilon(1.0)$	$\Upsilon(1.5)$	$\Upsilon(2.0)$
Si I	$^3P_2^o - {}^3P_1^o$	$8.072 + 5^{dw}$	$2.45 - 5^{ee}$	--	--	--	--
	$^1D_2 - {}^3P_0$	15875.8^{dm}	$4.70 - 7^{dm}$				
	$^1D_2 - {}^3P_1$	16068.3^{ee}	$9.75 - 4^{ee}$				
	$^1D_2 - {}^3P_2$	16454.5^{ee}	$2.71 - 3^{ee}$				
	$^1S_0 - {}^3P_1$	6526.8^{ee}	$3.55 - 2^{ee}$				
	$^1S_0 - {}^3P_2$	6591.4^{dm}	$9.02 - 4^{dm}$				
	$^1S_0 - {}^1D_2$	10991.4^{ee}	$7.96 - 1^{ee}$				
	$^3P_1 - {}^3P_0$	$1.297 + 6^{ee}$	$8.25 - 6^{ee}$				
	$^3P_2 - {}^3P_0$	$4.48 + 5^{dm}$	--				
	$^3P_2 - {}^3P_1$	$6.847 + 5^{ee}$	$4.20 - 5^{ee}$				
	$^5S_2^o - {}^3P_1$	3007.6^{dm}	--				
	$^5S_2^o - {}^3P_2$	3020.9^{dm}	--				
Si II	$^2P_{3/2}^o - {}^2P_{1/2}^o$	$3.48 + 5^{bb}$	$2.17 - 4^{bb}$	$5.59 + 0$	$5.70 + 0$	$5.78 + 0$	$5.77 + 0^{aa}$
	$^4P_{1/2} - {}^2P_{1/2}^o$	2335^{ac}	$4.55 + 3^{dp}$	$5.50 - 1$	$5.16 - 1$	$4.88 - 1$	$4.67 - 1^{aa}$
	$^4P_{1/2} - {}^2P_{3/2}^o$	2350^{cc}	$4.41 + 3^{dn}$	$4.33 - 1$	$4.02 - 1$	$3.81 - 1$	$3.65 - 1^{aa}$
	$^4P_{3/2} - {}^2P_{1/2}^o$	2329^{ac}	$1.32 + 1^{dp}$	$8.32 - 1$	$7.80 - 1$	$7.37 - 1$	$7.06 - 1^{aa}$
	$^4P_{3/2} - {}^2P_{3/2}^o$	2344^{cc}	$1.22 + 3^{dn}$	$1.13 + 0$	$1.05 + 0$	$9.97 - 1$	$9.56 - 1^{aa}$
	$^4P_{3/2} - {}^4P_{1/2}$	$9.23 + 5^{ac}$	--	$4.92 + 0$	$4.51 + 0$	$4.18 + 0$	$3.94 + 0^{aa}$
	$^4P_{5/2} - {}^2P_{1/2}^o$	2319.8^{ac}	--	$5.71 - 1$	$5.34 - 1$	$5.08 - 1$	$4.88 - 1^{aa}$
	$^4P_{5/2} - {}^2P_{3/2}^o$	2335^{cc}	$2.46 + 3^{dn}$	$2.33 + 0$	$2.19 + 0$	$2.08 + 0$	$1.99 + 0^{aa}$
	$^4P_{5/2} - {}^4P_{1/2}$	$3.53 + 5^{ac}$	--	$1.68 + 0$	$1.67 + 0$	$1.63 + 0$	$1.57 + 0^{aa}$
	$^4P_{5/2} - {}^4P_{3/2}$	$5.70 + 5^{ac}$	--	$7.36 + 0$	$6.94 + 0$	$6.58 + 0$	$6.32 + 0^{aa}$
Si III	$^3P_2^o - {}^1S_0$	1882.7^{ac}	$1.20 - 2^{ac}$	$6.96 + 0$	$5.46 + 0$	$4.82 + 0$	$4.41 + 0^{aa}$
	$^3P_1^o - {}^1S_0$	1892.0^{cc}	$1.67 + 4^{dq}$				
	$^3P_0^o - {}^1S_0$	1896.6^{ac}	--				
	$^1P_1^o - {}^1S_0$	1206.5^{dd}	$2.59 + 9^{dd}$	$5.30 + 0$	$5.60 + 0$	$5.93 + 0$	$6.22 + 0^{aa}$
	$^3P_1^o - {}^3P_0^o$	$7.78 + 5^{ee}$	$3.86 - 5^{ee}$	$1.78 + 0$	$1.81 + 0$	$1.83 + 0$	$1.83 + 0^{aa}$
	$^3P_2^o - {}^3P_0^o$	$2.56 + 5^{ac}$	$3.20 - 9^{ac}$	$3.66 + 0$	$3.62 + 0$	$3.53 + 0$	$3.43 + 0^{aa}$
	$^3P_2^o - {}^3P_1^o$	$3.82 + 5^{ee}$	$2.42 - 4^{ee}$	$1.04 + 1$	$1.04 + 1$	$1.02 + 1$	$1.00 + 1^{aa}$
Si IV	$^2P_{3/2}^o - {}^2S_{1/2}$	1393.8^{dd}	$7.73 + 8^{dd}$	$1.69 + 1$	$1.60 + 1$	$1.61 + 1$	$1.62 + 1^{dt}$
	$^2P_{1/2}^o - {}^2S_{1/2}$	1402.8^{dd}	$7.58 + 8^{dd}$				
Si VI	$^2P_{1/2}^o - {}^2P_{3/2}^o$	$1.964 + 4^{ee}$	$2.37 + 0^{ee}$		$2.42 - 1^{br}$		
S I	$^1D_2 - {}^3P_0$	11540.7^{ac}	$3.84 - 6^{ac}$				
	$^1D_2 - {}^3P_1$	11305.9^{ee}	$8.3 - 3^{ee}$				
	$^1D_2 - {}^3P_2$	10821.2^{ee}	$2.75 - 2^{ee}$				
	$^1S_0 - {}^3P_1$	4589.3^{ee}	$3.50 - 1^{ee}$				
	$^1S_0 - {}^3P_2$	4508.6^{ac}	$8.23 - 3^{ac}$				
	$^1S_0 - {}^1D_2$	7725.1^{ee}	$1.53 + 0^{ee}$				
	$^3P_0 - {}^3P_1$	$5.631 + 5^{ee}$	$3.02 - 4^{ee}$				
	$^3P_0 - {}^3P_2$	$1.74 + 5^{ac}$	$6.71 - 8^{ac}$				
	$^3P_1 - {}^3P_2$	$2.525 + 5^{ee}$	$1.40 - 3^{ee}$				
S II	$^2D_{5/2}^o - {}^4S_{3/2}^o$	6716.5^{ee}	$2.65 - 4^{ee}$	$4.90 + 0$	$4.66 + 0$	$4.44 + 0$	$4.26 + 0^{am}$
	$^2D_{3/2}^o - {}^4S_{3/2}^o$	6730.8^{ee}	$5.37 - 4^{ee}$	$3.27 + 0$	$3.11 + 0$	$2.97 + 0$	$2.84 + 0^{am}$
	$^2P_{3/2}^o - {}^4S_{3/2}^o$	4068.6^{ee}	$2.20 - 1^{ee}$	$1.11 + 0$	$1.38 + 0$	$1.52 + 0$	$1.59 + 0^{am}$
	$^2P_{1/2}^o - {}^4S_{3/2}^o$	4076.4^{ee}	$7.44 - 2^{ee}$	$5.54 - 1$	$6.90 - 1$	$7.59 - 1$	$7.94 - 1^{am}$
	$^2D_{5/2}^o - {}^2D_{3/2}^o$	$3.145 + 6^{ee}$	$3.46 - 7^{ee}$	$7.90 + 0$	$7.46 + 0$	$7.11 + 0$	$8.65 + 0^{am}$
	$^2P_{3/2}^o - {}^2P_{1/2}^o$	$2.14 + 6^{ee}$	$9.13 - 7^{ee}$	$1.55 + 0$	$1.95 + 0$	$2.13 + 0$	$2.22 + 0^{am}$
	$^2P_{3/2}^o - {}^2D_{5/2}^o$	10320.4^{ee}	$2.22 - 1^{ee}$	$4.56 + 0$	$4.77 + 0$	$4.75 + 0$	$4.68 + 0^{am}$
	$^2P_{3/2}^o - {}^2D_{3/2}^o$	10286.7^{ee}	$1.32 - 1^{ee}$	$2.62 + 0$	$2.74 + 0$	$2.74 + 0$	$2.71 + 0^{am}$
	$^2P_{1/2}^o - {}^2D_{5/2}^o$	10373.3^{dr}	$7.79 - 2^{dr}$	$1.90 + 0$	$1.99 + 0$	$1.99 + 0$	$1.97 + 0^{am}$
	$^2P_{1/2}^o - {}^2D_{3/2}^o$	10336.3^{ee}	$1.95 - 1^{ee}$	$1.69 + 0$	$1.76 + 0$	$1.76 + 0$	$1.73 + 0^{am}$

TABLE 1. (*Continued.*)

Ion	Transition	$\lambda(\text{Å})$	$A(\sec^{-1})$	$\Upsilon(0.5)$	$\Upsilon(1.0)$	$\Upsilon(1.5)$	$\Upsilon(2.0)$
S III	$^1D_2 - ^3P_0$	8833.9^{dm}	$5.82 - 6^{dm}$	$9.07 + 0$	$8.39 + 0$	$8.29 + 0$	$8.20 + 0^{ac}$
	$^1D_2 - ^3P_1$	9068.9^{ee}	$1.62 - 2^{ee}$				
	$^1D_2 - ^3P_2$	9531.0^{ee}	$9.40 - 2^{ee}$				
	$^1S_0 - ^3P_1$	3721.7^{ee}	$6.81 - 1^{ee}$	$1.16 + 0$	$1.19 + 0$	$1.21 + 0$	$1.24 + 0^{ac}$
	$^1S_0 - ^3P_2$	3797.8^{dm}	$1.05 - 2^{dm}$				
	$^1S_0 - ^1D_2$	6312.1^{ee}	$3.22 + 0^{ee}$	$1.42 + 0$	$1.88 + 0$	$2.02 + 0$	$2.08 + 0^{ac}$
	$^3P_1 - ^3P_0$	$3.347 + 5^{ee}$	$4.78 - 4^{ee}$	$2.64 + 0$	$2.59 + 0$	$2.38 + 0$	$2.20 + 0^{al}$
	$^3P_2 - ^3P_0$	$1.20 + 5^{dm}$	$4.61 - 8^{dm}$	$1.11 + 0$	$1.15 + 0$	$1.15 + 0$	$1.14 + 0^{al}$
	$^3P_2 - ^3P_1$	187129^{ee}	$2.06 - 3^{ee}$	$5.79 + 0$	$5.81 + 0$	$5.56 + 0$	$5.32 + 0^{al}$
	$^5S_2^o - ^3P_1$	1683.5^{dm}	$6.22 + 3^{eb}$	$--$	$3.8 + 0$	$3.7 + 0$	$3.6 + 0^{eb}$
	$^5S_2^o - ^3P_2$	$1698.86dm$	$1.70 + 4^{eb}$				
S IV	$^2P_{3/2}^o - ^2P_{1/2}^o$	$1.05 + 5^{ee}$	$7.73 - 3^{ee}$	$--$	$6.42 + 0$	$6.41 + 0$	$6.40 + 0^{ds}$
	$^4P_{1/2} - ^2P_{1/2}^o$	1404.9^{ds}	$5.50 + 4^{ds}$	$--$	$5.5 - 1$	$4.8 - 1$	$4.6 - 1^{ds}$
	$^4P_{1/2} - ^2P_{3/2}^o$	1423.9^{ds}	$3.39 + 4^{ds}$	$--$	$6.6 - 1$	$6.3 - 1$	$6.1 - 1^{ds}$
	$^4P_{3/2} - ^2P_{1/2}^o$	1398.1^{ds}	$1.40 + 2^{ds}$	$--$	$8.7 - 1$	$8.3 - 1$	$8.0 - 1^{ds}$
	$^4P_{3/2} - ^2P_{3/2}^o$	1017.0^{ds}	$1.95 + 4^{ds}$	$--$	$1.47 + 0$	$1.40 + 0$	$1.34 + 0^{ds}$
	$^4P_{3/2} - ^4P_{1/2}$	$2.91 + 5^{ds}$	$--$	$--$	$3.04 + 0$	$2.85 + 0$	$2.72 + 0^{ds}$
	$^4P_{5/2} - ^2P_{1/2}^o$	1387.5^{ds}	$--$	$--$	$9.5 - 1$	$9.1 - 1$	$8.8 - 1^{ds}$
	$^4P_{5/2} - ^2P_{3/2}^o$	1406.1^{ds}	$3.95 + 4$	$--$	$2.53 + 0$	$2.41 + 0$	$2.33 + 0^{ds}$
	$^4P_{5/2} - ^4P_{1/2}$	$1.12 + 5^{ds}$	$--$	$--$	$2.92 + 0$	$2.71 + 0$	$2.56 + 0^{ds}$
	$^4P_{5/2} - ^4P_{3/2}$	$1.85 + 5^{ds}$	$--$	$--$	$7.01 + 0$	$6.57 + 0$	$6.20 + 0^{ds}$
S V	$^3P_2^o - ^1S_0$	1188.3^{ac}	$6.59 - 2^{ac}$	$9.11 - 1$	$9.10 - 1$	$9.14 - 1$	$9.05 - 1^{ai}$
	$^3P_1^o - ^1S_0$	1199.1^{ac}	$1.26 + 5^{ac}$				
	$^3P_0^o - ^1S_0$	1204.5^{ac}	$--$				
	$^1P_1^o - ^1S_0$	786.48^{dd}	$5.25 + 9^{dd}$	$7.30 + 0$	$7.30 + 0$	$7.29 + 0$	$7.27 + 0^{ai}$
	$^3P_1^o - ^3P_0^o$	$2.71 + 5^{ee}$	$9.16 - 4^{ee}$	$2.72 - 1^{ag}$			
	$^3P_2^o - ^3P_0^o$	88401.7^{ac}	$--$	$4.00 - 1^{ag}$			
	$^3P_2^o - ^3P_1^o$	$1.312 + 5^{ee}$	$5.49 - 3^{ee}$	$1.24 + 0^{ag}$			
S VI	$^2P_{3/2}^o - ^2S_{1/2}$	933.38^{dd}	$1.7 + 9^{dd}$	$1.18 + 1$	$1.19 + 1$	$1.19 + 1$	$1.19 + 1^{ac}$
	$^2P_{1/2}^o - ^2S_{1/2}$	944.52^{dd}	$1.6 + 9^{dd}$				
	$^2P_{3/2}^o - ^2P_{3/2}^o$	$1.0846 + 5^{dz}$	$7.75 - 3^{dz}$	$5.845 + 0$	$6.67 + 0$	$7.095 + 0$	$7.27 + 0^{dz}$
Cl I	$^2P_{1/2}^o - ^2P_{3/2}^o$	$1.133 + 5^{ee}$	$1.24 - 2^{ee}$				
Cl II	$^1D_2 - ^3P_0$	9383.4^{ac}	$9.82 - 6^{ac}$			$3.86 + 0^{cg}$	
	$^1D_2 - ^3P_1$	9123.6^{ee}	$2.98 - 2^{ee}$				
	$^1D_2 - ^3P_2$	8578.7^{ee}	$1.07 - 1^{ee}$				
	$^1S_0 - ^3P_1$	3677.9^{ee}	$1.37 + 0^{ee}$			$4.56 - 1^{cg}$	
	$^1S_0 - ^3P_2$	3587.1	$1.79 - 2^{ac}$				
	$^1S_0 - ^1D_2$	6161.8^{ee}	$2.06 + 0^{ee}$			$1.15 + 0^{cg}$	
	$^3P_0 - ^3P_1$	$3.328 + 5^{ee}$	$1.50 - 3^{ee}$			$9.33 - 1^{cg}$	
	$^3P_0 - ^3P_2$	$1.004 + 5^{ac}$	$4.57 - 7^{ac}$			$4.43 - 1^{cg}$	
	$^3P_1 - ^3P_2$	$1.437 + 5^{ee}$	$7.50 - 3^{ee}$			$2.17 + 0^{cg}$	
Cl III	$^2D_{5/2}^o - ^4S_{3/2}^o$	5517.7^{ee}	$8.07 - 4^{ee}$	$1.941 + 0$	$2.046 + 0$	$2.044 + 0$	$2.035 + 0^{dv}$
	$^2D_{3/2}^o - ^4S_{3/2}^o$	5537.9^{ee}	$3.44 - 3^{ee}$	$1.292 + 0$	$1.359 + 0$	$1.358 + 0$	$1.352 + 0^{dv}$
	$^2P_{3/2}^o - ^4S_{3/2}^o$	3342.9^{ee}	$6.91 - 1^{ee}$	$7.69 - 1$	$8.37 - 1$	$8.88 - 1$	$9.20 - 1^{dv}$
	$^2P_{1/2}^o - ^4S_{3/2}^o$	3353.3^{ee}	$1.22 - 1^{ee}$	$3.85 - 1$	$4.18 - 1$	$4.44 - 1$	$4.61 - 1^{dv}$
	$^2D_{5/2}^o - ^2D_{3/2}^o$	$1.516 + 6^{ee}$	$3.08 - 6^{ee}$	$4.451 + 0$	$4.519 + 0$	$4.506 + 0$	$4.483 + 0^{dv}$
	$^2P_{3/2}^o - ^2P_{1/2}^o$	$1.081 + 6^{ee}$	$7.08 - 6^{ee}$	$1.729 + 0$	$1.755 + 0$	$1.806 + 0$	$1.858 + 0^{dv}$
	$^2P_{3/2}^o - ^2D_{5/2}^o$	8480.9^{ee}	$3.87 - 1^{ee}$	$3.747 + 0$	$4.203 + 0$	$4.331 + 0$	$4.324 + 0^{dv}$
	$^2P_{3/2}^o - ^2D_{3/2}^o$	8433.7^{ee}	$3.39 - 1^{ee}$	$2.009 + 0$	$2.192 + 0$	$2.338 + 0$	$2.252 + 0^{dv}$
	$^2P_{1/2}^o - ^2D_{5/2}^o$	8552.1^{ac}	$1.0 - 1^{ac}$	$1.444 + 0$	$1.563 + 0$	$1.599 + 0$	$1.604 + 0^{dv}$
	$^2P_{1/2}^o - ^2D_{3/2}^o$	8500.0^{ee}	$3.60 - 1^{ee}$	$1.452 + 0$	$1.653 + 0$	$1.712 + 0$	$1.716 + 0^{dv}$

TABLE 1. (*Continued.*)

Ion	Transition	$\lambda(\text{Å})$	$A(\text{sec}^{-1})$	$\Upsilon(0.5)$	$\Upsilon(1.0)$	$\Upsilon(1.5)$	$\Upsilon(2.0)$
Cl IV	$^1D_2 - {}^3P_0$	7263.4^{dm}	$1.54 - 5^{dm}$	$5.10 + 0$	$5.42 + 0$	$5.88 + 0$	$6.19 + 0^{cg}$
	$^1D_2 - {}^3P_1$	7529.9^{ee}	$5.57 - 2^{ee}$				
	$^1D_2 - {}^3P_2$	8045.6^{ee}	$2.08 - 1^{ee}$				
	$^1S_0 - {}^3P_1$	3118.6^{ee}	$2.19 + 0^{ee}$	$2.04 + 0$	$2.27 + 0$	$2.32 + 0$	$2.30 + 0^{cg}$
	$^1S_0 - {}^3P_2$	3204.5^{dm}	$2.62 - 2^{dm}$				
	$^1S_0 - {}^1D_2$	5323.3^{ee}	$4.14 + 0^{ee}$	$9.35 - 1$	$1.39 + 0$	$1.73 + 0$	$1.92 + 0^{cg}$
	$^3P_1 - {}^3P_0$	$2.035 + 5^{ee}$	$2.13 - 3^{ee}$	$4.75 - 1^{cg}$			
	$^3P_2 - {}^3P_0$	74521^{dm}	$2.70 - 7^{dm}$	$4.00 - 1^{cg}$			
	$^3P_2 - {}^3P_1$	$1.1741 + 5^{ee}$	$8.32 - 3^{ee}$	$1.50 + 0^{cg}$			
Cl V	$^2P^o_{3/2} - {}^2P^o_{1/2}$	67049^{ee}	$2.98 - 2^{ee}$			$1.05 + 0^{cg}$	
Ar I	$^2P^o_{1/2} - {}^2P^o_{3/2}$	69851.9^{ac}	$5.27 - 2^{ac}$			$6.35 - 1^{cg}$	
Ar III	$^1D_2 - {}^3P_0$	8038.7	$2.21 - 5^{ac}$			$4.74 + 0^{cg}$	
	$^1D_2 - {}^3P_1$	7751.1^{ee}	$8.44 - 2^{ee}$				
	$^1D_2 - {}^3P_2$	7135.8^{ee}	$3.24 - 1^{ee}$				
	$^1S_0 - {}^3P_1$	3109.1^{ee}	$4.09 + 0^{ee}$			$6.80 - 1^{cg}$	
	$^1S_0 - {}^3P_2$	3006.1	$4.17 - 2^{ac}$				
	$^1S_0 - {}^1D_2$	5191.8^{ee}	$2.59 + 0^{ee}$			$8.23 - 1^{cg}$	
	$^3P_0 - {}^3P_1$	$2.184 + 5^{ee}$	$5.31 - 3^{ee}$			$1.18 + 0^{cg}$	
	$^3P_0 - {}^3P_2$	63686.2	$2.37 - 6^{ac}$			$5.31 - 1^{cg}$	
	$^3P_1 - {}^3P_2$	89910^{ee}	$3.06 - 2^{ee}$			$2.24 + 0^{cg}$	
Ar IV	$^2D^o_{5/2} - {}^4S^o_{3/2}$	4711.3^{ee}	$2.07 - 3^{ee}$	$2.560 + 0$	$6.128 + 0$	$1.636 + 0$	$1.455 + 1^{du}$
	$^2D^o_{3/2} - {}^4S^o_{3/2}$	4740.2^{ee}	$1.72 - 1^{ee}$	$1.706 + 0$	$1.297 + 0$	$1.136 + 0$	$9.70 - 1^{du}$
	$^2P^o_{3/2} - {}^4S^o_{3/2}$	2853.7^{ee}	$1.88 + 0^{ee}$	$3.01 - 1$	$2.93 - 1$	$3.06 - 1$	$3.25 - 1^{du}$
	$^2P^o_{1/2} - {}^4S^o_{3/2}$	2868.2^{ee}	$7.60 - 1^{ee}$	$1.49 - 1$	$1.46 - 1$	$1.53 - 1$	$1.63 - 1^{du}$
	$^2D^o_{5/2} - {}^2D^o_{3/2}$	$7.741 + 5^{ee}$	$2.30 - 5^{ee}$	$6.353 + 0$	$6.128 + 0$	$6.025 + 0$	$5.926 + 0^{du}$
	$^2P^o_{3/2} - {}^2P^o_{1/2}$	564721^{ee}	$4.94 - 5^{ee}$	$2.242 + 0$	$2.33 + 0$	$2.53 + 0$	$2.72 + 0^{du}$
	$^2P^o_{3/2} - {}^2D^o_{5/2}$	7237.3^{ee}	$7.08 - 1^{ee}$	$4.29 + 0$	$4.437 + 0$	$4.402 + 0$	$4.338 + 0^{du}$
	$^2P^o_{3/2} - {}^2D^o_{3/2}$	7170.6^{ee}	$8.40 - 1^{ee}$	$2.446 + 0$	$2.472 + 0$	$2.435 + 0$	$2.387 + 0^{du}$
	$^2P^o_{1/2} - {}^2D^o_{5/2}$	7333.4^{dr}	$1.19 - 1^{dr}$	$1.779 + 0$	$1.788 + 0$	$1.757 + 0$	$1.718 + 0^{du}$
	$^2P^o_{1/2} - {}^2D^o_{3/2}$	7262.8^{ee}	$6.96 - 1^{ee}$	$1.614 + 0$	$1.688 + 0$	$1.682 + 0$	$1.661 + 0^{du}$
Ar V	$^1D_2 - {}^3P_0$	6135.2^{dm}	$3.50 - 5^{dm}$	$4.37 + 0$	$3.72 + 0$	$3.52 + 0$	$3.42 + 0^{ac}$
	$^1D_2 - {}^3P_1$	6435.1^{ee}	$1.61 - 1^{ee}$				
	$^1D_2 - {}^3P_2$	7005.7^{ee}	$4.70 - 1^{ee}$	$4.37 + 0$	$3.72 + 0$	$3.52 + 0$	$3.42 + 0^{ac}$
	$^1S_0 - {}^3P_1$	2691.0^{ee}	$5.89 + 0^{ee}$	$1.17 + 0$	$1.18 + 0$	$1.11 + 0$	$1.03 + 0^{ac}$
	$^1S_0 - {}^3P_2$	2686.8^{dm}	$5.69 - 2^{dm}$				
	$^1S_0 - {}^1D_2$	4625.5^{ee}	$5.18 + 0^{ee}$	$1.26 + 0$	$1.25 + 0$	$1.24 + 0$	$1.23 + 0^{ac}$
	$^3P_1 - {}^3P_0$	$1.307 + 5^{ee}$	$8.03 - 3^{ee}$	$2.57 - 1^{ac}$	$--$	$--$	$--$
	$^3P_2 - {}^3P_0$	49280.5^{dm}	$1.24 - 6^{dm}$	$3.20 - 1^{ac}$	$--$	$--$	$--$
	$^3P_2 - {}^3P_1$	79040^{ee}	$2.72 - 2^{ee}$	$1.04 + 0^{ac}$	$--$	$--$	$--$
Ar VI	$^2P^o_{3/2} - {}^2P^o_{1/2}$	45275^{ee}	$9.69 - 2^{ee}$			$7.98 - 2^{cg}$	
K III	$^2P^o_{1/2} - {}^2P^o_{3/2}$	46153.2^{ee}	$1.83 - 1^{ee}$			$1.78 + 0^{cg}$	
K IV	$^1D_2 - {}^3P_0$	7110.9^{ac}	$4.54 - 5^{ac}$			$1.90 + 0^{cg}$	
	$^1D_2 - {}^3P_1$	6795.0^{ee}	$2.03 - 1^{ee}$				
	$^1D_2 - {}^3P_2$	6101.8^{ee}	$8.38 - 1^{ee}$				
	$^1S_0 - {}^3P_1$	2711.1^{ee}	$1.05 + 1^{ee}$			$2.92 - 1^{cg}$	
	$^1S_0 - {}^3P_2$	2594.3^{ac}	$8.17 - 2^{ac}$				
	$^1S_0 - {}^1D_2$	4510.9^{ee}	$3.18 + 0^{ee}$			$7.98 - 1^{cg}$	
	$^3P_0 - {}^3P_1$	$1.539 + 5^{ee}$	$1.51 - 2^{ee}$			$4.21 - 1^{cg}$	
	$^3P_0 - {}^3P_2$	43081.2^{ac}	$1.01 - 5^{ac}$			$2.90 - 1^{cg}$	
	$^3P_1 - {}^3P_2$	59830.0^{ee}	$1.04 - 1^{ee}$			$1.16 + 0^{cg}$	

TABLE 1. (*Continued.*)

Ion	Transition	$\lambda(\text{Å})$	$A(\text{sec}^{-1})$	$\Upsilon(0.5)$	$\Upsilon(1.0)$	$\Upsilon(1.5)$	$\Upsilon(2.0)$
K v	$^2D^o_{5/2} - {}^4S^o_{3/2}$	4122.6^{ee}	$4.96 - 3^{ee}$	$9.25 - 1$	$8.51 - 1$	$8.24 - 1$	$8.18 - 1^{ci}$
	$^2D^o_{3/2} - {}^4S^o_{3/2}$	4163.3^{ee}	$8.06 - 2^{ee}$	$6.17 - 1$	$5.67 - 1$	$5.50 - 1$	$5.45 - 1^{ci}$
	$^2P^o_{3/2} - {}^4S^o_{3/2}$	2494.2^{ee}	$4.56 + 0^{ee}$	$1.49 - 1$	$3.68 - 1$	$4.94 - 1$	$5.47 - 1^{ci}$
	$^2P^o_{1/2} - {}^4S^o_{3/2}$	2514.5^{ee}	$1.90 + 0^{ee}$	$7.4 - 2$	$1.84 - 1$	$2.47 - 1$	$2.73 - 1^{ci}$
	$^2D^o_{5/2} - {}^2D^o_{3/2}$	$4.22 + 5^{ee}$	$1.41 - 4^{ee}$	$5.24 + 0$	$5.31 + 0$	$5.13 + 0$	$4.96 + 0^{ci}$
	$^2P^o_{3/2} - {}^2P^o_{1/2}$	$3.11 + 5^{ee}$	$2.94 - 4^{ee}$	$4.43 - 1$	$6.27 - 1$	$7.83 - 1$	$9.02 - 1^{ci}$
	$^2P^o_{3/2} - {}^2D^o_{5/2}$	6315.1^{ee}	$1.34 + 0^{ee}$	$2.56 + 0$	$3.07 + 0$	$3.31 + 0$	$3.40 + 0^{ci}$
	$^2P^o_{3/2} - {}^2D^o_{3/2}$	6221.9^{ee}	$1.97 + 0^{ee}$	$1.39 + 0$	$1.76 + 0$	$1.93 + 0$	$2.00 + 0^{ci}$
	$^2P^o_{1/2} - {}^2D^o_{5/2}$	6448.1^{ac}	$1.41 - 1^{ac}$				
	$^2P^o_{1/2} - {}^2D^o_{3/2}$	6349.2^{ee}	$1.37 + 0^{ee}$				
Ca II	$^2S_{1/2} - {}^2P^o_{3/2}$	3933.7^{dd}	$1.47 + 8^{dd}$	$1.56 + 1$	$1.75 + 1$	$1.92 + 1$	$2.08 + 1^{ac}$
	$^2S_{1/2} - {}^2P^o_{1/2}$	3968.5^{dd}	$1.4 + 8^{dd}$				
Ca IV	$^2P^o_{1/2} - {}^2P^o_{3/2}$	32061.9^{ee}	$5.46 - 1^{ee}$		$1.06 + 0^{cg}$		
Ca V	$^1D_2 - {}^3P_0$	6428.9	$8.42 - 5^{ac}$		$9.04 - 1^{cg}$		
	$^1D_2 - {}^3P_1$	6086.4^{ee}	$4.35 - 1^{ee}$				
	$^1D_2 - {}^3P_2$	5309.2^{ee}	$1.95 + 0^{ee}$				
	$^1S_0 - {}^3P_1$	2412.9^{ee}	$2.40 + 1^{ee}$		$1.16 - 1^{cg}$		
	$^1S_0 - {}^3P_2$	2281.2	$1.45 - 1^{ac}$				
	$^1S_0 - {}^1D_2$	3997.9^{ee}	$3.73 + 0^{ee}$		$7.93 - 1^{cg}$		
	$^3P_0 - {}^3P_1$	$1.1482 + 5^{ee}$	$3.62 - 2^{ee}$		$2.02 - 1^{cg}$		
	$^3P_0 - {}^3P_2$	30528.8	$3.67 - 5^{ac}$		$2.24 - 1^{cg}$		
	$^3P_1 - {}^3P_2$	41574.2^{ee}	$3.09 - 1^{ee}$		$7.60 - 1^{cg}$		
Fe II	$a^6D_{9/2} - a^6D_{7/2}$	$2.598 + 5^{ed}$	$2.13 - 3^{eh}$	$7.413 - 3$	$5.521 + 0$	$5.461 + 0$	$5.483 + 0^{ec}$
	$a^6D_{9/2} - a^6D_{5/2}$	$1.479 + 5^{ed}$		$6.142 + 0$	$1.491 + 0$	$1.550 + 0$	$1.642 + 0^{ec}$
	$a^6D_{9/2} - a^6D_{3/2}$	$1.159 + 5^{ed}$		$4.250 - 1$	$6.747 - 1$	$6.832 - 1$	$7.153 - 1^{ec}$
	$a^6D_{9/2} - a^6D_{1/2}$	$1.023 + 5^{ed}$		$1.627 - 1$	$2.839 - 1$	$2.836 - 1$	$2.947 - 1^{ec}$
	$a^6D_{9/2} - a^4F_{9/2}$	$5.339 + 4^{ed}$	$4.17 - 5^{ei}$	$4.769 - 2$	$3.595 + 0$	$3.192 + 0$	$2.894 + 0^{ec}$
	$a^6D_{9/2} - a^4D_{7/2}$	$1.257 + 4^{ed}$	$4.83 - 3^{ei}$	$3.096 + 0$	$1.098 + 1$	$9.651 + 0$	$8.772 + 0^{ec}$
	$a^6D_{5/2} - a^4D_{5/2}$	$1.294 + 4^{ed}$	$1.94 - 3^{ei}$	$3.460 + 0$	$2.847 + 0$	$2.556 + 0$	$2.361 + 0^{ec}$
	$a^6D_{3/2} - a^4D_{5/2}$	$1.328 + 4^{ed}$	$1.21 - 3^{ei}$	$7.110 - 1$	$6.264 - 1$	$6.048 - 1$	$5.866 - 1^{ec}$
	$a^6D_{1/2} - a^4D_{1/2}$	$1.270 + 4^{ed}$	$2.91 - 3^{ei}$	$2.056 + 0$	$1.649 + 0$	$1.437 + 0$	$1.299 + 0^{ec}$
	$a^4F_{9/2} - a^4D_{7/2}$	$1.644 + 4^{ed}$	$4.65 - 3^{ei}$	$2.283 + 0$	$2.106 + 0$	$2.136 + 0$	$2.179 + 0^{ec}$
	$a^4F_{9/2} - a^4D_{5/2}$	$1.533 + 4^{ed}$	$2.44 - 3^{ei}$	$7.773 - 1$	$7.248 - 1$	$7.544 - 1$	$7.829 - 1^{ec}$
	$a^4F_{9/2} - a^4P_{5/2}$	8617.0^{ed}	$2.73 - 2^{ei}$	$1.220 + 0$	$1.232 + 0$	$1.267 + 0$	$1.293 + 0^{ec}$
	$a^4F_{7/2} - a^4P_{3/2}$	8892.0^{ed}	$1.74 - 2^{ei}$	$5.532 + 0$	$5.541 - 1$	$5.638 - 1$	$5.710 + 0^{ec}$
	$a^4F_{9/2} - a^4H_{13/2}$	5157^{ef}	$4.4 - 1^{eh}$	$1.027 + 0$	$1.127 + 0$	$1.329 + 0$	$1.509 + 0^{ec}$
	$a^4F_{9/2} - a^4G_{11/2}$	4244^{ef}	$9.0 - 1^{eh}$	$6.450 - 1$	$8.237 - 1$	$1.021 + 0$	$1.158 + 0^{ec}$
	$a^4F_{7/2} - a^4H_{11/2}$	5262^{ef}	$3.1 - 1^{eh}$	$5.211 - 1$	$5.656 - 1$	$6.686 - 1$	$7.636 - 1^{ec}$
	$a^4F_{7/2} - a^4G_{9/2}$	4277^{ef}	$6.5 - 1^{eh}$	$3.667 - 1$	$4.471 - 1$	$5.354 - 1$	$5.971 - 1^{ec}$
	$a^4F_{5/2} - a^4H_{9/2}$	5334^{ef}	$2.6 - 1^{eh}$	$3.218 - 1$	$3.569 - 1$	$4.318 - 1$	$5.008 - 1^{ec}$
	$a^4F_{9/2} - b^4F_{9/2}$	4815^{ef}	$4.0 - 1^{eh}$	$1.061 + 0$	$1.207 + 0$	$1.422 + 0$	$1.604 + 0^{ec}$
Fe III	$^5D_4 - {}^5D_3$	229146^{ej}		$2.38 + 0$	$2.87 + 0$	$3.02 + 0$	$3.01 + 0^{ek}$
	$^5D_4 - {}^5D_2$	135513^{ej}		$9.70 - 1$	$1.23 + 0$	$1.31 + 0$	$1.32 + 0^{ek}$
	$^5D_4 - {}^5D_1$	107294^{ej}		$4.75 - 1$	$5.91 - 1$	$6.29 - 1$	$6.36 - 1^{ek}$
	$^5D_4 - {}^5D_0$	97436^{ej}		$1.43 - 1$	$1.78 - 1$	$1.90 - 1$	$1.94 - 1^{ek}$
	$^5D_3 - {}^5D_2$	331636^{ej}		$1.65 + 0$	$2.03 + 0$	$2.16 + 0$	$2.18 + 0^{ek}$
	$^5D_3 - {}^5D_1$	201769^{ej}		$6.12 - 1$	$7.94 - 1$	$8.45 - 1$	$8.46 - 1^{ek}$
	$^5D_3 - {}^5D_0$	169516^{ej}		$1.70 - 1$	$2.23 - 1$	$2.36 - 1$	$2.35 - 1^{ek}$
	$^5D_2 - {}^5D_1$	515254^{ej}		$1.04 + 0$	$1.28 + 0$	$1.36 + 0$	$1.36 + 0^{ek}$
	$^5D_2 - {}^5D_0$	346768^{ej}		$2.35 - 1$	$3.09 - 1$	$3.31 - 1$	$3.33 - 1^{ek}$
	$^5D_1 - {}^5D_0$	1060465^{ej}		$4.00 - 1$	$4.85 - 1$	$5.15 - 1$	$5.20 - 1^{ek}$

TABLE 1. (*Continued.*)

Ion	Transition	$\lambda(\text{\AA})$	$A(\text{sec}^{-1})$	$\Upsilon(0.5)$	$\Upsilon(1.0)$	$\Upsilon(1.5)$	$\Upsilon(2.0)$
	$^3H_6 - {}^3G_5$	22189.8^{ej}		$2.80+0$	$2.72+0$	$2.67+0$	$2.60+0^{ek}$
	$^3H_6 - {}^3G_4$	20448.4^{ej}		$1.18+0$	$1.20+0$	$1.19+0$	$1.16+0^{ek}$
	$^3H_6 - {}^3G_3$	19655.2^{ej}		$2.77-1$	$2.90-1$	$2.97-1$	$2.96-1^{ek}$
	$^3H_5 - {}^3G_5$	23505.2^{ej}		$1.26+0$	$1.28+0$	$1.26+0$	$1.23+0^{ek}$
	$^3H_5 - {}^3G_4$	21560.3^{ej}		$1.60+0$	$1.69+0$	$1.70+0$	$1.66+0^{ek}$
	$^3H_5 - {}^3G_3$	20680.3^{ej}		$1.07+0$	$1.12+0$	$1.12+0$	$1.10+0^{ek}$
	$^3H_4 - {}^3G_5$	24516.1^{ej}		$3.43-1$	$3.75-1$	$3.88-1$	$3.88-1^{ek}$
	$^3H_4 - {}^3G_4$	22407.9^{ej}		$1.18+0$	$1.23+0$	$1.23+0$	$1.21+0^{ek}$
	$^3H_4 - {}^3G_3$	21458.8^{ej}		$1.80+0$	$1.94+0$	$1.96+0$	$1.92+0^{ek}$

Note: References in Table 1

[aa]: Callaway Joseph, 1994, ADNDT, 57
[ab]: Sawey PM and Berrington KA, 1993, ADNDT, 55, 81
[ac]: Mendoza C, 1983, "planetary nebulae," edited by D. R. Flower, 143
[ad]: Aggarwal KM, Callaway J, Kingston AE and Unnikrishnan K, 1992, ApJS, 80, 473
[ae]: Johnson CT and Kingston AE, 1987, J. Phys. B, 20, 5757
[af]: Lang KR, 1980, "Astrophysical Formulae," p. 134
[ag]: Drake Gordon W. F. "Atomic Physics," Vol. 3, 269
[ah]: Kato T, 1989, ADNDT, 42, 313
[ai]: Giles K, 1980, Ph.D Thesis, U. of London.
[aj]: Zhang HL, Graziani M and Pradhan AK, 1994, Astron. Astrophys. 283, 319
[ak]: Aggarwal KM, 1983, ApJS, 52, 387
[al]: Johnson CT, Burke PG and Kingston AE, 1987, J. Phys. B, 20, 2553
[am]: Wei C and Pradhan AK, 1993, ApJS 88, 329
[an]: Mendoza C, 1987, MNRAS, 224, 7
[ao]: Giles K, 1981, MNRAS, 195, 63
[ap]: Butler K and Mendoza C, 1984, MNRAS, 208, 17
[ba]: Taylor PO, Gregory D, Dunn GH, Phaneuf RA and Crandall DH, 1977, Phys. Rev. Lett., 39, 1256
—— Gau JN and Henry RJW, 1977, Phys. Rev. A, 16, 986
[bb]: Wiese WL, Smith MW and Glennon BM, 1966, "Atomic Transition Probabilities," Vol. 1.
[bc]: Hata J and Grant IP, 1981, J. Phys. B, 14, 2111
[bd]: Drake GWF, 1979, Phys. Rev. A, 19, 1378
[be]: Pradhan AK, 1976, MNRAS, 177, 31
[bf]: Lin CD, Johnson WR and Dalgarno A, 1977, Phys. Rev. A, 15, 154
[bg]: Le Dourneuf M and Nesbet RK, 1976, J. Phys. B, 9, L241
[bh]: Nussbaumer H and Rusca C, 1979, Astron. Astrophys., 72, 129
[bi]: Pequignot D and Aldrovandi SMV, 1976, Asron. Astrophys., 50, 141
[bj]: Thomas LD and Nesbet RK, 1975, Phys. Rev. A, 12, 2387
[bk]: Nussbaumer H and Storey PJ, 1981, Astron. Astrophys., 96, 91
[bl]: Nussbaumer H and Storey PJ, 1978, Astron. Astrophys., 64, 139
[bm]: Dufton PL, Berrington KA, Burke PG and Kingston AE, 1978, Astron. Astrophys., 62, 111
[bn]: Zeippen CJ, 1982, MNRAS, 198, 111
[bo]: Berrington KA and Burke PG, 1981, Planet. Space Sci., 29, 263
[bp]: Dopita MA, Mason DJ and Robb WD, 1976, Astrophys. J., 207, 102
[bq]: Seaton MJ, 1975, MNRAS, 170, 475
[br]: Saraph HE, Seaton MJ and Shemming J, 1969, Phil. Trans. R. Soc. A, 264, 77
[bs]: Jackson ARG, 1973, MNRAS, 165, 53
[bt]: Nussbaumer H and Storey PJ, 1979, Astron. Astrophys., 71, L5
[bv]: Keenan FP, Berrington KA, Burke PG, Dufton PL and Kingston AE, 1986, Phys. Scr., 34, 216
[bu]: Nussbaumer H and Storey PJ, 1979, Astron. Astrophys., 74, 244
[bw]: Osterbrock DE and Wallace RK, 1977, APJL, 19, L11
—— Van Wyngaarden WL and Henry RJW, 1976, J. Phys. B, 9, 1461
[ca]: Fang, Kwong, Wang and Parkinson, 1993, Phys. Rev. A, 48, 1114

TABLE 1. (*Continued.*)

[cb]: Froese-Fischer C., 1994, Phys. Scr., 49, 323
[cc]: Parkinson WH and Smith PL, 1994, private communication
[cd]: Kwong, Fang, Gibbons, Parkinson and Smith, 1993, ApJ, 411, 431
[ce]: Hilbert A and Bates DR, 1981, Planet. Space Sci., 29, 263
[cf]: Fang, Kwong and Parkinson 1993, ApJL, 413, L141
[cg]: Krueger TK and Czyzak SJ, 1970, Proc. R. Soc. Lond. A, 318, 531
[ch]: Mendoza C, 1981, J. Phys. B, 14, 246
[ci]: Butler K, Zeippen CJ and Bourlot JL, 1988, A&A, 203, 189
[db]: Robert DB and Pradhan AK, 1992, ApJS, 80, 425
[dc]: Lennon DJ and Burke VM, 1994, A&AS, 103, 273
[dd]: Fuhr JR and Wiese WL, 1990, CRC Handbook of Chemistry and Phys. 71st Edition
[de]: Moore CE, 1985, Selected Tables of Atomic Spectra
[df]: Nussbaumer H and Storey PJ, 1981, A&A, 99, 177
[dh]: Hayes MA, 1982, MNRAS, 195, 63P.
[di]: Merkelis G, Vilkas MJ, Gaigalas G and Kisielius R, 1994, private communication
[dj]: Dufton PL, Doyle JG and Kingston AE, 1979, A&A, 78, 318; +Daresbury Laboratory Data Bank
[dk]: Tayal SS, Burke and Kingston AE, 1984, J. Phys. B, 17, 3864
[dl]: Clark REH, Magee NH, Mann JB and Merts AL, 1982, ApJ, 254, 412
—— Fabrikant II, 1974, J. Phys. B, 7, 91
[dm]: Mendoza C and Zeippen CJ, 1982, MNRAS, 199, 1025
[dn]: Calamai, Smith and Bergeson 1993, ApJL, 415, L59
[do]: Dufton PL, Keenan FP, Hibbert A, Stafford RP Byrne PB and Agnew D, 1991, MNRAS, 253, 474
[dp]: Nussbaumer H, 1977, A&A, 58, 291
[dq]: Ojha PC, Keenan FP and Hibbert A, 1988, J. Phys. B., 21, L395
[dr]: Mendoza C and Zeippen, 1982, MNRAS, 198, 127
[ds]: Dufton PL, Hibbert A, Kingston AE and Doschek GA, 1982, ApJ, 257, 338
[dt]: Dufton PL and Kingston AE, 1987, J. Phys. B, 20, 3899
[du]: Zeippen CJ, Butler K and Bourlot JL, 1987, A&A, 188, 251
[dv]: Butler K and Zeippen CJ, 1989, A&A, 208, 337
[dw]: Martin W and Zalubas Romuald, 1979, J. Phys. Chem. Ref. Data, V8, No. 3
[dz]: Johnson CT, Kingston AE and Dufton PL, 1986, MNRAS, 220, 155
[ea]: Tayal SS, Burke and Kingston AE, 1985, J. Phys. B, 18, 4321
[eb]: Hayes MA, 1986, J. Phys. B, 19, 1853
[ec]: Zhang HL and Pradhan AK, 1994, A&A, Submitted
[ed]: AK Pradhan and HL Zhang, 1993, ApJ, L77
[ee]: Kaufman V. and Sugar J., 1986, J. Phys. Chem. Ref. Data, 15, 321
[ef]: Osterbrock DE, Tran HD and Veilleux S, 1992, ApJ, 389, 305
[eg]: Kato T and Lang J, 1990, ADNDT, 44, 133
[eh]: Fuhr JR, Martin GA and Wiese WL, 1988, J. of Phys. and Chem. Ref. Data, 17, Supp. No. 4
[ei]: Nussbaumer N and Storey PJ, 1988, A&A, 193, 327
[ej]: Berrington KA, Zeippen CJ, Dourneuf ML, Eissner W and Burke PG, 1991, J. Phys. B, 24, 3467
[ek]: Zhang HL and Pradhan AK, 1994, to be published

TABLE 1. (*Continued.*)

Radiative Transfer

By D. G. HUMMER

Max-Planck-Institute for Astrophysics, Karl-Schwarzschild-Str. 1, 85740 Garching, Germany
and Institute for Astronomy and Astrophysics of the University of Munich, Scheinerstr. 1,
81679 Munich, Germany†

This brief overview for the non-specialist presents certain aspects of radiative transfer theory
important for the quantitative interpretation of astrophysical spectra.

1. Introduction

The theory of radiative transfer has made spectacular advances in the past decade,
both in the understanding of fundamentals and in computational techniques. However,
apart from the solar/stellar community, these important tools for the interpretation of
astrophysical spectra are neither recognized nor effectively used. It is hoped that this
brief overview will be useful in communicating the state of understanding and guiding
potential users to the appropriate literature. This paper is *not* intended as a review, but
as a discussion of two important developments related to Osterbrock (1962).

The role of radiative transfer theory in the quantitative interpretation of spectra seems
not to be widely understood. The crucial importance of radiative transfer processes as
the link between an astronomical object and the determination of its physical properties
is discussed in Sect. 2.

Although the necessity of treating radiation scattered in spectral lines as non-coherent,
i.e., experiencing slight shifts in frequency in each scattering, is well understood, the conditions under which one can employ the simplifying assumption of complete redistribution
are less well known. This issue is discussed in Sect. 3, starting from the discussion in
Osterbrock (1962). Sect. 4 contains a detailed comparison of numerical solutions of the
transfer equation with various assumptions concerning treatment of redistribution.

The solution of the combined radiative transfer and statistical equilibrium equations for
atomic models with a large number of levels, and in various geometrical configurations,
lies at the heart of the quantitative astrophysical spectroscopy. The class of practical
techniques for the solution of these problems, known as "Accelerated Lambda Iteration"
(ALI), has developed in the past decade to dominate the field. An overview of the
properties and the current state of these methods is given in Sect. 5, where some recent
applications are discussed briefly.

2. Role of radiative transfer

Radiative transfer processes play two distinct roles in astrophysics, the *constructive*
and the *diagnostic*. In the former, the degree of elaboration given to the treatment of
radiative transfer need be only commensurate with the importance of radiative processes.
For example, in a medium in which LTE is valid and the temperature is fixed by other
processes, there is no constructive aspect, although the diagnostic aspect, i.e., the calculation of the emergent spectrum is still necessary. In this case it consists only of a formal
solution of the radiative transfer equation. (In reality, the temperature of an LTE atmosphere is often determined by the condition of radiative equilibrium, which involves a

† present address: 313 Alder Lane, Boulder, CO 80304, USA

rather delicate radiative transfer problem requiring sophisticated methods.) Under some conditions an escape probability treatment of the radiative transfer will suffice, whereas under others, the full treatment will be necessary.

The diagnostic aspect, on the other hand, is always important if the spectrum of an optically thick medium is to be quantitatively interpreted. Because the *phenomena* of radiative transfer and of electron collisions with atoms (which couples the photons to the electron gas) *encodes* information concerning the physical conditions in the outgoing radiation, the *theory* of radiative transfer and the availability of the relevant atomic data are necessary to *decode* this information. Thus the disciplines of radiative transfer and atomic physics are links in the chain, along with telescope, spectrograph, and detector, which carry information on the physical conditions and processes within the object to the astrophysicist. The axiom that a chain is no stronger that its weakest link has the obvious corollary that the calculation of the observed spectrum from a model *must be at least as accurate* as any assumption or technique used to produce the model in the first place. Otherwise, one does not learn anything reliably and may, in fact, be seriously misled.

3. Partial redistribution in line formation

Photons experience slight frequency changes in the process of scattering. At low densities these arise primarily from the Doppler shift caused by the thermal motion of the atoms and from the natural widths of the levels, whereas at higher densities collisional broadening processes begin to play a role. The so-called *noncoherence* leads to radiation fields in spectral lines qualitatively different from those with coherent or monochromatic scattering (an abstract concept, which never occurs in nature). A very complete review of transfer phenomena with noncoherent scattering has been given by Nagirner (1987).

The *redistribution* functions that express the correlation between the incoming and outgoing frequencies are known for a variety for physical situations, as first systematically developed by Hummer (1962). A good review of subsequent work has been given by Hubeny (1985). Important later advances include the derivation of a general redistribution matrix for polarized light by Domke & Hubeny (1988), and a treatment of redistribution in hydrogen at higher densities where collisions play a role by Cooper, Ballagh & Hubeny (1989).

At low densities, the redistribution function for a resonance line is

$$R(x \to x') = \pi^{-3/2} \int_{\frac{1}{2}(\bar{x}-x)}^{\infty} du \, e^{-u^2} \left[\tan^{-1}\left(\frac{x+u}{a} \right) - \tan^{-1}\left(\frac{x-u}{a} \right) \right] du, \quad (3.1)$$

where $a = \gamma/\Delta$, γ and Δ being the natural and Doppler widths of the line, respectively. The form we give here has been averaged over the scattering angle of the photon, which is known to be a good approximation. The correlation of frequencies by such a complicated function greatly complicates the solution of the transfer equation.

In the approximation of *complete redistribution*, which was introduced in the early 1930's, the frequencies are uncorrelated and the distribution of emitted frequencies is proportional to the line absorption profile. This leads to a transfer equation, which can be solved numerically with relative ease and is the basis of nearly all astrophysical line transfer calculations. There are, however, important cases for which the true redistribution must be used, cf. Linsky (1985). In nebular situations the formation of the Lyman α lines of H I and He II in optically thick atmospheres and nebulae are of primary interest. We discuss the comparison of both forms of redistribution in two contexts, the mean number of scatterings $\langle N \rangle$ experienced by a resonance line photon, and the mean rate of photon absorption or scattering, which is proportional to

$$\bar{J} = \int_{-\infty}^{\infty} dx\, \phi(x) J(x). \tag{3.2}$$

As an unfortunate backformation from the expression "*complete* frequency redistribution" (CFR), scattering with correlation has acquired the label "*partial* frequency redistribution" (PFR). These abbreviations are used below. In the literature one finds also "CRD" and "PRD."

3.1. *Mean number of scattering*

Osterbrock (1962) used the mean escape probability for CFR given by Zanstra (1949) to infer that $\langle N \rangle \sim T$ (neglecting a logarithmic factor of order unity), where T is the mean optical thickness in the resonance line. For PFR Osterbrock simulated scattering of wing photons by an "almost random walk process" in frequency space, which is terminated when the photon reaches the frequency at which the optical distance to a surface is unity. This led to the result that $\langle N \rangle \sim T^2$. Shortly thereafter Hummer (1969) obtained accurate source functions by a numerical method for optical thicknesses large enough that the two approximations were qualitatively different. As $\langle N \rangle$ for PFR was smaller than with CFR, it seemed unlikely that Osterbrock's estimate was correct.

Adams (1972) showed that because of the simultaneous diffusion in space and frequency, photons can escape directly when $\mid x \mid > \tilde{x}$, where

$$\tilde{x} \approx (aT/\pi)^{1/3}, \tag{3.3}$$

where x is the frequency displacement from the line center in units of the thermal Doppler width. This led to a limiting form $\langle N \rangle \sim T$. This was verified by Harrington (1973), who solved the Fokker-Planck approximation to the transfer equation by means of eigenfunction expansions. Harrington's result was

$$\langle N \rangle = .3332368T, \quad T \to \infty \tag{3.4}$$

for an atmosphere with uniformly distributed true sources. For CFR, Hummer (1964) showed that

$$\langle N \rangle = T\left[\ln(0.2821T)\right], \quad T \to \infty. \tag{3.5}$$

The most detailed and accurate study to date is that of Hummer and Kunasz (1980), who obtained extensive numerical solutions with PFR from which both $\langle N \rangle$ and ρ, the ratio of the mean photon path length to the half thickness, were calculated. These results are in excellent agreement with Harrington's values for large optical thickness. Analytical methods have provided some useful insights and asymptotic results. A good review is given by H. Frisch (Kudritzki, Yorke, & Frisch 1988).

3.2. *Escape probability techniques with PFR*

Gayley (1992a,b,c) has recently developed a heuristic generalization of Adams's argument, which appears to yield good approximate solutions to transfer problems with partial redistribution. A second-order escape probability method appears to be quite useful in practical work. There appears to be some unidentified basic principle underlying these results, for the accuracy obtained is surprising given the ad hoc nature of the basic approximations.

4. Systematic comparison of solutions with partial and complete redistribution

There has been considerable confusion in the literature regarding the conditions under which it is necessary to employ partial redistribution. It is well established (see above) that for the first resonance line under nebular conditions PFR must be used when the optical thickness is so large that a significant amount of transfer occurs in the far line wings. This implies that when the number of scattering is limited by collisional processes or by branching, CFR should be adequate. By the same token, when the optical thickness is small or when photons are lost to continuous absorption, CFR is satisfactory.

Another source of confusion in this connection is the value of the Voigt parameter a to be used in the CFR approximation. Because of the large probability of branching to another transition, the mean number of scatterings in a resonance line other than the first is so small that relatively little transfer occurs in the line wings, with the consequence that one should set $a = 0$. Not to do so will lead to line profiles that are much too high in the wings. Only under conditions of high density for the pressure broadening is dominant should the true value of be used with CFR.

To provide clarification of these problems, we have computed the mean radiative excitation rate coefficient $\bar{J}$ throughout uniform, isothermal, collisionally excited planar atmospheres characterized by the mean optical thickness T and the loss probability per scattering ϵ. The quantity $\bar{J}$ is the average over the radiation field which is relevant to the determination of the atomic level populations. In situations in which the monochromatic radiation field itself, must be known, for example in pumping overlapping lines (Bowen mechanism!), the conditions for the use of CFR will be much more stringent.

Three approximations have been used: (1) partial frequency redistribution with $a = 10^{-3}$; (2) complete frequency redistribution with $a = 10^{-3}$; and (3) complete frequency redistribution with $a = 0$. The PFR calculations were carried out with a code kindly provided by Ivan Hubeny, and those with CFR were performed with a code developed by G. Rybicki and the author. Both codes employ some version of the Feautrier method. Note that the (true) source term was set equal to $T^{-1}\phi(x)$ to facilitate comparison of different cases.

In Figures 1–4 we plot the excitation rate factor $\bar{J}$ (written in figures as $\mathcal{J}$) vs. τ and the emergent flux profile $H(x)$ vs. x for $T = 10^2$, 10^4, 10^6 and 10^8, and for interesting values of ϵ between 0 and $10^{-1/2}$. For $T = 10^2$, all three approximations give essentially the same radiative rate factors $\bar{J}$, although for $\epsilon = 0$, some slight deviations are visible. The flux profiles behave in a similar way, except that the flux profile for CRD ($a = 10^{-3}$) begins to show very weak increases in the far wings for $\epsilon = 0$.

For $T = 10^4$, the three approximations for $\bar{J}$ are now quite different when $\epsilon = 0$, but otherwise are in essential agreement. The corresponding flux profiles show a similar pattern for $|x| \leq 3$, but in the wings are qualitatively different in each approximation. In particular, the very strong wings for CFR ($a = 10^{-3}$) are completely wrong as an approximation for the PRF cases. For collisional broadening they would represent enhanced energy loss. The situation for $T = 10^6$ is still more extreme. Now, the wing behavior of the fluxes in the three approximations are all different, although except for $\epsilon = 0$ the cores agree well.

Finally, for $T = 10^8$, we begin to see significant discrepancies in $\bar{J}$ for ϵ as large as 10^{-4}, which also appear in the line cores. Obviously, the three approximations are qualitatively different for $\epsilon = 0$—the classical nebular resonance line case.

The results for $\epsilon \sim 0.1$ are appropriate for the Lyman lines of HI and HeII with $n \geq 2$. This may be confirmed from Table 1 in Hummer & Storey (1992), where the

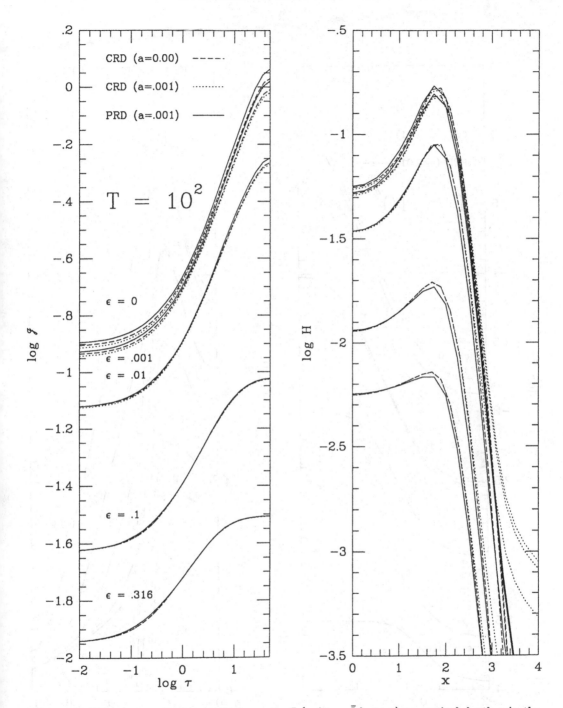

FIGURE 1. The mean radiative excitation rate $\mathcal{J}$ (written $\bar{J}$ in text) vs. optical depth τ in the left panel and the flux profile $H(x)$ vs. $x = (\nu - \nu_0)/n_0$ in the right panel. The mean optical depth is $T = 10^2$ and values of ϵ are given on each triplet of curves.

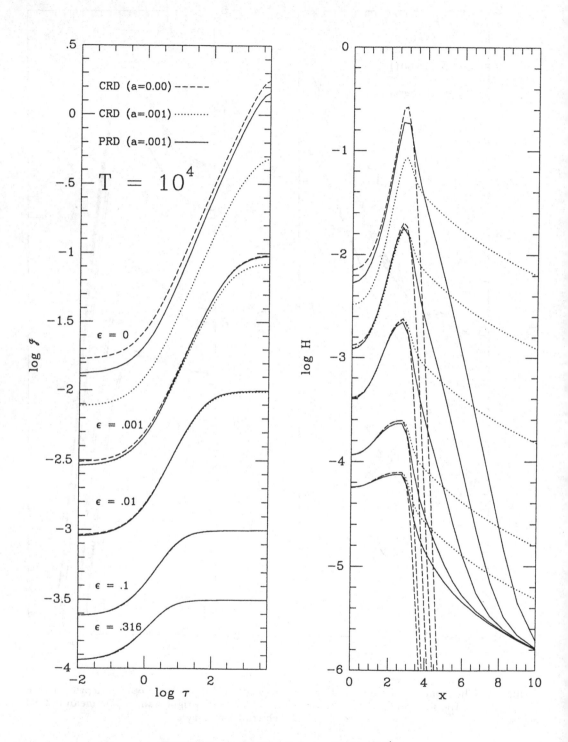

FIGURE 2. As for Fig. 1, but with $T = 10^4$.

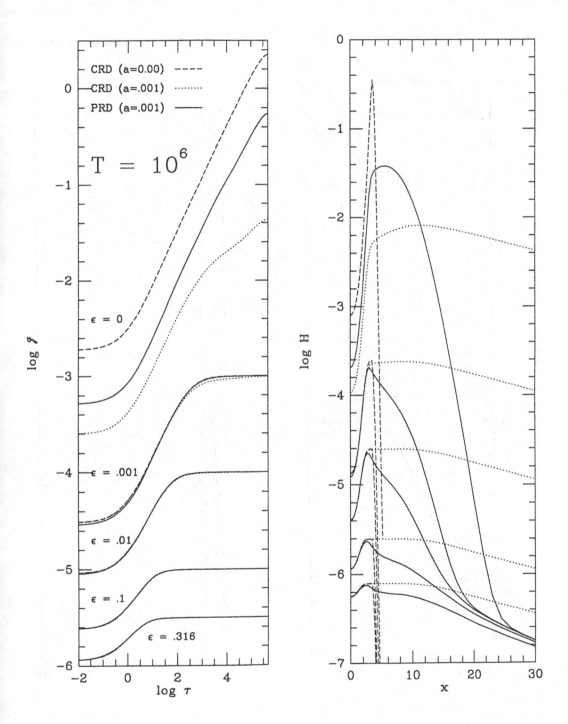

FIGURE 3. As for Fig. 1, but with $T = 10^6$.

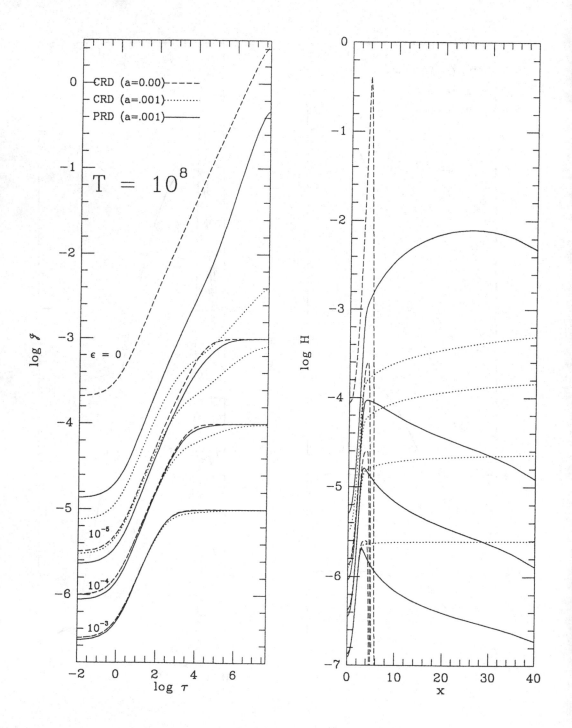

FIGURE 4. As for Fig. 1, but with $T = 10^8$.

excitation rates of the upper Lyman lines are calculated from both escape probability theory and from accurate numerical solutions. These results also show that the inclusion of continuous opacity in the Lyman line region improve the approximate results.

4.1. *Sobolev theory with partial and complete redistribution*

By setting the spatial derivatives to zero in the transfer equation for an expanding or contracting atmosphere, one obtains the basic equation for the Sobolev approximation. For uniform, spherically-symmetric expansion, this can be written as

$$\gamma \frac{dJ(x)}{dx} = \phi(x)J(x) - (1 - \epsilon) \int_{-\infty}^{\infty} J_{x'} R(x \to x') - G\phi(x); \quad J(\infty) = 0, \qquad (4.6)$$

where the dimensionless velocity gradient is

$$\gamma \equiv \frac{dV/dr}{v_{th}k}. \qquad (4.7)$$

For CPR, this can be solved exactly in terms of

$$y \equiv \int_{x}^{\infty} dx' \phi(x') \qquad (4.8)$$

and the escape probability

$$P(\gamma) = \gamma \left[1 - \exp(-1/\gamma)\right] \qquad (4.9)$$

as was shown by Sobolev (1957).

Numerical solutions for a variety of redistribution functions have be obtained and studied by Hummer & Rybicki (1992). Although the radiation fields obtained for PFR and CFR can be quite different, the relative excitation rate $\bar{J}$ differs by no more than 5 percent for $\gamma > 10^{-4}$.

5. Radiative transfer with statistical equilibrium

A central problem in the quantitative analysis of spectra is the simultaneous solution of the radiative transfer and statistical equilibrium equations, which are coupled because the intensities at each point depend on the level populations throughout the atmosphere, and the level populations depend on the intensities through the radiative excitation rates appearing in the statistical equilibrium equations. Thus the set of coupled equations are strongly non-linear and for three or more levels can be solved only by numerical procedures. In addition, it is often necessary to take into account more than one chemical species, which interact through the absorption and emission coefficients in the transfer equation.

Historically, the first procedure for solving these coupled equations was that of *lambda iteration* †, in which a set of level populations is assumed and the transfer equations solved to obtain the intensities. These intensities were used to evaluate the radiative excitation and ionization rates, from which a new set of improved level populations could be obtained. This procedure is found to converge very slowly except under certain, quite restrictive, circumstances. The main reason for this difficulty is that the atomic level populations are controlled by the mean intensity near line center, where the opacity is large and the flux is small. Because of the relatively small mean free path of line center photons, iterative corrections propagate very slowly.

---

† Originally, Λ iteration referred to the solution of the monochromatic transfer equation for scattering by iteration on the mean intensity J_ν

Rybicki (1972), in a seminal paper, eliminated this difficulty by solving the transfer equation only in the wings, assuming the radiation field in the line core was saturated there, i.e. that $I_\nu = S_\nu$ for frequencies lying within some interval γ about the line center. The core field was then corrected iteratively. This procedure was subsequently applied to multilevel problems by Rybicki and his collaborators (Flannery, Rybicki & Sarazan 1979, 1980).

The calculation of the intensity from the source function can be represented formally by the operator Λ, i.e.

$$I_{\nu,\mu} = \Lambda_{\nu,\mu}[S_\nu]. \tag{5.10}$$

Cannon (1973) introduced the operator splitting

$$\Lambda_{\nu,\mu} = \Lambda^*_{\nu,\mu} + (\Lambda_{\nu,\mu} - \Lambda^*_{\nu,\mu}), \tag{5.11}$$

where $\Lambda^*_{\nu,\mu}$ is an *approximate lambda operator* which is easily inverted. The intensity in an iterative procedure is then

$$I_{\nu,\mu} = \Lambda^*_{\nu,\mu}[S_{\nu,\mu}] + (\Lambda_{\nu,\mu} - \Lambda^*_{\nu,\mu})[S^\dagger_{\nu,\mu}], \tag{5.12}$$

where $S^\dagger$ is the source function from the previous iteration. As the iteration converges, $S^\dagger \to S$ and the intensity is exact.

Scharmer's (1981, 1984) synthesis of Rybicki's core saturation procedure with the idea of operator splitting was quickly adopted by the solar/stellar community, and many different choices of Λ^* and details of procedure were developed. Scharmer & Carlsson (1985) developed and circulated an ALI code "MULTI," which has become, in its many variants, one of the workhorses of the cool-star community. Werner & Husfeld (1985) generalized Scharmer's procedure to multi-level atoms; their procedure was further generalized to include the constraint of radiative equilibrium by Werner (1986), who then calculated non-LTE model atmospheres of hot stars.

5.1. *Multi-level approximate lambda iteration (MALI)*

Rybicki & Hummer (1991, 1992, 1994) have developed a variant (MALI) of the approximate lambda operator approach which is *intrinsically* multilevel in formulation, and is also computationally the most efficient, in that an internal iteration loop is avoided. To avoid unnecessary complexity, we consider here only the simplest case, in which only the lines are taken into account. Completely general cases are treated in the references just cited. The first of these gives also a complete summary of the development of ALI methods.

A fundamental and important feature of ALI methods in general is that the computational effort scales *linearly* with the number of frequency points, in contrast to the cubic dependence found in complete linearization techniques. The variant outlined here (MALI) also scales linearly in the number of depth points.

The expression for the Λ operator is readily derived. By taking the average of the forward and backward stream in each direction and defining

$$u_{\mu,\nu} \equiv (I_{\mu,\nu} + I_{-\mu,\nu})/2, \tag{5.13}$$

one can easily cast the transfer equation into second order form

$$d^2u/d\tau^2 = u - S, \tag{5.14}$$

where we have dropped the frequency and angle subscripts. If we now discretize the depth variable to the set $\tau_d, d = 1, 2, \ldots, ND$ we have

$$- A_d u_{d-1} + B_d u_d - C_d u_{d-1} = S_d, \quad d = 1, 2, \ldots, ND, \tag{5.15}$$

where the quantities A_d, B_d and C_d depend on the $\Delta\tau$'s and thus on the level populations. These equations can be written in matrix form as

$$\mathbf{T}u = S, \qquad (5.16)$$

where u and S are vectors of dimension ND. Solving for the vector u we have

$$u = \mathbf{T}^{-1}S \equiv \Lambda S. \qquad (5.17)$$

Although $\mathbf{T}$ is tridiagonal, $\Lambda \equiv \mathbf{T}^{-1}$ is a *full* matrix.

Following Olson, Auer & Buchler (1986), $\Lambda^\dagger$ is taken as the diagonal part of Λ. Rybicki and Hummer show how to calculate $\Lambda^\dagger$ in $O(ND)$ operations.

Let us now consider a system of L levels, labelled by $l = 1, 2, \ldots, L$, with energies E_l and statistical weights g_l, such that $E_{l+1} > E_l$. The opacity and emissivity of the $l \to l'$ transition are respectively

$$\chi_{ll'}(\nu) = (h\nu/4\pi)[n_{l'}B_{l'l} - n_l B_{ll'}]\phi_{ll'}(\nu), \quad l > l', \qquad (5.18)$$

and

$$\eta_{ll'}(\nu) = (h\nu/4\pi)n_l A_{ll'}\phi_{ll'}(\nu), \quad l > l', \qquad (5.19)$$

where we have assumed complete redistribution in writing η. The source function in this transition is

$$S_{ll'}(\nu) = n_l A_{ll'}/[n_{l'}B_{l'l} - n_l B_{ll'}], \quad l > l'. \qquad (5.20)$$

The total opacity, emissivity and source function at frequency ν are then

$$\chi_{\mu,\nu} = \sum_{l>l'} \chi_{ll'}, \qquad (5.21)$$

$$\eta_{\mu,\nu} = \sum_{l>l'} \eta_{ll'}, \qquad (5.22)$$

and

$$S_{\mu,\nu} = \eta_{\mu,\nu}/\chi_{\mu,\nu}. \qquad (5.23)$$

The equations of statistical equilibrium are

$$\sum_{l'<l}[n_l A_{ll'} - (n_{l'}B_{l'l} - n_l B_{ll'})\bar{J}_{ll'}] - \sum_{l'>l}[n_{l'}A_{l'l} - (n_l B_{ll'} - n_{l'}B_{l'l})\bar{J}_{ll'}]$$
$$+ \sum_{l'}[n_l C_{ll'} - n_{l'}C_{l'l}] = 0, \quad l = 1, 2, \ldots, L. \qquad (5.24)$$

Note that this set of equations is manifestly *bi-linear* in n and $\bar{J}$.

We now restrict ourselves to local approximate operator $\Lambda^*_{ll'}$, which is diagonal and hence multiplicative. The operator splitting introduced above gives us

$$I_{\mu,\nu} = \Lambda^*_{ll'}S_{ll'} + (\Lambda_{\mu,\nu} - \Lambda^*_{\mu,\nu})S^\dagger_{ll'}$$
$$= \Lambda^*_{ll'}S_{ll'} + I^\dagger_{\mu,\nu} - \Lambda^*_{\mu,\nu}S^\dagger_{ll'}$$
$$\equiv \Lambda^*_{ll'}S_{ll'} + I^{eff}_{\mu,\nu}, \qquad (5.25)$$

where the $\dagger$ denotes quantities from the previous iteration.

Substituting this expression into the definition of $\bar{J}$, we have

$$\bar{J}_{ll'} = \bar{\Lambda}^*_{ll'}S_{ll'} + \bar{J}^{eff}_{ll'} \qquad (5.26)$$

where

$$\bar{\Lambda}^*_{ll'} = \int d\Omega \int d\nu\, \phi_{ll'}\Lambda^* \qquad (5.27)$$

and

$$\bar{J}_{ll'}^{eff} = \int d\Omega \int d\nu \phi_{ll'} I_{\mu,\nu}^{eff} = \bar{J}_{ll'}^{\dagger} - \bar{\Lambda}_{ll'}^{*} S_{ll'}^{\dagger} \tag{5.28}$$

Substituting this expression for $\bar{J}_{ll'}$ into the statistical equilibrium equations gives the *preconditioned* form in which no reference to the core photons remains:

$$\sum_{l'<l}[n_l A_{ll'}(1 - \bar{\Lambda}_{ll'}^{*}) - (n_{l'}B_{l'l} - n_l B_{ll'})\bar{J}_{ll'}^{eff}] -$$
$$\sum_{l'>l}[n_{l'} A_{l'l}(1 - \bar{\Lambda}_{ll'}^{*}) - (n_l B_{ll'} - n_{l'}B_{l'l})\bar{J}_{ll'}^{eff}] +$$
$$\sum_{l'}[n_l C_{ll'} - n_{l'} C_{l'l}] = 0, \quad l = 1, 2, \ldots, L. \tag{5.29}$$

We note that this set of equations has the same form as the original but now includes the factor $(1 - \bar{\Lambda}_{ll'}^{*})$, which is an effective escape probability. Also, the n_l are "new," while all coefficients are known from the previous iteration. Moreover the equations are still linear, which eliminates the need for an internal iteration found in other multilevel formulations. Finally, it can be shown that all n_l are positive.

The procedure consists of the following steps:

(a) Assume a set of level populations n_l (for example, LTE values);

(b) Calculate $S_{ll'}$ and coefficients A_d, B_d and C_d in **T**:

(c) Calculate $u_{\mu,\nu}$ by *formal solution* of the transfer equation using $S_{ll'}$, frequency by frequency;

(d) Calculate $\Lambda_{d,d}$ by $O(ND)$ method given by Rybicki and Hummer (1991);

(e) Evaluate $\bar{J}_{ll'}^{eff}$;

(f) Solve preconditioned equations of statistical equilibrium for new n_l;

(g) Start next iteration, unless the maximum δn is smaller than some prespecified value, such as 10^{-4}.

Convergence of this process can be significantly improved by various numerical acceleration techniques, and by starting the process with large depth steps and halving them at appropriate stages of iteration. An optimum strategy for combining these techniques is given by Auer, Fabiano Bendicho & Trujillo Bueno (1994). A review of acceleration techniques in general is given by Auer (1991).

6. Applications of ALI/MALI

In addition to the many important applications which have been made to stellar atmospheres since the pioneering atmospheric model calculations of Werner (1986), these techniques have recently been applied with great success to the formation of molecular band spectra in planetary atmospheres. Moreover, generalizations to two and three dimensions, and to externally illuminated atmospheres have been made. Some of the more outstanding of these are outlined below.

6.1. *Molecular band spectra in planetary atmospheres*

The first application of MALI to molecular bands was given by Kutepov, Kunze, Hummer & Rybicki (1991), where restricted models of the 4.3μ band of CO_2 in the atmosphere of Venus and the 4.7 and 2.3 μ bands of CO in the earth's atmosphere were treated. Full calculations with very complete molecular models of CO, CO_2 and O_3 incorporating thousands of molecular levels and making use of newly available molecular parameters are being carried out by Kutepov and his collaborators in Munich and St. Petersburg.

6.2. *Stellar atmospheres and winds*

This field is now dominated by ALI calculations. As even a cursory summary of this work would run to many pages, reference is made to two recent reviews in which techniques as well as applications are discussed:

(a) "Stellar Atmospheres: Beyond Classical Models," eds. Crivellari, Hubeny & Hummer, (Kluwer: Dordrecht, 1991);

(b) "The Atmospheres of Early-Type Stars," eds. U. Heber & C. S. Jeffery, Lecture Notes in Physics, No. 401, (Springer Verlag: Berlin, 1992).

As an example of the kinds of models now in use, we mention the non-LTE models of sdO stars computed by Dreizler & Werner (1993), which treats non-LTE line blanketing including 130,000 lines of Iron-group elements by means of the frequency and multiplet regrouping technique of Anderson (1989). Extremely large non-LTE calculations with thousand of atomic levels and millions of transitions in non-static spherical winds and Wolf Rayet atmospheres are being carried out by groups in Munich (R. P. Kudritzki), in Kiel (W. R. Hamann) and in Pittsburgh (D. J. Hillier). Without ALI techniques these calculations would be impossible.

The MALI technique has been generalized by Heinzel (1995) to include radiation falling on the surface of the atmosphere.

Very recently, Hubeny & Lanz (1995) have developed a stellar atmosphere code combining ALI and complete linearization which appears to be even more powerful than ALI alone.

Obviously, this field is developing with great rapidity in both technique and in the understanding of the atmospheres and winds of hot stars.

6.3. *Two and three dimensions*

A generalization of MALI to two dimensions in a very effective form has been made by Auer, Fabiani Bendicho & Trujillo Bueno (1994). Partial redistribution in two spatial dimensions has been treated by Auer & Paletou (1994). In three dimensions, the radiation field in asymmetric accretion discs has been modelled using ALI techniques by Hummel (1994) and by Papkalla (1995).

Acknowledgements

For more than 35 years Mike Seaton has been my teacher, colleague and friend. It was he who aroused my lifelong interest in radiative transfer and atomic physics, the pillars of astrophysical quantitative spectroscopy. I regret that this paper must be such an imperfect reflection of the present state of quantitative spectroscopy. I am also indebted to my long-time collaborators and friends George Rybicki, Dimitri Mihalas, Rolf Kudritzki, Peter Storey and Ivan Hubeny, to name only a few, who have contributed so richly to this endeavor. Finally, I would like to thank the Fellows of JILA for access to their computing facilities and Ms. S. Rush for assistance in preparing the camera-ready copy.

REFERENCES

ADAMS, T. F. 1972 *ApJ* **168**, 575

ANDERSON, L. S. 1989 *ApJ* **339**, 558

AUER, L. 1991, in *Stellar Atmospheres: Beyond Classical Models* (ed. L. Crivellari, I. Hubeny & D. G. Hummer) Dordrecht: Kluwer, p. 9

AUER, L., FABIANI BENDICHO, P., & TRUJILLO BUENO, J. 1994 *A&A* **292**, 599

AUER, L. & PALETOU, F. 1994 *A&A* **285**, 675

CANNON, C. J. 1973 *ApJ* **185**, 621

COOPER, J., BALLAGH, R. J., & HUBENY, I. 1989 *ApJ* **344**, 949

DOMKE, H., & HUBENY, I. 1988 *ApJ* **334**, 527

DREIZLER, S., & WERNER, K. 1993 *A&A* **278**, 199

FLANNERY, B. P., RYBICKI, G. B. & SARAZIN, C. L. 1979 *ApJ* **229**, 1057

FLANNERY, B. P., RYBICKI, G. B. & SARAZIN, C. L. 1980 *ApJS* **44**, 539

GAYLEY, T. 1992a *ApJS* **78**, 549

GAYLEY, T. 1992b *ApJ* **390**, 573

GAYLEY, T. 1992c *ApJ* **390**, 583

HARRINGTON, J. P. 1973 *MNRAS* **162**, 43

HEINZEL, P. 1995 *A&A* in press.

HUBENY, I. 1985 in *Progress in Stellar Spectral Line Formation Theory* (ed. J. Beckman & L. Crivellari) Reidel: Dordrecht, p. 27

HUBENY, I., & LANZ, T. 1995 *ApJ* **439**, 875

HUMMEL, W. 1994 *A&A* **289**, 458

HUMMER, D. G. 1962 *MNRAS* **125**, 21

HUMMER, D. G. 1964 *ApJ* **140**, 276

HUMMER, D. G. 1969 *MNRAS* **145**, 313

HUMMER, D. G., & KUNASZ, P. B. 1980 *ApJ* **236**, 609

HUMMER, D. G., & STOREY, P. J. 1992 *MNRAS* **254**, 277

HUMMER, D. G., & RYBICKI, G. B. 1992 *ApJ* **387**, 248

KUDRITZKI, R. P., YORKE, H. W., & FRISCH, H. 1988, *Radiation In Moving Gaseous Media, 18th Advanced Saas-Fee Course* (eds. Y. Chmielski & T. Lanz) Sauverny-Versoix: Geneva Observatory

KUTEPOV, A. A., KUNZE, D., HUMMER, D. G., & RYBICKI, G. B. 1991 *JQSRT* **46**, 347

LINSKY, J. L. 1985 in *Progress in Stellar Spectral Line Formation Theory* (ed. J. Beckman & L. Crivellari) Reidel: Dordrecht, p. 1

NAGIRNER, D. I. 1987 *Astrophysics* **26**, 90

OLSON, G. L., AUER, L., & BUCHLER, J. R. 1987 *JQSRT* **38**, 325

OSTERBROCK, D. E. 1962 *ApJ* **135**, 195

PAPKALLA, R. 1995 *A&A* in press

RYBICKI, G. B. 1972 in *Line Formation in the Presence of Magnetic Fields* (eds. R. G. Athay, L. L. House, & G. Newkirk, Jr.) High Altitude Observatory, Boulder

RYBICKI, G. B. & HUMMER, D. G. 1991 *A&A* **245**, 171

RYBICKI, G. B. & HUMMER, D. G. 1992 *A&A* **262**, 209

RYBICKI, G. B. & HUMMER, D. G. 1994 *A&A* **290**, 553

SHARMER, G. B. 1983 *ApJ* **249**, 720

SHARMER, G. B. 1984 in *Methods in Radiative Transfer* (ed. W. Kalkofen) Cambridge: Cambridge University Press, p. 173.

SCHARMER, G. B., & CARLSSON, M. 1985 *J. Comp. Physics* **59**, 56

SOBOLEV, V. V. 1957 *Sov. Astr.* **1**, 678

WERNER, K. 1986 *A&A* **161**, 177

WERNER, K. & HUSFELD, D. 1985 *A&A* **148**, 417

ZANSTRA, H. 1949 *BAN* **11**, 1

Emission Lines from Winds

By JANET E. DREW

Department of Physics, Nuclear Physics Laboratory, Keble Road, Oxford OX1 3RH, U.K.

Some basic ideas about the origin of wind-formed line emission are presented. This is followed by three commentaries. The first focuses on the effects of clumped or inhomogeneous outflow on emission line formation, taking as examples the WR star HD 50896 and the Of star ζ Puppis. The second concerns wind-formed IR line emission: illustrations of the impact of overlapping continuum opacity on IR emission line spectra are presented, and the recent revolution in IR spectroscopy is demonstrated in the context of observations of highly-obscured luminous young stellar objects. Lastly, emission brought about by departures from spherical-symmetry is considered. Particular reference is made to the case of mass loss from the disk-accreting components located in cataclysmic binaries. HST data indicating a disk origin for the outflow are discussed briefly.

1. An overview

The classical spectroscopic signature of mass loss, first reviewed in the literature by Beals (1950), is the so-called P Cygni line profile. This label has come to be attached to the profile shape in which blueshifted absorption sits alongside redshifted emission. In truth, the practice of describing just this configuration as 'P Cygni' does little justice to the rich variety of profile forms that are to be found in this famous star's optical spectrum—Beals himself put the case for 4 different profile types characteristic of 'P Cygni stars.' Interestingly, from the perspective of this collection of papers on line emission, these other forgotten types include forms that emphasise emission rather than absorption. Indeed, those of us who have taken spectra of P Cygni itself are painfully aware of just how strong the strongest emission features (in Hα, He I λ5876) really are!

The narrowing of the term 'P Cygni line profile' is directly attributable to the key role played by space-borne ultraviolet astronomy in uncovering mass loss in a wide variety of stellar environments. The ultraviolet is host to many allowed ground-state transitions, while there are few to be found at optical wavelengths, or longward. The resonant scattering favoured for these transitions is as likely as not to lead to the development of an absorption component which, if due to an outflow, must be blueshifted. Unlike velocity-broadened line emission, there is absolutely no doubting the kinematic significance of blueshifted absorption—it *must* imply outflow. This has been seized upon gratefully, and has encouraged the occasional habit of describing even pure blueshifted absorption, unaccompanied by any emission, as 'P Cygni.'

At a time when infrared spectroscopy is coming of age and when echelle spectroscopy is promoting an avalanche of data at optical wavelengths, it is of some value to redress the balance. Many of the physical mechanisms well-known from nebular astrophysics are also at work in stellar winds and some, as in nebulae, will yield line emission. This article is mainly about the consequences of these processes and what can be learned from them. It also looks forward as much as it looks back. Guided by this author's experience, a significant fraction of the following concerns the top part of the H-R diagram occupied by stars, young and old, with $L \gtrsim 10^4$ L$_\odot$. All examples are derived from stellar or near-stellar, steady, ionized mass loss—transient phenomena and also cool, molecular outflows are dealt with elsewhere in this volume (e.g. the contributions by S. Pinto and A. Dalgarno).

The discussion begins below with some general ideas about the origin of wind-formed line emission. This is followed by three distinct commentaries that bring out a variety of physics relevant to different types of object and different wavelength ranges. Section 3 draws attention to the impact that outflow inhomogeneities may have on spectral line emission—an issue that assumes some practical importance because of the way in which emission line fluxes can be exploited to estimate mass loss rates and abundances. Section 4 shifts the focus onto infrared line emission (at $\lambda\lambda s > 1\mu m$), a phenomenon that has only recently become accessible to quantitative spectroscopy. The one departure from the top of the H-R diagram, in section 5, is to take in the example of non-spherical geometry encountered in the cataclysmic binaries—systems affording the best-constrained examples of accretion-driven mass loss.

2. Some physics of line emission from winds

The key differences between the radiative conditions in a wind and in a nebula rest in the gas density and the dilution of the ambient radiation field. In a 'typical' wind, densities are higher by several orders of magnitude, often spanning the range 10^6–10^{11} cm^{-3}. Whilst a working definition of a nebula might include the requirement that the exciting source is small enough, compared to the nebular volume, to allow it to be regarded as point-like, this is rarely a suitable approximation to apply in modelling an astrophysical wind. Winds are transitional entities that neither allow the full range of diffusive and thermodynamic approximations applicable to stellar atmospheres nor reduce entirely to the optically-thin statistical equilibrium often appropriate to nebulae. Winds are moving media capable of producing both optically-thick and optically-thin spectral lines. In short, complexity and variety are both attributes of wind-formed spectra.

2.1. *Optically-thin line emission*

Optically-thin line emission represents the simplest case. Transparency of the outflowing medium means that the observer can count all photons emitted in his or her direction, excepting those produced behind completely opaque obstacles such as (most commonly) the source of the outflow. In this circumstance, the emitted spectral line flux is additive to continuous emission produced by either the outflow source or the wind itself. Forbidden line emission is usually produced in this way. So also can permitted recombination line emission. A complication with respect to the nebular case is that account must be taken of density gradients and, in the case of forbidden lines, the possibility that gas densities above and below the critical density (at which collisional de-excitation matches spontaneous radiative decay) may both contribute significantly. Figure 1 provides an example of typically weak forbidden line emission in P Cygni's spectrum set amongst some minor classical P Cygni features. Uncomplicated by radiation transport effects, forbidden and other optically-thin features are of a great diagnostic value that, curiously, seems to be under-exploited.

2.2. *The optically-thick case*

To understand the outcome in the case of optically-thick wind emission, it helps to bring into use the concept of the line source function S_{line} (the line emissivity divided by its opacity—an intensity-like variable). At high optical depth ($\tau \gg 1$), the emergent flux in a spectral line can be shown from the formal solution of the transfer equation to obey the following proportionality:

$$F_{\text{line}} \propto S_{\text{line}} A_{\text{line}}. \tag{2.1}$$

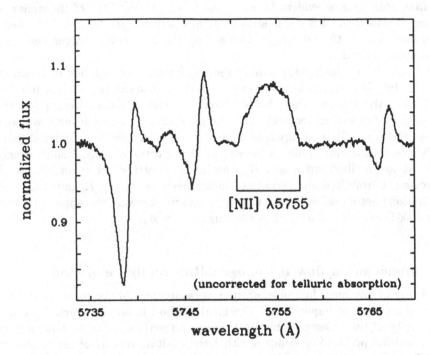

FIGURE 1. A fragment of an optical echelle spectrum of P Cygni in the vicinity of the [N II] 5755Å line. Like many of the forbidden lines seen in this star's spectrum this feature is weak relative to the continuum (cf. Stahl *et al.* 1991). Unlike many, its shape is rounded rather than square-topped, indicating that its zone of formation reaches into the accelerating portion of the outflow. Significantly, the critical density of this transition is relatively high.

Here A_{line} can be thought of as the size of the spectral line 'photosphere' (bounded by the $\tau = 1$ surface) projected onto the plane of the sky. Net line emission will be observed if:

$$S_{\text{line}}A_{\text{line}} > I_*A_*. \tag{2.2}$$

The right hand side of this inequality is the analogous quantity to that on the left calculated for the continuum emission produced primarily by the outflow source at the wavelength of the spectral line of interest.

From this inequality, it is easy to see that there are two main routes toward producing optically-thick net line emission from a wind. In most practical examples it is likely that $A_{\text{line}} \gtrsim A_*$. If, in fact, the projected size of the line-forming volume is not more than comparable with that of the continuum source (and covers it!), net line emission can only be observed if $S_{\text{line}} > I_*$. This may be thought of as the case of intrinsic line emission. It is a condition that is not so readily satisfied—especially at wavelengths close to or shortward of the Planck maximum of the continuum energy distribution emitted by the outflow source. It requires either the outflow to be hotter than its source (the coronal condition familiar to aficionados of solar-type stars) or the sustenance of a level population inversion. Radiative recombination is a physical process that can maintain the latter. However, whether or not this yields net line emission depends on the value of the ratio of the level departure coefficients, b_u/b_ℓ, and also on how the wind temperature compares with the continuum radiation temperature. For example, in radiation-driven stellar winds, the expectation must be that the wind temperature is significantly cooler than the stellar effective temperature (see Drew 1989). In such cases, recombination line emission may only become evident far along the Rayleigh-Jeans tail of the stellar energy distribution, where $B(\nu, T)$ is not so strongly temperature-dependent, since it then takes a less extreme value of the ratio b_u/b_ℓ to make up the difference between the wind and stellar Planck functions.

Where $S_{\mathrm{line}} \lesssim I_*$, inequality (2) implies the requirement for net line emission is that $A_{\mathrm{line}} >> A_*$. In this case, line emission can be attributed to extension of the line-forming region. This is the circumstance that explains the association between strengthening optical line emission and increasingly extreme mass loss well-known for main sequence and evolved OB stars. This picture is relevant also to the striking inclination-dependence of the UV resonance line profiles observed in the spectra of mass-losing cataclysmic variables: at low-inclinations where the continuum-emitting accretion disk is viewed almost face-on, blueshifted absorption dominates because $S_{\mathrm{line}} < I_*$ and $A_{\mathrm{line}} \sim A_*$; at high-inclinations net emission is seen instead, primarily because the apparent size of the accretion disk is greatly reduced, thus allowing $A_{\mathrm{line}} > A_*$.

3. The impact of outflow inhomogeneities on line emission

In the study of any mass loss phenomenon, measurement of the mass loss rate, $\dot{M}$, is bound to be an issue of importance. The ideal method is one in which there is negligible sensitivity of the observed quantity to other properties of the outflow in question. Methods requiring precise knowledge of either the excitation state of the outflow or the manner of its acceleration to terminal velocity are less attractive.

The favoured technique for OB star winds is the radio method (Wright & Barlow 1975; Panagia & Felli 1975). In essence this amounts to a measurement of the radio emissivity at a particular, large radius where it is safe to assume acceleration of the outflow is complete. Since the radio emission is in most instances attributable to free-free processes, the emissivity is proportional to ρ^2 (where ρ is gas density). As the emitters are free electrons, the factors introduced to relate the measured radio flux to the mass loss rate (more strictly, the ratio $\dot{M}/v_\infty$) are typically quite easily determined. The snag with this method is that not so many stars are detected by existing radio telescopes. By contrast there are many more stars for which it is possible to measure optical/IR hydrogen line emission (for highly-evolved stars, like Wolf-Rayets, helium lines become the better mass loss tracers). Since this emission is due primarily to radiative recombination, the emissivity and hence also the total line flux are, in the optically-thin limit proportional to ρ^2. Again, conversion of an observed line flux into a mass loss rate proceeds via relatively straightforward abundance and ionization corrections. A drawback is that optical and infrared line emission from OB stars originates from quite close to the stellar surface, thereby requiring consideration of the wind velocity law.

An implicit assumption of both methods is that the wind density profile is a smooth, monotonic function of position. Noting that the mean squared density $< \rho^2 >$ is not formally equivalent to the square of the mean density $< \rho >^2$, it can be appreciated the presence of density inhomogeneities may cause the radio and recombination line emissivities to mimic those of a smooth, but denser outflow. If there is reason to suppose that outflow 'clumping' is significant, then there is also reason to worry about the impact this might have on diagnostic emission line fluxes.

In recent years this issue has begun to be taken seriously in the context of hot star mass loss, in parallel with the growing acceptance that there is no avoiding the instability of the radiation-driving mechanism. A particularly compelling discussion of this problem was

	A	MODEL 'shell'	B
$M(M_\odot \text{ yr}^{-1})$:	5×10^{-5}	2.9×10^{-5}	2.9×10^{-5}
density profile:	monotonic	sinusoidally modulated	monotonic
		Line EWs in Å	
He I $\lambda10830$	67	86	33
He I $\lambda5876$	3.7	4.5	1.6
He II $\lambda1640$	226	146	147
He II $\lambda4686$	373	264	325

TABLE 1. Emission line equivalent widths calculated from WN5 wind models
(from Hillier 1991)

presented by Hillier (1991). He took the case of the WN5 star HD 50896 and calculated wind models for it based both on monotonic and clumped density profiles. To give a taste for the results obtained, some of derived He I and He II equivalent widths are set out here in Table 1. The wind of HD 50896 is one in which He II lines are markedly stronger than those of He I. This implies that He^+ is much more abundant than the neutral species and that the neutral lines are due to $He^+ \rightarrow He^0$ recombination. Hillier calculated 3 related models. Model A (monotonic density profile) and the 'shell' model (modulated density profile, emulating clumping) differ in mass loss rate, but share the same volume-integrated recombination rate, whereas the mass loss rates for the 'shell' model and model B (monotonic density profile) are the same.

An important point to note from Table 1 is that the He I line equivalent widths of the 'shell' model are very similar to those of model A, yet distinctly higher than those of model B. This is directly attributable to the ρ^2 dependence of the recombination-dominated emissivities of the He I lines. The integrated recombination rates of models A and 'shell' are the same, implying that the He I line fluxes will be similar also. On the other hand, the the 'shell' model agrees better with model B in the matter of the He II equivalent widths. These lines are most likely to be radiatively excited and will accordingly exhibit a more nearly linear density dependence in their emissivities. Hence the sensitivity of the emergent line fluxes to wind clumping can be seen to depend on the way in which the line emission is excited.

A further example of sensitivity to wind clumping may be drawn from the spectra of the less extreme O stars. It was shown by Käufl (1993) that the strength of Brα emission in the spectrum of ζ Puppis (O4f) was a factor of a few higher than the prediction of the 'unified' wind model of Gabler *et al.* (1989). An independent observation of this line is shown in Figure 2 and compared with a profile synthesised from a model calculated using the simpler method of Drew (1989). In keeping with the result of Käufl (1993) it was necessary to apply a vertical stretch of ×2.4 to the calculated emission line profile in order to achieve reasonable agreement with the observed profile. The mass loss rate of the model used here and that of the Gabler *et al.* wind model are practically the same, and consistent with the radio determination for this star ($M = 5\times10^{-6}$ $M_\odot$ yr^{-1}, Bieging *et al.* 1989). Both models are built on smooth monotonic wind density profiles. The likely explanation for this striking discrepancy is that density inhomogeneities in ζ Puppis' wind cause an enhancement of the effective Brα emissivity—once again, this line is due mainly to recombination, since neutral hydrogen is very much the minority species.

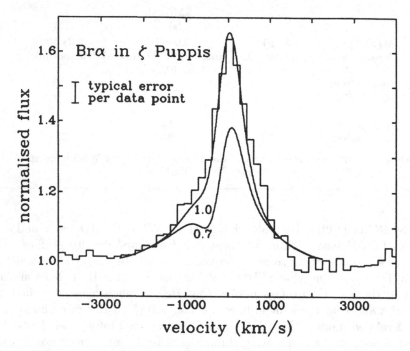

FIGURE 2. Brα in the spectrum of ζ Puppis (O4f) compared with theoretical profiles synthesised for two choices of wind velocity law ($\beta = 0.7, 1.0$ where β is the acceleration index of the commonly-used Castor & Lamers (1979) law). Both theoretical profiles have been stretched relative to the continuum by a factor of 2.4. For both, the adopted mass loss rate and terminal wind speed were 5×10^{-6} M$_\odot$ yr^{-1} and 2400 km s^{-1} respectively. From Bunn (1992).

At Hα some discrepancy between prediction and observation persists. However, in the case of ζ Puppis, it is reduced to an observed flux 'excess' of around 50% (depending on choice of velocity law, see Drew 1990). This reflects the fact that this transition is excited radiatively as well as by recombination and is partially opaque. Given that Hα flux measurements are used as a means of estimating mass loss rates, differences of this magnitude can be troublesome if measures are not taken to correct for them. Fortunately, the scaling of Leitherer's (1988) $\dot{M}$—L(Hα) relation for O-star winds to the empirical relation of Garmany & Conti (1984), based on UV and radio measurements, ensures that a mean correction is included. However, precisely for this reason, the fact that Leitherer's relation brings Hα-derived mass loss rates into agreement with radio determinations is not a basis for deducing there is no enhancement of O-star Hα emission due to the effects of wind-shocking (contrary to the claim of Lamers & Leitherer 1993).

The conclusion is thus that wind-formed line emission can be used to determine mass loss rates, but it should not be presumed that other than great care is necessary with the calibration.

4. Winds and infrared line emission

The wavelength decade from ~1μm up to ~10μm must be the spectral range of choice for the stellar observer seeking wind-formed line emission, unsullied as it is by much absorption. As noted in section 2, the weakening sensitivity of the Planck flunction to temperature in the Rayleigh-Jeans tail has much to do with this. It also has to do with the fact that most of the prominent line transitions at these wavelengths either involve excited levels well above ground (e.g. the hydrogen and helium recombination lines), or are spectroscopically forbidden.

Within the past few years, the introduction of area detectors to infrared astronomy has meant that it has at last become possible to obtain infrared spectra of a quality approaching that which has long been routine in the visible. And the technology continues to improve. One might ask what, uniquely, spectral line emission in the infrared can tell us about stellar winds. First, there is the obvious point that there are whole classes of highly-reddened mass-losing objects that simply cannot be studied effectively at shorter wavelengths. Secondly, the emphasis in the physics and the spatial scales sampled can change. Last, and not to be discounted, we stand to be surprised by what we find on entering new territory. There are examples of this already. Among the more striking discoveries is the cluster of He I emission line stars in the Galactic Center (Krabbe *et al.* 1991). This has begun to be followed up by quantitative spectroscopy (e.g. Najarro *et al.* 1994) and soon we shall know better whether these stars are evolved O stars or something more exotic (Morris 1993).

To mark this new opportunity, the following sub-sections deal with a physical effect of particular relevance to the infrared, and with a class of objects that come into their own longward of $\sim 1\mu$m.

4.1. *The effects of overlapping continuous opacity*

While line opacity is often cause for concern in understanding the properties of stellar winds, it is less common that significant continuous opacity has to be directly accounted for, especially within the accessible UV and optical wavelength ranges. This norm changes somewhat on moving up into the infrared, because of the rising opacity due mainly to free-free processes (bremsstrahlung). Bound-free opacity contributes at these wavelengths also, but is often the minority partner. In those objects whose spectra exhibit pronounced permitted line emission longward of 1μm, the suspicion must be that continuous opacity could have a significant role to play. Unsurprisingly, the most graphic illustrations of the effects of overlapping continuous opacity at IR wavelengths are to be derived from the WR stars—stars with winds so extreme that the observed continuum at every wavelength forms at a different depth in an expanding rather than a static atmosphere.

Since free-free opacity rises roughly as λ^2, the possibility exists that the observed spectrum at a shorter wavelength may probe a deeper zone in a stellar wind than that obtained at a longer wavelength. A nice demonstration that this can occur is to be found in the study of the WC8 component of γ Vel by Barlow, Roche & Aitken (1988). Observations of this source spanning the 1.4–4.1μm range provided emission line fluxes that could be used, with recombination line theory, to estimate the degree of helium ionization: Hummer, Barlow & Storey (1982) derived $n(He^+)/n(He^{2+}) = 3.8$ this way, while Smith & Hummer (1988) obtained the slightly lower ratio of 2.9. Barlow *et al.* (1988) carried out a similar exercise based on observations in the wavelength range 12.3–13.3μm. They derived $n(He^+)/[n(He^{2+}) + n(C^{2+})] = 6.4(+4.6, -2.5)$ (the appearance of the abundance of C^{2+} here is made necessary by an unresolved blend of C III and He II emission). In this result we see that the two wavelength ranges sample different volumes and, because we know the increasing opacity hides more of the inner wind at longer wavelengths, that the ionization of helium must decline with increasing radius in the WC8 wind.

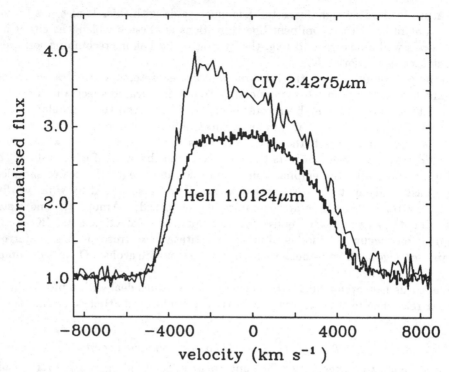

FIGURE 3. Two IR line profiles in the spectrum of WR 142, a WO star, obtained by Eenens & Williams (1991). A similar asymmetry is apparent in all of the IR transitions observed, and is likely to be due to greater erosion of red wing flux resulting from overlapping continuous opacity.

A more subtle effect than this is possible—one that can alter observed line profile shapes. Eenens & Williams (1991) have obtained observations of the He II 1.0124μm and C IV 2.4275μm emission lines in the spectrum of WR 142, one of the select group of WO stars. These are reproduced in Figure 3. Both profiles are asymmetric, and give the impression that some opacity source has intervened to preferentially erode the red wing flux. Significantly more erosion has occurred in the longer wavelength C IV line than in the He II line. Assuming that free-free opacity is the main culprit, this difference of effect could simply be due to its $\sim \lambda^2$ dependence.

How is it that overlapping continuous opacity might achieve this effect? Naively one might imagine that it just takes the concentration of continuous opacity at the heart of an expanding outflow to occult a part of the retreating hemisphere of line-emitting gas. Bunn (1992, see also Bunn & Drew 1991) explored this by means of line profile synthesis incorporating both line and continuous opacity, and showed that this purely geometric effect was insufficient. The missing ingredient turned out to be a requirement that the line source function departs significantly from the LTE source function relevant to the free-free process. The synthesised line profiles shown in Figure 4 illustrate this point. It can be seen that where the line source function exceeds the continuum source function, the effect of overlapping continuous opacity is to preferentially depress the red wing of the line. Both the emission lines shown in Figure 3 can be presumed to be due to recombination and hence may be characterised by $S_{\mathrm{line}} > B$ (where B is the Planck function). An amusing prospect is that overlapping continuous opacity could, in principle, lead to the opposite profile asymmetry. Examples of this have yet to be identified. In truth, the data of Eenens & Williams (1991) presently seems to stand alone in showing

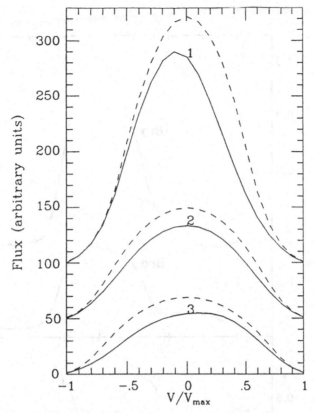

FIGURE 4. Schematic profiles illustrating the combined effect of departures from LTE and overlapping free-free opacity (Bunn 1992). Dashed lines show profiles calculated allowing for line opacity only. The solid lines are those calculated when continuum opacity is also accounted for. In 1, the line source function is raised above its LTE value; in 2, $S = B$; and in 3, the line source function is depressed by the same amount as it was increased for case 1.

red wing erosion, but then the number of published high-quality well-resolved IR emission line profiles is yet very limited.

4.2. *The nature of BN-type embedded stellar sources*

A curious feature of the evolution of stars is that they begin and approach the end of their lives in a shrouded condition that, in both instances, is accompanied by a phase of dramatic mass loss. A corollary of this is that it can be difficult to tell the young apart from the old (criteria designed to pick out YSOs not infrequently turn up post-AGB objects!). To study the inner ionized outflows of such objects, infrared spectroscopy is the appropriate tool.

The first IR spectrometers of over a decade ago began to probe regions of massive star formation and revealed in them embedded point sources with luminosities commensurate with those of normal field OB stars. The prototype of this class of presumably young objects is the Becklin-Neugebauer Object (BN) in OMC-1. The problem posed by these sources is that they exhibit signs of mass loss that are quite unlike those of their unobscured main-sequence counterparts: mass loss rates deduced from radio flux measurements are typically too high by an order of magnitude ($\dot{M} \sim 10^{-6}$ M$_\odot$ yr^{-1}), and the expansion velocities measured from H I recombination line emission are too low by a similar amount ($100 \lesssim v_{max} \lesssim 300$ km s^{-1}). So what is going on? Could these sources

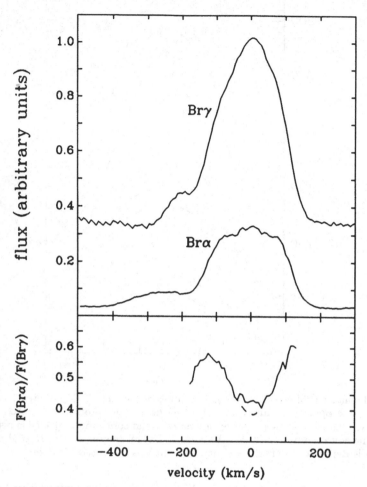

FIGURE 5. The Brα and Brγ line profiles in the spectrum of P Cygni (upper panel). Both transitions are blended in the blue wing with He I line emission. Also shown is the F(Brα)/F(Brγ) flux ratio as a function of velocity across the line profile (lower panel). The blending with He I emission is the cause of the down-turn in the flux ratio at the most negative velocity offsets shown.

still be accreting some matter, or are the phenomena merely evidence of a circumstellar clear-out?

Whilst wondering about answers to these questions, we might also consider whether we have much grasp of the IR spectral properties of normal OB stars. As yet, we do not. This needs to change, if only to provide a better context for understanding highly reddened sources. To assist the ensuing discussion, Figure 5 provides examples of the Brα and Brγ lines in the spectrum of P Cygni—an object whose wind can be described, to a first approximation, as 'understood.' These data show how the situation has been transformed in that IR line profile shapes, as well as integrated line fluxes can be used to challenge models.

Moving on, Figure 6a contains data for the same transitions observed in the spectrum of GL 989, a BN-type object. The Brα profile does not have the look of a line formed within a single body of gas—there seems to be a narrow emission feature superposed

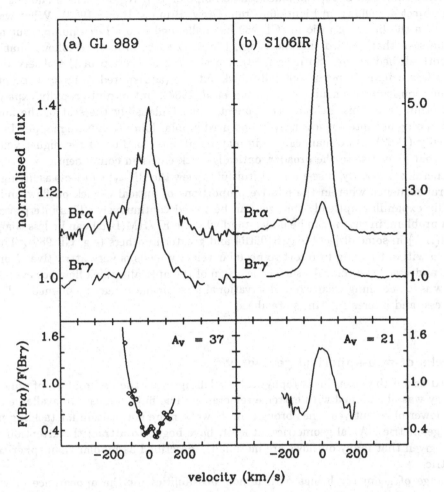

FIGURE 6. Brα and Brγ profiles in the spectra of the BN-type objects (a) GL 989, (b) S106IR. For clarity, a constant offset has been added to the Brα data. The lower panels show the flux ratios after correction for extinctions derived using methods independent of the H I line ratios. For GL 989, the points are values of the ratio derived from the profiles shown, while the solid line was obtained by dividing profile fits. The ratio plotted for S106IR is corrected for the He I feature blended into the blue wing of Brα.

on a broader component. The impression that something strange is going on close to line centre is reinforced by the velocity-dependent line flux ratio, also shown. For the case of accelerating outflow one might expect a simple 'dished' ratio, as provided by the example of P Cygni. The wings in GL 989 conform with this, but at line centre the ratio is inflected. This behaviour could be hinting at a hybrid origin for the line emission: perhaps the narrow central component showing a steeper decrement is due to optically-thin circumstellar gas, while most of the line flux arises in an optically-thick outflow of the sort targeted by the modelling of the 1980s (e.g. Höflich & Wehrse 1987, and references therein).

If correctly so described, the optically-thin component in GL 989 is little more than a minor contaminant. Intriguingly, it appears that a contaminant in one source can become the main source of emission in another. An example of this is to be found in S106IR, the exciting source of S106, the much-studied bipolar H II region. The analogous data for this object are shown in Figure 6b (from Drew, Bunn & Hoare 1993). What was an inflection in the line flux ratio for GL 989 has ballooned into a large hump. But again it can be seen that toward the line wings, the ratio turns around to follow what may be a gently-dished shape. So it seems that a significant fraction of the observed Brα and Brγ flux originates not in optically-thick outflow (as required to make sense of the radio emission from this source—see Simon *et al.* 1983), but in quite rapidly expanding optically-thin gas. This 'nebular' component is most plausibly the seat of the ionized flow expanding out into the less heavily-obscured bipolar lobes that were mapped by Solf & Carsenty (1982). Its dynamical origin may be in mass-loading of the higher velocity outflow that gives rise to the broader, optically-thick emission component.

To generalize instantly, it seems most fruitful to view the BN-type objects as fitting into a spectral sequence wherein the relative proportions of optically-thick outflow and not so rapidly expanding optically-thin gas can be varied continuously. This interpretation solves a problem, in that it can be understood why the FWHM(Brα) can be less than the FWHM(Brγ) in some objects (e.g. S106IR) and greater in others (e.g. GL 989). That it can do so without having to resort to peculiar velocity fields is very attractive. None of this has addressed the central issue of the origin of the optically-thick flow however—but only now is it becoming clear what observational phenomena it may be blamed for! This is progress, and interesting times are ahead.

5. Tricks of non-spherical geometry

It used to be the case, in astrophysical modelling, that the assumption of spherical symmetry was abandoned with extreme reluctance. The huge increase in available computing power of recent years has brought in its wake more enthusiasm for tackling more general geometries. Axial geometries, at least, have begun to attract the attention they deserve, given that many outflows are more fairly described as bipolar than spherically-symmetric.

A change of geometry brings with it new possibilities for the appearance of wind-formed spectral line profiles. An amusing and undoubtedly extreme case to note in this context is that of SS 433 with its precessing jets and consequently highly mobile optical hydrogen line emission (e.g. Margon & Anderson 1989). Continuing the infrared theme of the previous section for a moment, an interesting example of the intrusion of geometry upon line profile form has been put forward by Oudmaijer *et al.* (1994). They have used geometry to explain curious differences between the appearance of IR H I line profiles and the form of the Hα line in the spectrum of IRC+10420 (F8I⁺). Specifically, both blueshifted and redshifted emission is detected at Hα but the redshifted component is missing from the IR lines. The suggested cause of this is that the emission arises in a bipolar flow oriented such that geometric occultation of the receding emission occurs in all but the strongest line, Hα, emanating from the largest volume. Is this is a bipolar planetary nebula in the making?

Accretion disks have become commonplace in astrophysics. A happy feature of the accretion disks and associated outflows in binary systems is that the binarity routinely affords us a truly independent estimate of our viewing angle. For this and a list of reasons, these systems are without question the 'laboratories' of choice for testing our understanding of accretion physics and related phenomena. It is interesting to note that already much effort has gone into developing wind models for the accretion disks associated with star formation. Some thought about how such models would apply to

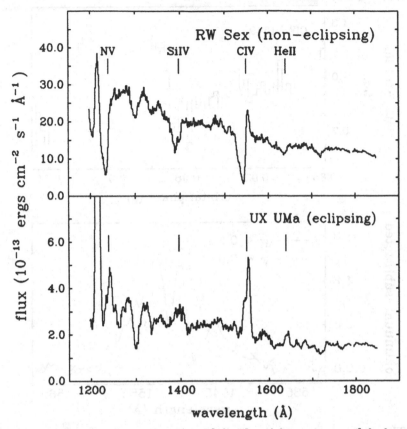

FIGURE 7. The upper panel shows a segment of the ultraviolet spectrum of the low-inclination (non-eclipsing) nova-like variable SW Sex. Note the well-developed blueshifted resonance line absorption. The lower panel shows the same wavelength range in the spectrum of the high-inclination nova-like variable UX UMa. The change of orientation throws those same resonance transitions seen mainly in absorption in lowly-inclined systems into emission.

the winds of cataclysmic variables could be richly rewarded by the opportunity to test them with a rigour unthinkable in the context of star formation.

It has been plain for some time that a change of aspect can completely transform the spectral signatures of outflow seen in the spectra of cataclysmic variables. The ultraviolet resonance lines seen primarily as blueshifted absorption features in non-eclipsing disk-accreting close binaries, become apparently unabsorbed broad emissions in their eclipsing relatives. Figure 7 provides a comparison between the spectra of a non-eclipsing and an eclipsing nova-like variable. The fact that the spectral transformation is so complete has been shown, quantitatively, to favour bipolar outflow (Drew 1987). The essential reasoning underpinning this may be expressed in the language of section 2. The apparent area of a bipolar structure is larger when viewed from the side than down its axis— conversely a flattened disk (the continuum source in the present case) appears larger face-on than edge-on. Putting these attributes together, it may be seen that it is easy to arrange $A_{line} < A_{disk}$, and hence net line absorption, in non-eclipsing systems at the same time as $A_{line} > A_{disk}$ and net emission in the eclipsing case.

The presence of accretion disk winds in eclipsing binaries can be exploited in a yet more discriminating way. Time series of ultraviolet spectra spanning primary eclipse can be obtained in order to follow changes in the wind-formed lines that are presumably dictated

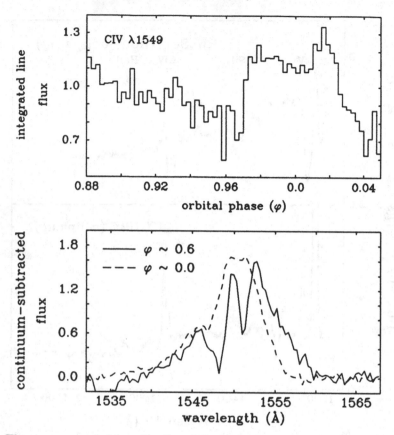

FIGURE 8. The upper panel shows the C IV λ1549 line flux eclipse light curve (obtained by merging data from three eclipse sequences, and normalising to the out-of-eclipse line flux). The lower panel compares the out-of-eclipse line profile, obtained around orbital phase 0.6, with the mid-eclipse profile. The net line flux increase during continuum eclipse is due to the disappearance of the superimposed narrow absorption features seen away from eclipse—this more than offsets the occultation of some redshifted line flux.

by the outflow geometry and kinematics. By following lines of differing ionization, it should be possible to learn something of the ionization structure as well.

A first experiment of this kind has been undertaken using the GHRS on board the Hubble Space Telescope (Mason *et al.* 1995). The target selected was UX UMa, the eclipsing system whose spectrum is shown in Figure 7. Sequences of spectra were obtained of the C IV λ1549 and He II λ1640 emission profiles at an effective spectral resolution of ∼0.3Å and time resolution better than a minute. Some aspects of what was found had been correctly anticipated (e.g. the form of the line flux eclipse light curve, the appearance of the mid-eclipse line profile: see Figure 8 and Drew 1987). The surprise came in the form of the superposed absorption component observed. In Figure 8 it can be seen that the out-of-eclipse C IV profile is indented near line centre by two quite sharp absorption components (the components of the doublet). Observations obtained using the *International Ultraviolet Explorer* are of insufficient S/N and spectral resolution to reveal them. It is not yet clear whether these features should be attributed to the outflow or to disk-edge structure. If due to the outflow, their presence is likely to imply very tight collimation (Knigge, Woods & Drew 1994) and the timing of their eclipse would certainly indicate that the mass loss begins in the accretion disk, at least to a radius of order 5 white-dwarf radii, rather than in either the boundary layer or white dwarf.

This is a beginning. There is plenty of scope both for further observations and further numerical modelling. It may be deemed ironic that it should happen to be the emission form of the resonance lines, rather than the 'P Cygni' profile, that offers greatest promise of unravelling the structure of these flows.

6. Postscript

It is a misleading but not uncommon impression that the primary theoretical tool of mass loss studies is an inscrutable mix of radiation transport theory and numerical simulation. In truth, direct physical insight remains the commodity of highest value. This property, which this contribution has endeavoured to keep to the fore, is undoubtedly shared with the study of nebulae. Were this not so, this author could hardly be so endebted to a former association of several years with Professor Seaton's group at University College London.

Acknowledgments

JED is presently in receipt of an Advanced Fellowship funded by the Particle Physics & Astronomy Research Council of the United Kingdom. The ζ Puppis data (Figure 2) were obtained as a service observation on the United Kingdom Infrared Telescope, also funded by the Particle Physics & Astronomy Research Council.

REFERENCES

BARLOW, M. J., ROCHE, P. F., & AITKEN, D. 1988 *MNRAS* **232**, 821

BEALS, C. 1950 *Publ. Dom. Astrophys. Obs. Victoria* **9**, 1

BIEGING, J. H., ABBOTT, D. C., & CHURCHWELL, E. 1989 *ApJ* **340**, 518

BUNN, J. C. 1992 D. Phil. Thesis, Oxford, England

BUNN, J. C., & DREW, J. E. 1992 *MNRAS* **255**, 449

BUNN, J. C., HOARE, M. G., & DREW, J. E. 1994 *MNRAS*, in press

CASTOR, J. I., & LAMERS, H. J. G. L. M. 1979 *ApJS* **39**, 481

DREW, J. E. 1987 *MNRAS* **224**, 595

DREW, J. E. 1989 *ApJS* **71**, 267

DREW, J. E. 1990 *ApJ* **357**, 573

DREW, J. E., BUNN, J. C., & HOARE, M. G. 1993 *MNRAS* **265**, 12

EENENS, P. R., & WILLIAMS, P. M. 1991 in *The infrared spectral region of stars* (eds. C. Jaschek & Y. Andrillat) CUP, England, p. 158

GABLER, R., GABLER, A., KUDRITZKI, R., PULS, J., & PAULDRACH, A. 1989 *A&A* **226**, 162

GARMANY, C. D., & CONTI, P. S. 1984 *ApJ* **284**, 705

HILLIER, D. J. 1991 *A&A* **247**, 455

HÖFLICH, P., & WEHRSE, R. 1987 *A&A* **185**, 107

HUMMER, D. G., BARLOW, M. J., & STOREY, P. J. 1982 in *Wolf-Rayet Stars*, IAU Sym. No. 99, (eds. C. de Loore & A. Willis) Reidel: Dordrecht, Holland, p. 149

KÄUFL, H. U. 1993 *A&A* **272**, 452

KNIGGE, C., WOODS, J. A., & DREW, J. E. 1994 *MNRAS* in press

KRABBE, A., GENZEL, R., DRAPATZ, S., & ROTACIUC, V. 1991 *ApJ* **382**, L19

LAMERS, H. J. G. L. M., & LEITHERER, C. 1993 *ApJ* **412**, 771

LEITHERER, C. 1988 *ApJ* **326**, 356

MARGON, B., & ANDERSON, S. F. 1989 *ApJ* **347**, 448

MASON, K. O., DREW, J. E., CORDOVA, F. A., HORNE, K., HILDITCH, R., KNIGGE C., LANZ, T., & MEYLAN, T. 1995 submitted to *MNRAS*

MORRIS, M. 1993 *ApJ* **408**, 496

NAJARRO, F., HILLIER, D. J., KUDRITZKI, R. P., KRABBE, A., GENZEL, R., LUTZ, D., DRAPATZ, S., & GEBALLE, T. R. 1994 *A&A* **285**, 573

OUDMAIJER, R. D., GEBALLE, T. R., WATERS, L. B. F. M., & SAHU, K. C. 1994 *A&A* **281**, L33

PANAGIA, N., & FELLI, M. 1975 *A&A* **39**, 1

SIMON, M., FELLI, M., CASSAR, L., FISCHER, J., & MASSI, M. 1983 *ApJ* **266**, 623

SMITH, L. F., & HUMMER, D. G. 1988 *MNRAS* **230**, 511

SOLF, J., & CARSENTY, U. 1982 *A&A* **113**, 142

STAHL, O., MANDEL, H., SZEIFERT, T., WOLF, B., & ZHAO, F. 1991 *A&A* **244**, 467

WRIGHT, A. E., & BARLOW, M. J. 1975 *MNRAS* **170**, 41

Photoionising Shocks

By MICHAEL A. DOPITA

Mount Stromlo and Siding Spring Observatory, Institute of Advanced Studies,
The Australian National University

A high velocity radiative shock, or one moving into high-metallicity gas provides an efficient means to generate a strong UV photon field. If there is gas available in the pre- or post-shock zones to absorb this, then the optical emission from the shock and precursor region can be dominated by photoionisation, rather than by cooling and the optical + UV emission scales as the mechanical energy flux through the shock. The diagnostic characteristics of such shocks are discussed for the cases of supernova remnants, NLRs, LINERs and cooling flows.

---

1. Introduction

Let me preface this paper to say how honoured I am to have been given this opportunity to pay tribute simultaneously to two of the principal sources of scientific inspiration of my career. In my attempts over the past twenty years to understand and to analyse the optical and UV spectra of shock-excited plasmas, Don's books (1974, 1988) have been invaluable to both myself and to my students. In Australia we used to refer to the *Physics of Gaseous Nebulae* somewhat irreverently as "the new testament" to distinguish it from the earlier work by Aller! The famous diagnostic diagrams of Baldwin, Phillips & Terlevich (1981), of Veilleux & Osterbrock (1987), and of Osterbrock, Tran & Veilleux (1992) provide both an inspiration, and a powerful means of distinguishing between various excitation mechanisms. The name of the photoionisation/shock emission code which Luc Binette, Ralph Sutherland and I jointly developed (MAPPINGS), was chosen to convey the sense of an analytic tool to map out uncharted reaches of such diagnostic space. Of course, the accuracy and utility of all such codes is limited entirely by the quality and quantity of the atomic data, and an understanding of the relevant physical processes to apply which is largely due to Mike's work over at least the last thirty five years (*e.g.* Seaton 1958, 1959), the contributions of his students and the work of his London group.

When preparing this tribute, I was tempted to provide a review of how the subject of optical shock diagnostics has developed since the pioneering work of Don Cox (1972), John Raymond (1976) and myself (1976). However, such a review would probably seem too turgid for the experts and too compressed for the neophytes. In any event, excellent reviews covering the developments over that period have already been supplied by McKee & Hollenbach (1980) and by Draine & McKee (1993), and it would be hard to improve on them. Since the work of both Don and Mike is orientated towards the future rather than being rooted in the past, I have indulged instead in an exposition on a relatively new subject that I believe provides a key to a proper understanding of shock phenomena in young supernova remnants and in active galaxies, that of "photoionising shocks." Much of the work described is from the thesis of my ex-student Ralph Sutherland (1993), published in Sutherland & Dopita (1993, 1994a,b) and in Sutherland, Bicknell & Dopita (1993).

2. Photoionising shocks

A radiative shock in interstellar space has long been recognised as supplying a potential source of hydrogen ionising UV photons. These photons, produced in the hot, post-shock cooling plasma can diffuse both upstream, to generate a photoionised precursor; and downstream, to influence the ionisation and temperature structure of the recombination region of shock. The first reference to the photoionisation effects of precursors to shocks was in Daltabuit & Cox (1972) in the context of ~ 1000 km s^{-1} cloud-cloud collisions in the broad-line regions of AGN. For lower velocity radiative shocks ($v < 150$ km s^{-1}), the shock models of Raymond (1976) and Dopita *et al.* (1984a) allowed for photoionisation both up- and down-stream. However, since the photoionising fields are weak in these low velocity shocks, the emission from the precursor region is negligible, and the photons serve merely to pre-ionise the gas coming into the shock. This in turn radically affects the emission line spectrum of the shock allowing sensitive shock diagnostics to be made. Shocks with velocities lower than about 100 km s^{-1} are unable to pre-ionise the precursor region, and these produce an enhanced two-photon continuum through collisional excitation of the neutral hydrogen, an effect first recognised by Dopita, Binette & Schwartz (1982).

As the shock velocity increases, the shape of the cooling curve ensures that the cooling plasma cools in a thermally unstable fashion. Innes, Giddings & Falle (1987a,b) claimed that this invalidated all shock diagnostics. However further work by Innes (1992) has established rather that the plasma undergoes limit cycle oscillations, and that the time averaged mean spectrum is more like the steady-flow solution. In fully three dimensional modelling, it is clear that secondary shocks driven by the development of pressure differences between adjacent cooling regions may have an important contribution to make to the resultant optical/UV spectrum, but as yet such models have not been made. An early attempt to investigate the effects of radiative transfer in thermally unstable cooling flows or radiative shocks was by Binette, Dopita & Tuohy (1985). The hot plasma was modelled on the assumption of collisional ionisation equilibrium (CIE) cooling using the Raymond & Smith (1977) code, and the effect of thermal instabilities was approximated by assuming the formation of photoionised dense cloudlets in pressure equilibrium with the hot plasma. The opposite extreme, of plane parallel sheets cooling in thermal stability was also treated.

A fully self-consistent model for high velocity shocks cooling in a thermally stable fashion had to await the development of codes capable of dealing with the non-CIE cooling, which include X-ray emission processes and radiative transfer. Recently, this was accomplished in the code MAPPINGS II (Sutherland & Dopita 1993) and similar results have been produced by Schmutzler & Tscharnuter (1993). However, the proper treatment of secondary shocks still remains to be done. The results of this paper have been generated neglecting these secondary shocks, and dealing with thermal instabilities only through the limits which these place on the geometrical parameters of the photoionised plasma with respect to the diffuse photoionisation source which is generated by the hot plasma.

A radiative shock radiates effectively all the available enthalpy. At high velocities, the predominant emission mechanisms are electron thermal bremsstrahlung and line emission. The non-equilibrium ionisation in the post-shock plasma enhances line emission enormously over the equilibrium case, particularly for the hydrogen-like and helium-like ionic species, which undergo many collisional excitations of their principal resonance lines before they are ionised. The output spectra for a number of shock velocities is shown in Figure 1. Note that the bulk of the radiation occurs in the FUV region, with an exponential bremsstrahlung tail extending to thermal X-ray wavelengths for the higher

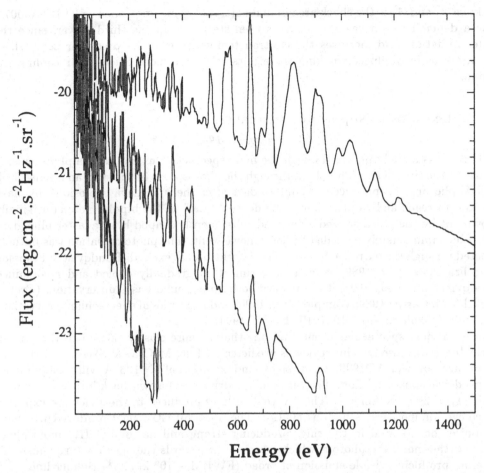

FIGURE 1. The EUV/soft X-Ray fluxes generated by fully radiative shocks with velocities 200 km s^{-1} (lower curve), 300 km s^{-1} (middle) and 500 km s^{-1} (upper). Note the rich resonance line spectrum, and the underlying exponential thermal Bremstrahlung continuum.

velocity shocks (> 300 km s^{-1}). The total EUV/X-ray radiative flux is closely coupled to the total dissipation of mechanical energy through the shock. For a plane-parallel shock the upstream ionising flux is given by:

$$L_{FUV} = 4.08 \times 10^{-3} n_o V_{200}{}^{3.0} \text{erg cm}^{-2}\text{ s}^{-1} \qquad (2.1)$$

A substantial proportion of this radiation field is fluorescently converted through photoionisation to photons at optical wavelengths. This suggests that the shock luminosity in the optical lines should also scale as the area of the shock, and as the cube of the shock velocity. Globally, this expectation is borne out by detailed models and this scaling also holds for individual emission lines which are mainly collisionally excited such as [N II], [S II] or the UV resonance and intercombination lines. However, in the case of the recombination lines of hydrogen, the decrease in recombination coefficient at high temperature tends to decrease the efficiency of production of Balmer photons at higher shock velocities. For Hβ, the total shock + precursor luminosity is:

$$L_{H\beta} = 6.9 \times 10^{38} n_o V_{200}{}^{2.44} A_6 \text{erg s}^{-1} \qquad (2.2)$$

where n_o is the pre-shock density (cm^{-3}), V_{200} is the velocity of the fast shock in units of

200 km s^{-1}, and A_6 is the shock area in units of kpc^2. Some lines, such as [O III] $\lambda5007$Å show a dependence on velocity which is even steeper than the third power, since the harder radiation field increases the electron temperature in the precursor zone where they are mostly produced, making these lines relatively more efficient for cooling the plasma.

3. Applications to supernova remnants

3.1. *Supernova 1987A*

SN 1987A has already provided one of the more spectacular examples of a photoionising shock. At the time of shock breakout through the photosphere of the star, an intense flash of EUV photons was produced. About 70 days after the supernova explosion, this flash was seen to reach and to photoionise the dense "waist" of the bipolar nebula originally thrown out by the precursor red giant, and subsequently shaped by the faster blue giant wind which immediately preceded the supernova event. The photoionisation was initially detected through the narrow UV emission lines which were excited (Lundqvist & Fransson 1987; Fransson *et al.* 1989), and, as the ring itself gradually cooled and recombined (Lundqvist & Fransson 1991) it was imaged both from ground-based observations (Crotts, Kunkel & McCarthy 1989; Wampler *et al.* 1990) and in a series of spectacular observations with HST (Jakobsen *et al.* 1991; Plait *et al.* 1994).

An even more spectacular event, and one that is more closely related to the type of shocks being discussed in this review is predicted by Luo, McCray & Slavin (1994). This will occur in or near AD1999, when the expanding ejecta of SN 1987A will finally impact on this dense equatorial ring. The collision will drive a radiative shock into the ring with a velocity of 200–400 km s^{-1}. The UV cooling lines produced in the shock are expected to achieve luminosities as high as 800L$_\odot$. The powerful EUV field produced will both pre-ionise the material in the ring, producing strong and narrow [O III] and Balmer lines. Furthermore, the photons which escape back towards the ejecta will re-ionise this material, producing visible emission of broad (FWHM $\sim 10^4$ km s^{-1}) Balmer lines.

3.2. *Oxygen rich supernova remnants*

The young supernova remnants (SNRs) are offer an key to test theories of nucleosynthesis in massive stars, since only in these objects do we find material from the stellar interiors exposed to direct investigation. A small class of SNRs show fast moving (> 1000 km s^{-1}) knots of material having lines of oxygen, neon, and other heavy elements, but neither hydrogen or helium, consistent with their origin from within the helium-burnt layers of a massive (> 12 M$_\odot$) progenitor star. In our Galaxy, Cas A is the prototype of this class (see Chevalier & Kirshner 1979). The other members are the galactic objects G292+1.8 and Puppis A (Goss *et al.* 1979; Winkler & Kirshner 1985), three remnants in the Magellanic Clouds (Lasker 1978; Dopita *et al.* 1981; Kirshner *et al.* 1989), and two unresolved objects in more distant galaxies M83 and NGC 4449 (Long *et al.* 1989; Blair *et al.* 1983).

A fundamental barrier to a quantitative interpretation of the spectra from these objects has been our inability to produce a plausible model for the excitation of these knots. It is clear that the knots are formed by an early instability in the supernova ejecta, and are excited at the time that they pass through the reverse shock, which is propagated back into the low-density phase of the ejecta as this interacts with the surrounding ISM. Logically, therefore, the knots are excited by slow shocks resulting from the increase in external pressure. However, in oxygen-rich material, the structure of these shocks is quite different from those which occur in material of normal composition. In particular,

Itoh (1981a, b) and Dopita, Binette & Tuohy (1984) demonstrated that the very high metallicity causes the cooling time to remain shorter than the recombination time until the shocked plasma has cooled to a few hundred degrees K. As a consequence, the optical [O III] lines are predicted to be bright, but lines of species of lower ionisation are weak, unless the shock velocity is less than ~ 100 km s^{-1}. However, in these remnants, lines of both [O I] and of [O II] are observed to be bright. Itoh (1981b) argued that the powerful EUV field generated in the cooling zone could escape upstream to produce an extensive warm photoionised precursor region. This idea has the difficulty that the extent of this region would have to be greater than the observed size of the knots. Furthermore, the temperature could only be maintained at a temperature high enough to excite optical forbidden lines in the case that it was composed of pure oxygen. The efficient cooling in the fine-structure lines of any other elements present would normally force the photoionisation equilibrium temperature in this zone to lie close to 300K, suppressing the optical cooling.

Dopita (1987) suggested that a non-equilibrium solution for this precursor region may be closer to the mark. The buildup of the strong precursor radiation field occurs over a cooling timescale, allowing an ionisation front to be propagated into the cloud. For an R-Type ionisation front driven by a strong enough photoionising field, the heating effect of the photon field dominates at the leading edge of the front, and the gas is strongly superheated, producing optical forbidden lines of a range of ionisation states. As the gas approaches its photoionisation equilibrium, the temperature of the plasma approaches its equilibrium value, and optical lines are suppressed in the low temperature, high ionisation plasma.

This idea was fully developed in Sutherland's thesis (1993); Sutherland & Dopita (1994b). He showed that the cloud shock by itself could drive the R-Type ionisation front, rather than the bow-shock emission which had been invoked by Dopita (1987). The buildup of the EUV emission is sufficiently rapid to allow an R-Type ionisation to be detached from the shock in $\sim 1/10$ of the cooling timescale; about $3/n_o$ years, where n_o is the pre-shock density in units of cm^{-3}. The optical emission is produced in a brief period while and after the cloud shock becomes radiative, and as the thin superheated zone close to the ionisation front sweeps through the cloud. The ionisation front velocity is comparable with the dynamical timescale, the time need for the cloud to fully enter the reverse shock, and this timescale is itself a few times shorter than the cloud crushing timescale. For this reason, the lifetime of an individual knot is short, and theory shows that, for cloud densities ~ 100 cm^{-3}, it is comparable with the 25 year e-folding lifetimes observed for the Cas A knots by Kamper & van den Bergh (1976). The combination of the (steady-flow) cloud shock emission and the contribution from the R-Type precursor gives a good description to the diagnostics derived from the various ratios that can be formed from [O I] $\lambda 6300$Å, [O II] $\lambda 3727,9$Å and $\lambda 7316,24$Å, and the [O III] $\lambda 4363$Å and $\lambda 5007$Å lines. An example from Sutherland (1993) is given in Figure 2. Each of the line ratios indicate a cloud shock velocity of 100–200 km s^{-1} is appropriate for the best observed objects. However, UV diagnostics from observations such as those by Blair *et al.* (1989) are required in order to obtain useful abundance data, especially for elements such as C, Mg and Si. We look forward to HST to provide such data.

4. Applications to AGN

All classes of AGN show some evidence of "narrow Line" emission in the [O III] lines. Seyferts of Classes 1 and 2 (Khachikian & Weedman 1974) both show a rich spectrum of narrow lines, as do the narrow line radio galaxies (NLRGs). The universal characteristic

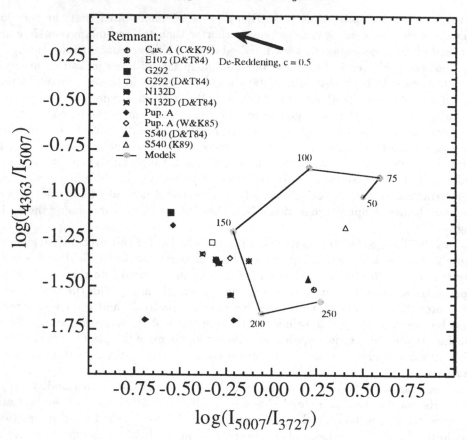

FIGURE 2. Oxygen line shock diagnostic diagram for oxygen-rich SNR. The theoretical ratios for the shock + precursor region are shown by a line marked with the shock velocity. Shock velocities in the range 100–200 km s^{-1} are indicated by these models.

of these narrow line regions is a very large [O III] λ5007/Hβ ratio ($\sim$7–20), the presence of He II and strong lines of [O I], [N II] and [S II]. Although the lines are termed "narrow," this is something of a misnomer, since the typical velocity widths are 200 < FWHM < 500 km s^{-1}, and a few objects even exceed 1000 km s^{-1} (Veilleux 1991a,b).

In the last fifteen years, the hypothesis that these narrow line regions (NLRs) of AGN are excited by a hard power-law spectrum of UV photons has gained almost universal acceptance. In such a model, the NLRs would be excited by a fairly flat UV power law, or else a truncated power law with an ionisation parameter (U) typically of order 10^{-2} (Koski 1978; Stasinska 1984; Veilleux & Osterbrock 1987; Osterbrock, Tran & Veilleux 1992; Baum, Heckman & van Breugel 1992). The source of these photons is assumed to be the nucleus, but the exact means whereby the non-thermal UV spectrum is generated remains obscure. Generally what is done is to simply join the observed soft X-ray to the observed FUV point, or to extrapolate the FUV slope with an exponential cutoff to match the soft X-ray. Neither of these procedures properly addresses the physical processes producing the emission, and there remains a serious shortfall in the estimated number of ionising photons based on an extrapolation of the observed UV spectrum (Binette, Parker & Fosbury 1993).

The development of so called "unified models" of AGN (Barthel 1989) has been driven fairly heavily by the observation of cones of highly ionised narrow-line gas extending from the nucleus (Pogge 1988a,b, 1989; Tadhunter & Tsvetanov 1989; Fosbury 1989). In the standard picture, the nuclear radiation is assumed to be directionally beamed (Morganti *et al.* 1991, 1992). Despite a deal of success in describing the emission line spectrum, the weaknesses of such models are that a number of parameters remain arbitrary. In particular, the model fails to give an explanation for the relatively small range of ionisation parameter over a very large radial scale range, and it fails to even address the question of the observed velocity dispersion of the gas.

The other major class of narrow-line AGNs is that of the low-ionisation nuclear emission-line regions, LINERs (Heckman 1980). These fall into a distinct region of excitation space when classified according to their emission line ratios (Baldwin, Phillips & Terlevich 1982; Veilleux & Osterbrock 1987). If LINERs are photoionised, then again, the ionising spectrum has to be fairly flat, and the ionisation parameter is low, $U \sim 10^{-4}$ (Ferland & Netzer 1983). LINERs, or objects having LINER-like spectra are found in the nuclei of spiral galaxies lacking nuclear starbursts (Keel 1983a, b), in as many as 50% of elliptical galaxy nuclei (Phillips *et al.* 1986), and in the "cooling flows" associated with first-ranked elliptical galaxies in clusters (Heckman 1981; Cowie *et al.* 1983; Johnstone, Fabian & Nulsen 1987; Heckman *et al.* 1989). Again, typical velocity dispersions observed in all these objects are a few hundred km s^{-1}.

A promising model for all these classes of object is based on the input of mechanical energy, rather than on photoionisation from a central source. The source of the mechanical energy may be either turbulent motions of clouds in the potential of the galaxy, or direct energy deposition from an outflowing jet of relativistic electrons. It should be remarked that this idea is by no means new. Osterbrock (1971) argued for shocks as a means to set up the multi-phase structure of AGN, and Daltabuit & Cox (1972) pointed out how such high velocity shocks in a turbulent medium could of themselves provide an ample supply of ionising photons.

4.1. *Jet excited NLRs*

We cannot consider the interaction of jets with the surrounding interstellar medium without going past the beautiful and instructive observational series of papers by Morse and his colleagues. They have studied a number of Herbig-Haro (HH) jets ejected from pre-main sequence stars, which demonstrate all the hydrodynamic phenomena that had been predicted, including bow shocks and Mach disks (Morse *et al.* 1993a,b), jet instabilities, entrainment, and wall shocks (Hartigan *et al.* 1993; Morse *et al.* 1994; Raymond *et al.* 1994).

The analogy of these phenomena with AGN is not perfect, since the HH jets are cool and heavy with respect to their surrounding medium, but the AGN case is one of interaction of a light hot or even relativistic plasma with a surrounding relatively dense and cool medium. Nonetheless all of these phenomena also occur, as shown in the hydrodynamic calculations of Bicknell (1990; 1994). In these, a mildly supersonic jet confined by a denser medium will strongly interact with the surrounding medium through the operation of the Kelvin-Helmoltz instability at the boundary. This generates slow shocks in the surrounding medium, and turbulent vortices pull off dense clouds of gas embedded in a lower-density phase which is partially entrained in the jet. The turbulent collisions of clouds generate fast shocks, and slower shocks are driven into them.

In an entraining jet-driven flow of this type, we would expect to find a number of global and local correlations peculiar to the shock-excited model. Because the hot plasma is confined by the cooler, we would expect to find the optical emission surrounding any radio lobes in regions where the wall density is sufficient to allow the shocks to become radiative. Because the brightness of the radiative shock and its precursor scales as the rate of dissipation of mechanical energy through it, we would expect to find a strong correlation between local velocity dispersion and local surface brightness. Since the driving pressure is the ram pressure, the luminosity per unit area will scale with the local density ρ and the pressure P as $P^{3/2}\rho^{-1/2}$. This implies that, for an isothermal matter distribution in the ISM, the local surface brightness will decline as $1/r^2$, exactly the same functional variation as expected in the case of photoionisation from a central source. However, the jet-driven flow will be distinguished by a global outflow velocity in the lobes, superimposed on the local rotation curve, whereas the photoionised case will show only rotation. In the shock-excited case the excitation of the narrow-line region will be a strong function of the local velocity dispersion and/or outflow velocity, highly shocked regions having higher excitation, and showing stronger forbidden lines with respect to the Balmer lines. Since we would expect the shocks to be driven by the pressure in the relativistic plasma, we would expect to find similar pressures in the relativistic and NLR gas. Since radiative shocks in general are efficient producers of forbidden lines then, to the extent that the radio emission is a measure of the gas pressure in the lobes, we would expect to see a correlation of the flux in these lines with the radio luminosity.

In fact, all of the properties of entraining jet-driven flows listed above are encountered in various well-observed NLRs. However, what is not yet so clear is that these properties are universal to all NLRs. The papers that could be referenced here are legion, so the sample I now quote should be taken as simply giving examples of each phenomenon. Close correlations between the forbidden line regions and the radio continuum regions are found in the sample of nearby Seyferts studied in the extensive survey by Tsvetanov, Fosbury & Tadhunter (1994). In particular, there is often an excellent alignment with inner radio jets (Unger *et al.* 1987; Wilson & Tsvetanov 1994). Higher redshift and more luminous sources likewise show strong alignments between the NLR and the radio (McCarthy *et al.* 1987; Baum *et al.* 1988; Meisenheimer *et al.* 1994). Evidence of balance between the thermal pressure in the NLR gas and the relativistic and magnetic pressure in the lobes was found by De Bruyn & Wilson (1978) and Unger *et al.* (1987). Correlations of line flux with radio luminosity were found by Baum & Heckman (1989; [N II] + Hα), Forbes & Ward (1993; [Fe II]), and Morganti *et al.* (1994; [O III]). A dynamical correlation between [O III] line width and the radio luminosity was found for high luminosity sources by Wilson & Willis (1980) and Whittle (1992, 1994). Evidence for global outflow and for line splitting and broadening of several hundred km s^{-1} has been found in many objects 9se review by Whittle 1989). To give a few particular examples, 3C120 (Meurs *et al.* 1989), NGC 1068 (Antonucci & Miller 1985; Cecil & Bland 1989), NGC 3516 (Mulchaey *et al.* 1992; Veilleux *et al.* 1993), N2992 (Tsvetanov, Dopita & Allen 1994, in prep), and NGC 5252 (Morse *et al.*, this conference). A relationship between the surface brightness and level of excitation of the NLER was discussed by Fosbury (1989) and Prieto *et al.* (1989), and Koekemoer *et al.* (1994) showed that a flux : dynamical relationship of a form similar to Eq. 2.2 exists for the NLERs of luminous radio galaxies.

The nearest narrow-line AGN, Cen A, offers a perfect test case in which to study these phenomena in detail. Here, direct dynamical evidence for entrainment has been discovered in the filaments associated with the NW radio lobe (Bicknell, Dopita & Sutherland 1994). There is a remarkably strong correlation between the spatial position and the dynamics, in that regions of highest velocity are found furthest in the direction of the radio lobe. The optical spectrum of these filaments has been self-consistently modelled as the result of a collision between two clouds of unequal density by Sutherland, Bicknell & Dopita (1993), using our code MAPPINGS II. The advantage of this model is that the shocked region itself produces significant amounts of optical radiation, particularly in respect of the [O III] λ4363Å line. As a result, the electron temperature inferred for the "global" spectrum (summed over the low and high velocity cloud shocks and the photoionised precursor region) is predicted to be a good deal higher (∼17000K) than is expected with simple photoionisation models (∼11000K). The high temperature inferred from the shock model is much closer to what is observed (Morganti *et al.* 1991). The low temperature given by photoionised models, compared with observation, is a problem which is common to many other NLRs, for which [O III] electron temperatures are observed to range up to ∼22000K (Tadhunter, Robinson & Morganti 1989).

The paper by Sutherland, Bicknell & Dopita (1993), emphasised how thermal instabilities and shock structure can influence optical line ratios such as [O III]/Hβ. In addition, shock curvature can have a profound influence on the excitation in the precursor zone. These effects can be illustrated using diagnostic diagrams of Baldwin, Phillips & Terlevich (1981) or Veilleux & Osterbrock (1987), which provide a powerful means of distinguishing between the various excitation mechanisms. As an example, in Figure 1, we show the observational results compared with the computed plane-parallel, and spherical shocks and with the computed H II region sequences in the [O I]/Hα vs [O III]/Hβ plane. It is clear that plane parallel geometries place us in the region of the NLRs. On the other hand, spherical shocks provide a good description of the LINER spectra (see next section). Furthermore, the well-known correlations of the [O I], [N II], and the [S II] flux ratios with respect to H-Alpha (Keele 1983; Veilleux & Osterbrock 1987), are also found to be reproduced by simple sequences of shock velocities.

The shock and the photoionised models both produce a good description of the optical spectral diagnostics, with the exception of the temperature sensitive line ratios. However, the parameters which control the mapping onto the diagnostic planes are different. In the photoionisation models, the parameters are the spectral index and the ionisation parameter. In the shock models the corresponding parameters are shock velocity and shock geometry with respect to the photoionised precursor regions. To achieve a definitive distinction between the two models from emission line diagnostics alone, we have to make observations in the UV. Those resonance lines which are predominantly produced in the cooling zone of the shock are predicted to be much stronger than in a photoionisation model that would give a similar optical spectrum. In particular, the UV line ratios that are temperature sensitive are expected to give a much higher temperature than the [O III] line ratio. In this respect the HUT (Kriss *et al.* 1992) on NGC 1068 provide an unambiguous diagnostic. In this object the C III λλ(1907+9)/977Å intensity ratio gives a temperature in excess of 27000K, and the N III λλ1750/991Å ratio suggests temperatures in excess of 24000K. In addition, the absolute intensity of the C IV line is much higher than would be predicted from a photoionised model, even accounting from the flux in the broad wings, which are probably derived from scattering from the hidden central object. In NGC 1068, there seems little doubt that shocks are of paramount importance in understanding the emission line spectrum.

4.2. *LINER galaxies*

A number of lines of evidence suggest that the relationship between LINERS and other types of narrow-line AGN is quite intimate. LINERS are characterised by large [O II]/ [O III], strong [O I] [S II] and [N II] lines in the red, and typically display velocity dispersions of 100–400 km s^{-1}. They continue the trend of decreasing velocity dispersion and Balmer luminosities towards lower excitation. A survey of "normal" elliptical galaxies by Phillips *et al.* (1986) showed that low-level LINER activity is found in an appreciable fraction of these. In the spiral galaxies, LINER activity appears to be found in virtually all nuclei which do not show nuclear starburst activity (Keel 1983a,b). Elliptical LINER

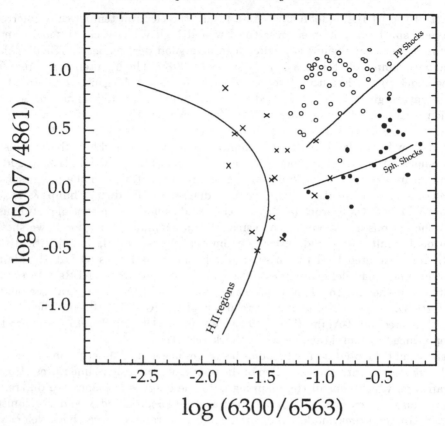

FIGURE 3. The observational results for Seyfert 2, NLRGs (open circles) and other objects with narrow emission line nuclei (crosses) from Veilleux & Osterbrock (1987); and for spiral nuclei with LINER emission from Keele (1983b) plotted on the oxygen excitation diagram. The H II region sequence is shown, as well as the sequence for plane-parallel shocks between 150–450 km s^{-1}, and for spherical shocks between 150–350 km s^{-1}. It is clear that these three sequences can provide an adequate description of the excitation of all classes of narrow-line emission nuclei.

galaxies cover a range of radio properties, including radio quiet objects, weak nuclear sources, and Fanaroff-Riley (FR) Class I sources. However, LINERs are **never** found in the high-power FR II radio sources. In terms of the correlation between the optical and the radio luminosities, a similar trend emerges. There is a good correlation for the FR II class, but this disappears smoothly at the FR I-II transition. These correlations suggest that the optical emission makes a transition from predominantly jet-excited to some other excitation mechanism at the FR I-II transition.

LINERS show a variety of internal motions. Generally speaking, the lowest excitation objects show a more ordered rotation, whereas the higher excitation AGN are more chaotic systems (Baum, Heckman & van Breugel 1992). This strongly suggests that mechanical energy input and excitation are correlated. This is consistent with the expectations of shock model for the excitation, first suggested in the case of NGC 1052 by Fosbury *et al.* (1978).

The recent HST observations of the central disk of M87 by Ford *et al.* (1994) show that, in this object, the disk within one arc second of the nucleus is in rapid rotation (~ 500 km s^{-1}). Optically, this disk displays a classic LINER spectrum, and the velocity dispersion across the 0.3 arc sec aperture is not very much greater than would be expected on the basis of the velocity shear of the rotation across the aperture. Appreciable fluctuations in surface brightness occur across the disk, whilst outside of the central bright region, a series of spiral features emanate from it with the sense of winding that is suggested by the sense of rotation. Clearly, the dissipation in the disk is being driven by some kind of fluting instability, giving rise to shocks of relatively low velocity. In this sense, the shocked accretion flow of the M87 disk could well be a prototype for the class of LINERS in general.

Even where the shock velocities are high and effects of photoionisation are important, a LINER-like spectrum can still be generated. The relative strength of the lines produced in the photoionised precursors is strongly dependent upon the geometry of the shock. The photoionised region has a considerable thickness compared with that of the radiative shock region. For example, a shock having a velocity of 300 km s^{-1} has a thickness of $\sim (2.5/n_o)$ pc, whereas its photoionised precursor has a thickness of $\sim (366/n_o{}^2)$ pc in the case of plane-parallel geometry. Thus, in the case that the high-velocity shock is produced as a bow-shock around a dense, gravitationally-confined cloud, the precursor H II region will be characterised by spherical geometry rather than plane parallel geometry provided that the cloud is smaller than about $\sim (100/n_o{}^2)$ pc. The consequence of this is that the mean ionisation parameter in the H II region drops considerably, from $U \sim 10^{-2}$ in the plane-parallel case typical of NLER galaxies, QSOs and Seyferts to of order 10^{-4} for bow shocks around clouds which are small in comparison to their shock precursor H II regions. Such low ionisation parameters in the precursor region have the effect of suppressing the lines of high ionisation, and enhancing the lines of lower ionisation, relative to the Balmer lines.

The condition that a dense cloud can survive in its passage through the shock-heated intercloud medium is that the gravitational pressure in the cloud should exceed the ram pressure of the intercloud medium. Thus;

$$\sigma_c > (\rho V_s^2/G)^{1/2} = 580 P_{-9}^{1/2} \ M_\odot \ \mathrm{pc}^{-2} \tag{4.3}$$

where σ_c is the surface density of the cloud, V_s is the velocity of the cloud relative to the intercloud medium, and P_{-9} is the ram pressure measured in units of 10^{-9} dynes cm^{-2}. This pressure should correspond fairly closely to the pressure given by the [S II] $\lambda 6717/6731$Å line ratio. Since many LINERs have $P_{-9} \sim 1$, the surface density of the clouds which can survive is somewhat higher than that which characterises the molecular clouds in our solar neighbourhood. If we identify the clouds with dense molecular clouds being turbulently accreted to the nucleus, then presumably the intercloud medium through which they are moving derives from those (less bound) portions of the molecular clouds which have been ablated in previous cloud-intercloud or cloud-cloud encounters. In either the shocked accretion disk model, or the accreting cloudlet model, the luminosity of the LINER will be determined by the rate of change of gravitational binding energy of the accreting gas, and so can be used as a measure of this accretion.

4.3. *Cooling flows*

Since their discovery, the emission line regions associated with the first-ranked ellipticals in X-ray bright clusters have usually been interpreted as being the result of cooling of the X-ray plasma in the deepest part of the potential well, where cooling times are shortest (Ford & Butcher 1979). However, this simple interpretation encountered a serious problem as a result of recognition of the "recombination problem" (Johnstone, Fabian & Nulsen 1987; Heckman *et al.* 1989; hereafter HBvM). It was shown that each hydrogen atom would have to make between 10 and 1000 recombinations in order to account for the observed luminosity assuming that the cooling flow has a mass flux equal to that derived from consideration of the X-ray properties (*e.g.* Arnaud 1988). Two suggestions have emerged which attempt to address this question. First, that repressurising shocks occur as large parcels of gas cool isochorically, and drop out of pressure equilibrium, and second, that emission is generated in mixing layers between the hot and the cold medium (Voit & Donahue 1990; Donahue & Voit 1991). Neither of these models gives sufficient additional recombinations to solve the energy problem. Crawford & Fabian (1992; hereafter CF) have pointed out that the EUV photons in the mixing-layer hypothesis determine the emission spectrum, and have argued that emission from self-absorbed mixing layers can explain the observed line ratios of the HBvM Class I sources, such as the M87 filaments associated with the radio lobes in M87 (Ford & Butcher 1979; Sparks, Ford & Kinney 1993). These are distinguished from the Class II sources by having stronger [N II] and [S II] lines relative to their [O I]. This may be an abundance effect, since the central accretion disk of M87 is also characterised by relatively strong [N II] lines, but it may also be related to photoionisation by a Synchrotron spectrum.

The Class II sources have line ratios which put them in exactly the same part of the diagnostic plots as the LINERS, and this strongly suggests that the same excitation mechanism, radiative autoionising shocks around small, dense clouds applies to Class II cooling flows as well. The idea that these are shock-excited was argued by Crawford & Fabian (1992), who presented very direct dynamical evidence that this is indeed the case. They found a relationship between the local velocity dispersion and the local surface brightness of the emission which follows that expected from equation (1), but applied to Hα. This is shown in Figure 4, where we have normalised each galaxy to the mean of A496 to reduce the scatter caused by the different fn_o, the product of the area covering factor of shocks and the pre-shock density. The values derived for this quantity are given in Table 1.

The excitation line diagnostics given in HBvM have been used to derive the shock parameters. It appears clear that, in cooling flows, the N/O ratio must be enhanced over solar values to between $0.15 < N/O < 0.35$ by number if we are to successfully account for the absolute strength of the [N II] $\lambda 6584$Å line. This result should not be too unexpected, since we believe that the gas in elliptical galaxies is derived mainly from planetary nebulae, which are both theoretically expected to be (Becker & Iben 1980), and observed to be (Dopita & Meatheringham 1991) enhanced in N and He as a consequence of the dredge-up processes. In the particular excitation plot from HBvM shown in Figure 3, it can be seen that individual cooling flows are characterised by a unique value of N/O, as they should be, but that different points in the cooling flows lie along a shock sequence for this value of the abundance ratio. The shock velocities indicated lie between ~ 270 km s^{-1} in the inner regions, falling to as little as 160 km s^{-1} in the outer parts. This range agrees well with the shock velocity estimated from the dynamics (taken as 0.42 times the velocity width in the HBvM and the CF objects). The two Type II cooling flows of CF appear to have shock velocities in the same range, again judged from both their line diagnostics and their velocity widths. However, the Type I flows (Sersic 159-03 in CF, and M87, A262 and A2052 in HBvM) do not fall on these diagnostic plots. Not only do the line ratios fail to give consistent shock velocities, but their line widths are unusually small, and inconsistent with the autoionising shock hypothesis. In the case of M87, it is clear that the emission line material is intimately associated with the radio lobes (Sparks, Ford & Kinney 1993), so it seems likely that an additional excitation mechanism is a work in Type I flows, possibly the self-absorbed mixing layers advocated by CF.

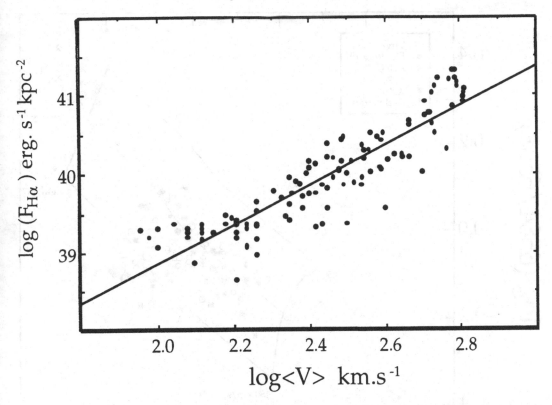

FIGURE 4. The velocity dispersion: surface brightness relation for the cooling flows studied by Crawford & Fabian (1992), normalised to A496. The line is a least squares fit to the data with the slope constrained at 2.4, the value given by shock theory (*c.f.* Eqn. 1).

The luminosity of the Type II cooling flow is the energy flux across the shock:

$$dE/dt = 8.6 \times 10^{40} f n_o V_{200}^3 A_6 \text{ erg s}^{-1} \tag{4.4}$$

where the symbols have the same meaning as equation (2.2) and f is the area covering factor of the shocks. All these parameters can be estimated from observation. Ideally, we would like to relate the energy dissipation rate to the mass inflow rate of dense clouds, assuming that the luminosity represents the turbulent dissipation of gravitational potential energy. We can do this (to order of magnitude) by assuming that the dense clouds are derived from the cooling X-ray gas, and that these release potential energy at the rate $L_{cf} = 3 \dot{M}_x \sigma_{mean}^2 / 2$, where $\dot{M}_x$ is the cooling flow mass flux, and σ_{mean} is the mean line of sight velocity dispersion. The properties of the well-observed Type II flows are summarised in Table 1. We find, as did HvBM, that these mechanical luminosities agree to order of magnitude. Interestingly enough, the turbulent pressure of the intercloud gas, given by $P_{IC} = 3 \mu m_H f n_o \sigma_{mean}^2 / 2$ is in the same ratio to the X-ray gas pressure as L_{tot}/L_{cf} for the cases in which we can estimate it. This would then imply that the intercloud medium is kept in turbulent support with the X-ray plasma by being repeatedly shocked and partially entrained in the passage of the dense clouds.

These estimates of the properties of cooling flows allow us to construct a self consistent scenario of the way in which the cooling flow feeds the nuclear regions. In the outer regions of the flow, clouds which form by cooling are moving at subsonic velocities with respect to the surrounding hot medium, and coalesce into dense clouds. How exactly self-gravitating

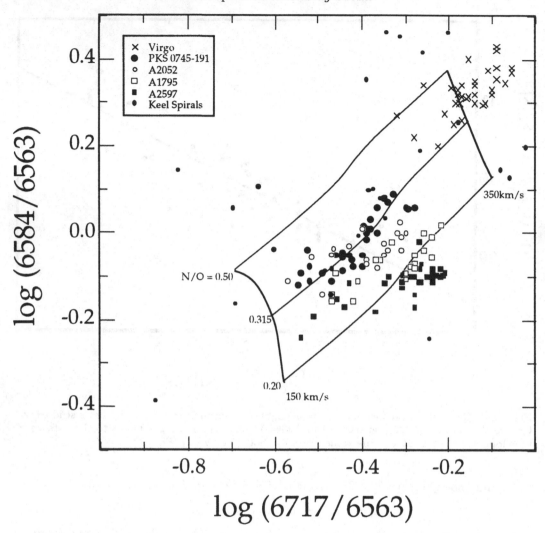

FIGURE 5. The [N II]/Hα vs [S II]/Hα diagnostic plot for cooling flows from HBvM. A grid of theoretical models are shown having shock velocities between 150 and 350 km s^{-1}, and N/O ratios between 0.2 and 0.5. The Type II cooling flows appear to characterised by shock velocities in the range 140–270 km s^{-1}.

Cooling Flow	$L_{H\alpha}$ erg s^{-1}	σ_{mean} km s^{-1}	σ_{cen} km s^{-1}	fn_o cm^{-3}	L_{tot} erg s^{-1}	$\dot{M}_x$ M$_\odot$ yr^{-1}	L_{cf} erg s^{-1}
NGC 1275	4.5×10^{41}	177	265	1.7	1.7×10^{43}	76	2.2×10^{42}
PKS 0745-191	7.5×10^{41}	180	210	7.1	3.0×10^{43}	310	1.0×10^{43}
3C 218	5.0×10^{40}	210	265	8.0	2.2×10^{42}	267	1.1×10^{43}
A 496	1.5×10^{40}	140	190	2.0	5.2×10^{41}	63	1.2×10^{42}
A 1795	2.0×10^{41}	200	265	0.8	8.5×10^{42}	144	5.7×10^{42}
A 2597	4.4×10^{41}	250	270	4-8	2.2×10^{43}	164	1.0×10^{43}

TABLE 1. Core properties of Cooling Flows

clouds can form in this process is unclear, but the implication of the observations is that they exist. Falling in under gravity, these clouds enter the core region where the cooling time of the X-ray plasma is short compared with the dynamical time. Here they move at supersonic velocities with respect to the intercloud, generating a bow shock, entraining shocked intercloud in their wake, and losing energy as a result. If the X-ray halo has little net rotational support, then the infalling clouds will plunge in on radial orbits. This would ensure the increase in velocity dispersion seen in the cores of these systems, and the corresponding increase in surface brightness in the flow, without the need to pack the intercloud medium near the centre of the flow, or necessarily have a much greater density of clouds near the centre of the galaxy. As long as this accretion process is chaotic, the intercloud medium will not acquire any net rotational support, but will be maintained as a turbulent 'froth' presumably embedded in a hotter coronal phase. The existence of high velocity bow shocks ensures that a radio non-thermal component will exist in close association with the optical emission. This would provide a qualitative explanation for the observed correlation between the radio power and $\dot{M}_x$ found by HvBM.

REFERENCES

ANTONUCCI, R. R. J. & MILLER, J. S. 1985 *ApJ* **297**, 621

ARNAUD, K. A. In *NATO Workshop on Cooling Flows in Galaxies and Clusters* (ed. A. C. Fabian) Dordrecht: Reidel

BARTHEL, P. 1989 *ApJ* **336**, 606

BALDWIN, J. A., PHILLIPS, M. M., & TERLEVICH, R. 1981 *PASP* **93**, 5

BAUM, S. A., *et al.* 1988 *ApJS* **68**, 643

BAUM, S. A., & HECKMAN, T. M. 1989 *ApJ* **336**, 702

BAUM, S. A., HECKMAN, T. M., & VAN BREUGEL, W. 1992 *ApJ* **389**, 208

BERTOLA *et al.* 1991 *ApJ* **373**, 369

BICKNELL, G. V. 1990 *ApJ* **354**, 98

BICKNELL, G. V. 1993 In *The Physics of Active Galaxies* (eds. G. V. Bicknell, M. A. Dopita & P. J. Quinn) *PASP Conf. Ser* **54**, 231

BECKER, S. A. & IBEN, I. JR. 1980 *ApJ* **237**, 111

BINETTE, L., DOPITA, M. A., & TUOHY, I. R. 1985 *ApJ* **297**, 476

BINETTE, L. 1985 *A&A* **143**, 334

BINETTE, L., PARKER, D. & FOSBURY, R. A. E. 1993 (preprint)

BLAIR, W. P., KIRSHNER, R. P., & WINKLER, P. F. 1983 *ApJ* **272**, 84

BLAIR, W. P., RAYMOND, J. C., DANZIGER, J., & MATTEUCCI, F. 1989 *ApJ* **338**, 812

CECIL, G., & BLAND, J. 1989 In *IAU Symp. #134, Active Galactic Nuclei* (eds. D. E. Osterbrock & J. S. Miller) Kluwer: Dordrecht, p. 345

CHEVALIER, R. A., & KIRSHNER, R. P. 1979 *ApJ* **233**, 154

COWIE, L. L., HU, E. M., JENKINS, E. B., & YORK, D. G. 1983 *ApJ* **272**, 29

COX, D. P. 1972 *ApJ* **172**, 143

CRAWFORD, C. S., & FABIAN, A. C. 1992 *MNRAS* **259**, 262

CROTTS, A. P. S., KUNKEL, W. E., & MCCARTHY, P. J. 1989 *ApJ* **347**, L61

DALTABUIT, E. & COX, D. 1972 *ApJ* **173**, L13

DE BRUYN, A. G. & WILSON, A. S. 1978 *AA* **64**, 433

DONAHUE, M. & VOIT, G. M. 1991 *ApJ* **381**, 361

DOPITA, M. A. 1976 *ApJ* **207**, 394

DOPITA, M. A. 1987 *Aus. J. Phys.* **40**, 789

DOPITA, M. A., BINETTE, L., D'ODORICO, S. & BENVENUTI, P. 1984a *ApJ* **276**, 653

DOPITA, M. A., BINETTE, L. & SCHWARTZ, R. D. 1982 *ApJ* **261**, 183

DOPITA, M. A., BINETTE, L. & TUOHY, I. R. 1984b *ApJ* **282**, 142

DOPITA, M. A. & MEATHERINGHAM, S. J. 1991 *ApJ* **377**, 480

DOPITA, M. A., TUOHY, I. R. & MATHEWSON, D. S. 1981 *ApJ* **248**, L105

DRAINE, B. T. & McKEE, C. F. 1993, *Ann Rev A&A* **31**, 373

FERLAND, G. J. & NETZER, H. 1983 *ApJ* **264**, 105

FORBES, D. A. & WARD, M. J. 1993 *ApJ* **416**, 150

FORD, H. C. *et al.* 1994 ST ScI Press Release, PR94-23

FORD, H. C. & BUTCHER, H. 1978 *ApJS* **41**, 147

FOSBURY, R. A. E. 1989 In *Extranuclear Activity in Galaxies* (eds. E. J. A. Meurs & R. A. E. Fosbury) ESO: Garching, p. 169

FOSBURY, R. A. E., MEBOLD, U., GOSS, W. M. & DOPITA, M. A. 1978 *MNRAS* **183**, 549

FRANSSON, C. *et al.* 1989 *ApJ* **347**, L61

GOSS, W. M. *et al.* 1979 *MNRAS* **188**, 357

HARTIGAN, P., MORSE, J., HEATHCOTE, S. & CECIL, G. 1993 *ApJL* **414**, L121

HECKMAN, T. M. 1980 *A&A* **87**, 142

HECKMAN, T. M. 1981 *ApJ* **250**, L59

HECKMAN, T. M., BAUM, S. A., VAN BREUGEL, W. J. M. & McCARTHY, P. 1989 *ApJ* **338**, 48

HO, L. C., FILIPPENKO, A. V. & SARGENT, W. L. 1993 *ApJ* **417**, 63

INNES, D. E. 1992 *A&A* **256**, 660

INNES, D. E., GIDDINGS, J. R. & FALLE, S. A. E. G. 1987a *MNRAS* **224**, 179

INNES, D. E., GIDDINGS, J. R. & FALLE, S. A. E. G. 1987a *MNRAS* **226**, 67

ITOH, H. 1981a *PAS Japan* **33**, 121

ITOH, H. 1981b *PAS Japan* **33**, 521

JACOBSEN, P. *et al.* 1991 *ApJ* **369**, 63

JOHNSTONE, R. M., FABIAN, A. C. & NULSEN, P. E. J. 1987 *MNRAS* **224**, 74

KAMPER, K. W. & VAN DEN BERGH, S. 1976 *ApJS* **32**, 351

KEELE, W. C. 1983a *ApJS* **52**, 229

KEELE, W. C. 1983b *ApJ* **269**, 466

KHACHIKIAN, E. YE. & WEEDMAN, D. W. 1974 *ApJ* **192**, 581

KIRSHNER, R. P. *et al.* 1989 *ApJ* **342**, 260

KOSKI, A. T. 1978 *ApJ* **223**, 56

KOEKEMOER, A. M., BICKNELL, G. V. & DOPITA, M. A. 1994 In *The Physics of Active Galaxies* (eds. G. V. Bicknell, M. A. Dopita & P. J. Quinn) *PASP Conf. Ser* **54**, 371

KRISS, G. A. *et al.* 1992 *ApJ* **394**, L37

LASKER, B. M. 1978 *ApJ* **223**, 109

LONG, K. S., HELFAND, D. J. & GRABELSKY, D. A. 1981 *ApJ* **248**, 925

LUNDQVIST, P. & FRANSSON, C. 1987 In *Proc ESO/EIPC Workshop, SN1987A and Other Supernovae* (eds. I. J. Danziger & K. Kjär) Garching: ESO, p. 607

LUNDQVIST, P. & FRANSSON, C. 1991 *ApJ* **380**, 575

LUO, D, McCRAY, R. & SLAVIN, J. 1994 *ApJ* (in press)

McCARTHY, P. J., VAN BREUGEL, W., SPINRAD, H. & DJORGOVSKI, S. 1987 *ApJ* **321**, L29

McKEE, C. F. & HOLLENBACH, D. J. 1980, *Ann Rev A&A* **18**, 219

MEISENHEIMER, K., HIPPELEIN, H. & NEESER, M. 1994 In *The Physics of Active Galaxies* (eds. G. V. Bicknell, M. A. Dopita & P. J. Quinn), *PASP Conf. Ser* **54**, 397

MEURS, E. J. A. *et al.* 1989 In *Extranuclear Activity in Galaxies*, (eds. E. J. A. Meurs & R. A. E. Fosbury) ESO: Garching, p. 249

MORGANTI, R. *et al.* 1991 *MNRAS* **249**, 91

MORGANTI, R. *et al.* 1992 *MNRAS* **256**, 1P

MORGANTI, R., FOSBURY, R. A. E., OOSTERLOO, T. A. & TADHUNTER, C. N. 1994 In *The Physics of Active Galaxies* (eds. G. V. Bicknell, M. A. Dopita & P.J. Quinn) *PASP Conf. Ser* **54**, 267

MORSE, J. A., HEATHCOTE, S., HARTIGAN, P. & CECIL, G. 1993a *AJ* **106**, 1133

MORSE, J. A. *et al.* 1993b *ApJ* **410**, 764

MORSE, J. A. *et al.* 1994 *ApJ* (in press)

MULCHAEY, J. S., TSVETANOV, Z., WILSON, A. S. & PÉREZ-FOURNON, I. 1992 *ApJ* **394**, 91

OSTERBROCK, D. E. 1971 *Pontificiae Academiae Scientiarum, Scripta Varia* (ed. D. J. K. O'Connell) Amsterdam No. 35, p. 151

OSTERBROCK, D. E. 1974 *The Physics of Gaseous Nebulae*, Freeman & Co., San Francisco

OSTERBROCK, D. E. 1989 *Astrophysics of Gaseous Nebulae and Active Galactic Nuclei*, University Science Books, Mill Valley: Calif

OSTERBROCK, D. E., TRAN, H. D., & VEILLEUX, S. 1992 *ApJ* **389**, 196

PHILLIPS, M. M. *et al.* 1986 *AJ* **1**, 1062

PLAIT, P. C., LUNDQVIST, P., CHEVALIER, R. A. & KIRSHNER, R. P. 1994 *ApJ* (in press)

POGGE, R. W. 1988a *ApJ* **328**, 519

POGGE, R. W. 1988b *ApJ* **332**, 702

POGGE, R. W. 1989 *ApJ* **345**, 730

PRIETO, A., DI SEREGO ALIGHIERI, S. & FOSBURY, R. A. E. 1989 In *Extranuclear Activity in Galaxies* (eds. E. J. A. Meurs & R. A. E. Fosbury) ESO: Garching, p. 31

RAYMOND, J. C. 1976, PhD Thesis, University of Wisconsin

RAYMOND, J. C. & SMITH, B. W. 1977 *ApJS* **35**, 419

RAYMOND, J. C., MORSE, J. A. *et al.* 1994 *ApJ* (in press)

SCHMUTZLER, T. & TSCHARNUTER, W. M. 1993 *A&A* **273**, 318

SEATON, M. J. 1958 *Rev. Mod. Phys.* **30**, 979

SEATON, M. J. 1959 *MNRAS* **119**, 81

SPARKS, W. B., FORD, H. C. & KINNEY, A. L. 1993 *ApJ* (in press)

STASINSKA, G. 1984 *AA* **135**, 341

SUTHERLAND, R. S. 1993, PhD Thesis, The Australian National University

SUTHERLAND, R. S., BICKNELL, G. V. & DOPITA, M. A. 1993 *ApJ* **414**, 510

SUTHERLAND, R. S. & DOPITA, M. A. 1993 *ApJS* **88**, 253

SUTHERLAND, R. S. & DOPITA, M. A. 1994a,b *ApJ* (in press)

TADHUNTER, C. N., ROBINSON, A. & MORGANTI, R. 1989 In *Extranuclear Activity in Galaxies* (eds. E. J. A. Meurs & R. A. E. Fosbury) ESO: Garching, p. 293

TADHUNTER, C. N. & TSVETANOV, Z. I. 1989 *Nature* **341**, 422

TSVETANOV, Z. I., FOSBURY, R. A. E. & TADHUNTER, C. N. 1994 *ApJ* (in press)

UNGER, S. W. *et al.* 1987 *MNRAS* **228**, 671

VEILLEUX, S. 1991a *ApJS* **75**, 357

VEILLEUX, S. 1991b *ApJS* **75**, 383

VEILLEUX, S. & OSTERBROCK, D. E. 1987 *ApJS* **63**, 295

VEILLEUX, S., TULLY, R. B. & BLAND-HAWTHORN, J. 1993 *AJ* **105**, 1318

VOIT, G. M. & DONAHUE, M. 1990 *ApJ* **360**, L15

WAMPLER, E. *et al.* 1990 *ApJ* **362**, L13

WHITTLE, M. 1989 In *Extranuclear Activity in Galaxies* (eds. E. J. A. Meurs & R. A. E. Fosbury)
     ESO: Garching, p. 199

WHITTLE, M. 1992 *ApJS* **79**, 49

WHITTLE, M. 1994 *ApJS* **91**, 491

WILSON, A. S. & TSVETANOV, Z. I. 1994 *AJ* **107**, 1227

WILSON, A. S. & WILLIS, A. G. 1980 *ApJ* **240**, 429

WINKLER, P. F. & KIRSHNER, R. P. 1985 *ApJ* **299**, 981

The Lexington Benchmarks for Numerical Simulations of Nebulae

By G. FERLAND,[1] L. BINETTE,[2] M. CONTINI,[3]
J. HARRINGTON,[4] T. KALLMAN,[5] H. NETZER,[3]
D. PÉQUIGNOT,[6] J. RAYMOND,[7] R. RUBIN,[8]
G. SHIELDS,[9] R. SUTHERLAND,[10] S. VIEGAS[11]

[1]Physics and Astronomy, University of Kentucky, Lexington, KY 40506,
gary@cloud9.pa.uky.edu

[2]ESO, D-85748, Garching bei Muenchen, Germany, lbinette@eso.org

[3]School of Physics and Astronomy, Tel Aviv University, 69978 Tel Aviv, Israel,
netzer@wise.tau.ac.il, contini@ccsg.tau.ac.il

[4]Astronomy, U of Maryland, College Park, MD 20742, jph@astro.umd.edu

[5]Code 665, NASA Goddard SFC, Greenbelt, MD 20771, tim@xstar.gsfc.nasa.gov

[6]Observatoire de Paris, Meudon F-92195, Meudon Principal Cedex, France,
pequignot@obspm.fr

[7]CfA, 60 Garden St., Cambridge, MA 02138, raymond@cfassp8.harvard.edu

[8]NASA/Ames Research Center, MS 245-6, Moffett Field, CA 94035-1000,
rubin@cygnus.arc.nasa.gov

[9]Astronomy, University of Texas, Austin, TX 78712, shields@astro.as.utexas.edu

[10]JILA, University of Colorado, Boulder, CO 80309-0440, ralph@zwicky.colorado.edu

[11]IAGUSP, Av. Miguel Stefano 4200, 04301 São Paulo, S.P., Brazil, viegas@iag.usp.ansp.br

We present the results of a meeting on numerical simulations of ionized nebulae held at the University of Kentucky in conjunction with the celebration of the 70th birthdays of Profs. Donald Osterbrock and Michael Seaton.

1. Introduction

Numerical simulations of emission line regions, whether photo or shock ionized, are a vital tool in the analysis and interpretation of spectroscopic observations. Models can determine characteristics of the central source of ionizing radiation, the composition and conditions within the emitting gas, or, for shocks, the shock velocity. Osterbrock (1989) and Draine & McKee (1993) review the basic physical processes in these environments.

Although numerical simulations are a powerful tool, this capability is somewhat mitigated by the complexity of the calculations. There will always be underlying questions regarding the astronomical environment (i.e., the shape of the ionizing continuum, inhomogeneities, or the composition of the gas) and uncertainties introduced by the evolving atomic/molecular data base. On top of this, however, the numerical approximations, assumptions, and the complexity of the simulations themselves introduce an uncertainty that cannot be judged from a single calculation.

With these questions in mind Daniel Péquignot held a meeting on model nebulae in Meudon, France, in 1985. This provided a forum where investigators could carefully compare model predictions and identify methods, assumptions, or atomic data which led to significant differences in results. The results of this meeting were a set of benchmarks

83

Parameter/ Table	2	3	4	5	6	7	8
$Q(H)$ or $\phi(H)$	1(49)	4.26(49)	1(13)	5.42(47)	3.853(47)	3.853(47)	3(12)
R_{inner} (cm)	3(18)	3(18)	PP	1(17)	1.5(17)	1.5(17)	PP
TBB (°K)	20,000	40,000	40,000	150,000	75,000	75,000	power law
Geometry	closed	closed	closed	closed	closed	closed	open
Stopping criteria	H-front	H-front	H-front	H-front	7.5E17 cm	7.5E17 cm	1E22 cm^{-2}
n_H (cm^{-3})	100	100	1(4)	3000	500	15,810	10,000
He	0.10	0.10	0.095	0.10	0.10	0.10	0.10
C	2.2(-4)	2.2(-4)	3.0(-4)	3.0(-4)	2.0(-4)	2.0(-4)	3.0(-4)
N	4.0(-5)	4.0(-5)	7.0(-5)	1.0(-4)	6.0(-5)	6.0(-5)	1.0(-4)
O	3.3(-4)	3.3(-4)	4.0(-4)	6.0(-4)	3.0(-4)	3.0(-4)	8.0(-4)
Ne	5.0(-5)	5.0(-5)	1.1(-4)	1.5(-4)	6.0(-5)	6.0(-5)	1.0(-4)
Mg				3.0(-5)	1.0(-5)	1.0(-5)	3.0(-5)
Si				3.0(-5)	1.0(-5)	1.0(-5)	3.0(-5)
S	9.0(-6)	9.0(-6)	1.0(-5)	1.5(-5)	1.0(-5)	1.0(-5)	1.5(-5)
Ar			3.0(-6)				

TABLE 1. Parameters for Photoionized Nebulae

which represented the state of the art at that time (Péquignot 1986). A second meeting on model nebulae was held in Lexington in association with the celebration of the seventieth birthdays of Professors Osterbrock and Seaton, who both defined the field and took an active part in the deliberations. This is a summary of the results of that workshop.

2. The codes

Brief summaries of the codes follow: the lead author(s) of the code is followed by the identification number used in the following tables.

Binette: (1) MAPPINGS I is a combination of a shock code developed by Mike Dopita and a photoionization code developed by Luc Binette. Work started in 1980 and a first version is described in Binette *et al.* (1985). Outward-only continuum transfer is assumed. Recent revisions include a 5-level hydrogen atom which allows calculations at much higher densities and the effects of dust scattering and absorption on the emergent line spectrum. A description of the dust transfer and of most recent modifications can be found in Binette *et al.* (1993a, b). Instructions for obtaining the code via anonymous ftp may be obtained from the author.

Ferland: (2) Work on Cloudy began at the Institute of Astronomy, Cambridge, in 1978. Version 87 was used to generate the results presented here. A recent publication is Ferland, Fabian, & Johnston (1994). The code is fully described in the document *Hazy, a Brief Introduction to Cloudy*. Instructions for obtaining both the documentation and the code via anonymous ftp may be obtained from the author. For the models presented here outward-only continuum transfer was used, and the guesses of the dielectronic recombination rate coefficients described by Ali *et al.* (1991) were turned off.

Harrington: (3) These codes are discussed in Harrington (1968), Harrington *et al.* (1982), and Clegg *et al.* (1987). Most versions are for PNe; the radiation extends only to 200 eV. A version developed by S. Kraemer (1986) for AGNs goes to 3 keV. Most work has spherical geometry, though plane-parallel versions exist (Bregman & Harrington 1986). The spherical code used here (except for Table 4) starts with the outward-only

		Mean	1	2	3	5	6	8	9	10	11
L(Hβ)	E36	4.93	4.99	4.98	4.93	4.85	4.83	4.93	4.94	5.01	4.91
[NII]	6584+	0.85	0.82	0.91	0.82	0.97	0.82	0.84	0.84	0.83	0.83
[OII]	3727+	1.18	1.11	1.16	1.22	1.32	1.14	1.21	1.24	1.14	1.11
[NeII]	12.8 μ	0.31	0.36	0.35	0.29	0.29	0.29	0.29	0.35	0.29	0.29
[SII]	6720+	0.57	0.69	0.64	0.55	0.61	0.52	0.52	0.60	0.45	0.58
[SIII]	18.7 μ	0.32	0.26	0.27	0.36	0.17	0.37	0.37	0.33	0.40	0.30
[SIII]	34 μ	0.52	0.43	0.47	0.60	0.27	0.61	0.62	0.54	0.67	0.51
[SIII]	9532+	0.55	0.40	0.48	0.55	0.64	0.60	0.56	0.49	0.62	0.58
L(total)	E36	21.2	20.3	21.3	21.7	20.7	21.0	21.8	21.7	22.1	20.6
T(in)		6793	6860	6952	6749	6980	6870	6747	6230	6912	6838
T(H+)		6744	6690	6740	6742	6950	6660	6742	6770	6720	6681
⟨He+⟩/⟨H+⟩		0.054		0.041	0.044	0.068	0.048	0.034	0.055	0.090	
R(out)	E18	8.96	9.00	8.93	8.94	9.00	8.93	9.00	8.87	9.02	8.97

TABLE 2. Cool HII Region

approximation, followed by iterations in which the diffuse field is computed by Feautrier techniques from the opacity and source functions of the previous model. Convergence may take ~ 20 iterations for cases like the Meudon PN model (i.e., high optical depth and a hot star). Codes are not available as they are not documented and exist in many versions. A version ("nebmod") is in the UK Starlink collection, but documentation is minimal.

Kallman: (4) The XSTAR code dates from work at the University of Colorado by R. McCray and students: J. Buff, S. Hatchett, C. Wright, and R. Ross. The code was originally developed primarily for the prediction of X-Ray line emission from gas in X-Ray binaries, and later modified to treat optical and UV line emission from broad line clouds in quasars. A comprehensive description and collection of results were published in Kallman & McCray (1982). A more up-to-date description is included in the user's manual which, together with the source code, is available by anonymous ftp from legacy.gsfc.nasa.gov.

Netzer: (5) ION was born in 1976 and reached some maturity in 1978. The earlier versions of the code are described in Netzer & Ferland (1984), Rees, Netzer & Ferland (1989) and Netzer (1993). ION is designed to cover the full range of densities from extremely low up to 10^{13} cm^{-3}. The HI, HeI and HeII solutions include a large number of levels and a full transfer treatment. Line transfer is calculated with the local escape probability and continuum transfer with either modified on-the-spot or outward only approximations. The present calculations assume outward-only and low temperature dielectronic recombination to magnesium and sulfur were not included. ION is available upon request from the author.

Péquignot: (6) NEBU is a descendant of a photoionization code whose main features were established by 1978 in a collaboration with G. Stasinska and S. Viegas at Meudon and São Paulo. Continuum and lines are treated outward only along 20 directions in spherical symmetry. NEBU is a combined photoionization and planar-shock code taking into account time-dependent ionization. The shock version was used by Stasinska and Péquignot at the Meudon meeting. Non-collisional excitation of forbidden lines was implemented in 1989 (Petitjean, Boisson & Péquignot 1990; this includes a short description of the code). The codes AANGABA (Viegas) and PHOTO (Stasinska) diverged from NEBU in 1985 and 1988 respectively.

Raymond: (7) The shock code described by Raymond (1979) grew out of the one developed by Cox (1972), who gives a clear description of the physical processes and the

		Meu	Lex	1	2	3	4	5	6	8	9	10	11
L(Hβ)	E37	2.06	2.03	1.96	2.06	2.04	1.86	2.02	2.02	2.05	2.10	2.11	2.09
HeI	5876	0.116	0.116	0.125	0.109	0.119	0.110	0.101	0.116		0.125	0.115	0.120
CII	2326+	0.17	0.16	0.07	0.19	0.17	0.16	0.16	0.14	0.18	0.28	0.12	0.14
CIII]	1909+	0.051	0.06	0.050	0.059	0.059	0.027	0.078	0.065	0.076	0.082	0.077	0.071
[NII]	122 μ		0.031	0.032	0.033				0.036	0.031	0.030	0.037	0.034
[NII]	6584+	0.73	0.79	0.61	0.88	0.74	0.94	0.87	0.78	0.73	0.78	0.81	0.75
[NIII]	57 μ	0.30	0.27	0.16	0.27	0.29		0.26	0.30	0.30	0.17	0.27	0.39
[OII]	3727+	2.01	2.16	1.50	2.19	2.14	2.56	2.3	2.11	2.26	2.41	2.20	1.95
[OIII]	51.8 μ	1.10	1.07	1.10	1.04	1.11	1.04	0.99	1.08	1.08	1.23	1.04	0.97
[OIII]	88.4 μ	1.20	1.23	1.30	1.07	1.28		1.16	1.25	1.26	1.42	1.20	1.14
[OIII]	5007+	2.03	2.06	2.30	1.93	1.96	1.47	2.29	2.17	2.10	2.23	2.22	1.89
[NeII]	12.8 μ	0.21	0.22	0.26	0.23	0.19	0.23	0.22	0.20	0.20	0.22	0.22	0.20
[NeIII]	15.5 μ	0.44	0.38	0.37	0.43	0.43	0.47	0.37	0.42	0.42	0.22	0.34	0.38
[NeIII]	3869+	0.096	0.086	0.085	0.103	0.086	0.071	0.100	0.079	0.087	0.081	0.087	0.078
[SII]	6720+	0.14	0.20	0.24	0.23	0.16	0.25	0.22	0.17	0.13	0.21	0.15	0.21
[SIII]	18.7 μ	0.55	0.55	0.56	0.48	0.56	0.53	0.5	0.55	0.58	0.58	0.58	0.55
[SIII]	34 μ	0.93	0.89	0.91	0.82	0.89		0.81	0.88	0.94	0.92	0.92	0.91
[SIII]	9532+	1.25	1.29	1.16	1.27	1.23	1.15	1.48	1.27	1.30	1.31	1.32	1.46
[SIV]	10.5 μ	0.39	0.34	0.22	0.37	0.42	0.35	0.36	0.41	0.33	0.26	0.38	0.27
L(total)	E37	24.1	24.2	21.7	24.1	24.1	17.4	24.8	24.3	24.6	26.4	25.5	24.1
T(in)		7378	7552	7630	7815	7741	8057	7670	7650	7399	6530	7582	7445
T(H+)		7992	8034	7880	8064	8047	7879	8000	8060	8087	8220	8191	7913
⟨He+⟩/⟨H+⟩			0.77		0.71	0.77	0.69	0.76	0.75	0.83	0.86	0.79	0.77
R(out)	E18	1.45	1.48	1.43	1.46	1.46	1.61	1.47	1.46	1.46	1.46	1.49	1.47

TABLE 3. Meudon HII Region

		Mean	1	2	3	4	5	6	8	9	10	11
I(Hβ)		4.62	4.60	4.59	4.81	3.89	4.69	4.67	4.70	4.85	4.58	4.78
HeI	5876	0.12	0.12	0.13	0.11	0.11	0.12	0.12		0.12	0.11	0.12
CII	2326+	0.18	0.06	0.14	0.20	0.30	0.10	0.15	0.23	0.35	0.11	0.16
CII	1335+	0.09	.002	0.17	0.14	0.02	0.13	0.16		0.01	0.02	0.13
CIII]	1909+	0.17	0.13	0.22	0.17	0.08	0.18	0.15	0.20	0.25	0.14	0.23
[NII]	6584+	0.87	0.67	0.58	0.94	1.48	0.74	0.90	0.87	0.92	0.82	0.83
[NIII]	57 μ	.031	.032	.035	.033		.033	.032	.034	.014	.032	.033
[OII]	7330+	0.12	0.06	0.10	0.13	0.19	0.09	0.12	0.14	0.15	0.08	0.10
[OII]	3727+	0.88	0.53	0.73	0.98	1.39	0.69	0.86	1.04	1.04	0.73	0.86
[OIII]	51.8 μ	0.29	0.29	0.31	0.29	0.26	0.28	0.28	0.28	0.32	0.28	0.27
[OIII]	5007+	4.13	4.50	4.74	3.90	3.28	4.40	3.90	3.96	4.51	4.16	3.98
[NeII]	12.8 μ	0.36	0.45	0.32	0.33	0.44	0.35	0.33	0.35	0.36	0.37	0.35
[NeIII]	15.5 μ	0.98	0.93	1.24	1.07	1.09	0.96	1.04	1.00	0.59	0.92	0.97
[NeIII]	3869+	0.33	0.33	0.48	0.32	0.31	0.35	0.26	0.29	0.35	0.31	0.31
[SIII]	18.7 μ	0.35	0.37	0.31	0.34	0.37	0.31	0.33	0.35	0.37	0.34	0.39
[SIII]	9532+	1.53	1.52	1.41	1.46	1.62	1.51	1.42	1.53	1.61	1.42	1.82
[SIV]	10.5 μ	0.46	0.26	0.54	0.52	0.42	0.51	0.53	0.43	0.36	0.50	0.51
I(total)		50.3	47.1	52.6	52.4	44.2	50.4	49.4	50.3	54.9	47.4	52.9
T(in)		7989	8300	8206	7582		8200	8200	7366	7740	8122	8189
T(H+)		8263	8170	8324	8351		8310	8200	8328	8220	8217	8250
⟨He+⟩/⟨H+⟩		0.85		0.94	0.78		0.93	0.79	0.84	0.86	0.85	0.84
ΔR	E17	2.96	2.90	2.88	3.08		2.93	2.98	3.09	3.10	2.67	3.03

TABLE 4. Blister HII Region

numerical methods. Updated atomic rates are described in Cox & Raymond (1985), and an extensive grid of models is given in Hartigan, Raymond & Hartmann (1987). The code assumes steady flow, which simplifies the hydrodynamics to a calculation of the radiative cooling rate and the resulting compression. It computes the time-dependent ionization balance including photoionization and the associated heating.

Rubin: (8) This code "NEBULA" has a detailed treatment for the Lyman continuum radiative transport. The gas structure (ionization and thermal equilibrium) and transfer of the diffuse ionizing radiation are solved iteratively. For the example H II region models here, uniform convergence is obtained in 15–20 iterations. A publication that uses a recent version of the 1-dimensional (spherical) code is by Simpson *et al.* (1995). A 2-dimensional, axisymmetric version, which treats the diffuse radiation in a less precise fashion, has been applied to model the Orion Nebula (Rubin *et al.* 1991; 1993).

Shields: (9) This code, called NEBULA, was first developed at Caltech in 1971 for work on Seyfert galaxies. It was later streamlined to include only physics relevant to H II regions and planetary nebulae. Some description is contained in Garnett & Shields (1987), and further details may be obtained from Shields. The models here were computed with OTS for the Lyman continua of H I and He II, outward-only for the Lyman continuum of He I, and outward-only for the helium Lyman lines.

Sutherland: (10) The MAPPINGS code was started in 1976 by Dopita to model plane parallel steady shocks. In 1990 a rewrite was performed to extend MAPPINGS into the high temperature regime (Sutherland & Dopita 1993). MAPPINGS II, does not supersede the Binette MAPPINGS code but is rather an extension in a different direction. MAPPINGS II v1.0.5x was used for the results presented here. The diffuse field is calculated and integrated in the outward only approximation. Dielectronic recombination rates for sulfur and argon based on the Opacity Project cross-sections have been used for Ar I-V and S I-V. Inquiries about the availability of MAPPINGS II should be made to the author.

Viegas: (11) AANGABA (which means "image" in the Brazilian native language) is a descendant of a photoionization code whose main features were established by 1978 in a collaboration with G. Stasinska and D. Péquignot at Meudon and São Paulo. The diffuse radiation field is calculated using the outward-only approximation along 20 directions in spherical symmetry. A brief description of the code is given in Gruenwald & Viegas (1992). The codes NEBU (Péquignot) and PHOTO (Stasinska) diverged from AANGABA in 1985.

Viegas and Contini: (12) The first version of SUMA, which accounts for the coupled effects of photoionization and shocks, appeared in 1982. The code has been revised in 1992, and the photoionization calculation is in full agreement with AANGABA. The last publication refers to results obtained after updating the code and including the effect of dust (Viegas & Contini 1994).

3. The model nebulae

These benchmarks are designed to be simple yet still test the numerical approximations used in the simulations. Blackbodies are used instead of stellar atmospheres; the compositions have low gas phase abundances for refractory elements but do not include the grains themselves; and all have constant density. For shocks, the gradual liberation of refractory elements in the course of grain destruction (see Jones *et al.* 1994) and departures from steady flow (especially for $v_s \geq 150$ km/s; see Innes 1992) are ignored, as are the photoionization precursors which contribute significantly to the optical emission lines if one observes an entire SNR rather than a single bright filament (see Shull 1983).

		Meu	Lex	1	2	3	4	5	6	10	11
L(Hβ)	E35	2.60	2.53	2.06	2.63	2.68	2.35	2.73	2.68	2.30	2.80
He I	5876	0.11	0.09	0.09	0.11	0.10	0.05	0.10	0.11	0.09	0.11
He II	4686	0.33	0.41	0.40	0.32	0.33	0.81	0.35	0.32	0.43	0.34
C II]	2326+	0.38	0.27	0.20	0.33	0.43	0.12	0.27	0.30	0.22	0.32
C III]	1909+	1.70	2.14	2.40	1.82	1.66	2.92	1.72	1.87	3.14	1.63
C IV	1549+	1.64	2.51	2.60	2.44	2.05	1.44	2.66	2.18	4.74	1.94
[N II]	6584+	1.44	1.49	1.43	1.59	1.45	1.69	1.47	1.44	1.47	1.38
N III]	1749+	0.11	0.12	0.16	0.13	0.13	0.01	0.11	0.13	0.16	0.16
[NIII]	57 μ		0.13	0.11	0.12	0.13		0.13	0.13	0.13	0.14
N IV]	1487+	0.12	0.20	0.22	0.20	0.15	0.20	0.21	0.19	0.26	0.17
N V	1240+	0.09	0.21	0.17	0.18	0.12	0.34	0.23	0.15	0.38	0.15
[O I]	6300+	0.15	0.14	0.17	0.15	0.12	0.15	0.14	0.14	0.16	0.13
[O II]	3727+	2.23	2.28	2.35	2.23	2.27		2.31	2.18	2.50	2.14
[O III]	5007+	20.9	20.7	21.8	21.1	21.4	19.8	19.4	21.1	20.2	20.9
[O III]	4363	0.16	0.16	0.18	0.16	0.16	0.19	0.14	0.16	0.16	0.15
[O III]	52 μ	1.43	1.34	1.39	1.42	1.44	0.96	1.40	1.46	1.26	1.41
[O IV]	26 μ	3.62	3.92	3.90	3.52	3.98	4.48	3.32	3.86	5.01	3.33
O IV]	1403+	0.13	0.27	0.36	0.20	0.23	0.18	0.26	0.33	0.41	0.15
O V]	1218+	0.09	0.24		0.20	0.11	0.35	0.29	0.19	0.33	
[Ne III]	15.5 μ	2.51	2.49	2.67	2.75	2.76	0.72	2.80	2.81	2.71	2.74
[Ne III]	3869+	2.59	2.63	3.20	3.33	2.27	0.88	2.74	2.44	3.35	2.86
Ne IV]	2423+	0.56	0.95	1.05	0.72	0.74	1.64	0.91	0.74	1.19	0.63
[Ne V]	3426+	0.73	0.90	0.79	0.74	0.60	2.29	0.73	0.61	0.81	0.63
[Ne V]	24.2 μ	1.67	0.88	1.20	0.94	0.76	1.16	0.81	0.99	0.25	0.95
Mg II	2798+	1.48	1.56	2.50	2.33	1.60	0.63	1.22	1.17	1.15	1.92
[Mg IV]	4.5 μ	0.09	0.12	0.11	0.12	0.13			0.12	0.14	0.12
[Si II]	34.8 μ	0.13	0.18	0.14	0.16	0.26		0.19	0.17	0.15	0.16
Si II]	2335+	0.11	0.23	0.23	0.15		0.53	0.16	0.16		0.15
Si III]	1892+	0.20	0.68	0.79	0.39	0.32	1.95	0.46	0.45		0.40
Si IV	1397+	0.15	0.15	0.10	0.20	0.15	0.03	0.21	0.17		0.16
[S II]	6720+	0.39	0.33	0.24	0.21	0.45	0.08	0.33	0.43	0.41	0.51
[S III]	18.7 μ	0.49	0.53	0.60	0.48	0.49		0.46	0.49	0.60	0.55
[S III]	9532+	2.09	1.91	2.31	2.04	1.89	0.36	2.05	1.87	2.34	2.42
[S IV]	10.5 μ	1.92	1.84	1.58	1.92	2.21	0.93	1.81	1.98	2.36	1.94
L(total)	E35	129	132	114	139	136	105	135	136	130	142
T(in)	E4		1.80		1.83	1.78	1.63	1.84	1.78	1.95	1.81
T(H+)	E4		1.26		1.22	1.21	1.32	1.35	1.21	1.29	1.20
⟨He+⟩/⟨H+⟩			0.69		0.74	0.74		0.71	0.71	0.60	0.68
R(out)	E17		4.02		4.04	4.04		4.07	4.07	3.83	4.08

TABLE 5. Meudon Planetary Nebula

3.1. *Cloud parameters and tabulated quantities*

Table 1 lists the parameters for the benchmarks presented here. The first row gives the table number for the models, and following rows list parameters. Standard notation is used (Osterbrock 1989).

The luminosity or intensity of the continuum is specified by a) for the spherical geometry, Q(H), the total number of hydrogen-ionizing photons emitted per second into 4π sr; or b) for the plane-parallel case, the flux of hydrogen ionizing photons at the illuminated face of the cloud. The inner radius is given for spherical geometries, or the notation "PP" if the geometry is plane parallel. The black body temperature follows, except for model 8 which uses the power-law described below. Most models assumed a "closed" geometry,

		Mean	1	2	3	4	5	6	10	11
L(Hβ)	E34	5.85	5.67	6.05	5.96	6.02	5.65	5.74	5.72	6.02
He I	5876	0.12	0.12	0.13	0.13	0.10	0.12	0.13	0.13	0.13
He II	4686	0.081	0.096	0.080	0.087	0.039	0.085	0.092	0.090	0.083
C III]	1909+	0.83	0.90	0.60	0.60	0.89	0.99	0.89	1.03	0.74
C IV	1549+	0.34	0.24	0.35	0.29	0.45	0.40	0.37	0.32	0.29
[N II]	6584+	0.12	0.12	0.11	0.11	0.14	0.15	0.12	0.12	0.12
[NIII]	57 μ	0.39	0.27	0.37			0.40	0.41	0.40	0.48
[O II]	3727+	0.29	0.32	0.22	0.24	0.35	0.35	0.26	0.32	0.27
[O III]	5007+	11.5	12.1	10.0	10.1	12.7	12.2	11.7	11.9	11.2
[O III]	52 μ	2.02	2.03	1.88	1.96	2.39	1.95	2.02	2.02	1.94
[O IV]	26 μ	0.79	0.76	0.68	0.80	1.09	0.71	0.86	0.77	0.67
[Ne III]	15.5 μ	1.35	1.35	1.30	1.32	1.55	1.30	1.35	1.34	1.31
[Ne III]	3869+	1.03	1.15	1.02	0.92	1.02	1.13	0.89	1.11	1.00
Mg II	2798+	0.13	0.34	0.10	0.07	0.14	0.05	0.10	0.10	0.11
Si III]	1892+	0.15	0.10	0.09	0.10	0.37	0.15	0.13		0.11
[S III]	18.7 μ	0.34	0.45	0.24	0.32		0.26	0.28	0.36	0.49
[S III]	9532+	1.13	1.40	0.81	0.92		1.02	0.85	1.10	1.77
[S IV]	10.5 μ	2.05	1.66	2.19	2.21	2.10	2.11	2.35	2.31	1.46
L(total)	E34	133	133	122	120	140	132	131	134	134
T(in)	E4	1.48	1.45	1.48	1.42	1.83	1.40	1.45	1.36	1.44
T(H+)	E4	1.05	1.07	1.01	1.01	1.03	1.14	1.05	1.06	1.03
$\langle$He+$\rangle$/$\langle$H+$\rangle$		0.92		0.92	0.92		0.92	0.91	0.92	0.92

TABLE 6. High Ionization PN

one in which radiation escaping from the illuminated face of the cloud does not escape the system. An open geometry is one in which such radiation freely escapes.

The stopping criteria follow. There are either the hydrogen ionization front ("H-front"), a certain thickness ΔR, or a column density N_H. The hydrogen density n_H is listed, followed by abundances by number relative to hydrogen.

3.2. *Results*

The tables present several types of results for the models. The most important (observationally) are the intensities of lines—these are presented relative to Hβ. For multiplets the total intensities of all lines are listed; and this is indicated by the "+" after the wavelength of the multiplet. The total intensity or luminosity ("I(total)" or "L(total)") is the intensity or luminosity of Hβ (the first line) multiplied by the sum of the listed intensities. This is a measure of the energy conservation of the codes. The electron temperature at the illuminated face of the cloud is next, followed by the mean temperature T(H+), and the mean He$^+$/H$^+$:

$$T(H^+) = \frac{\int n_e n_p T_e dV}{\int n_e n_p dV}, \qquad \frac{\langle He^+ \rangle}{\langle H^+ \rangle} = \frac{n(H)}{n(He)} \frac{\int n_e n(He^+) dV}{\int n_e n_p dV}.$$

Table 2. This is an HII region ionized by a very cool star. This is the simplest model since helium is predominantly neutral.

Table 3. This was one of the original Meudon tests. It is a spherical HII region with a star warm enough to nearly fully fill the H$^+$ Strömgren sphere with He$^+$. Atomic helium radiative transfer is important, but the complexities introduced by fully ionized helium are not yet present. In this and the following tables the entries "Meu" and "Lex" indicate means for the Meudon and Lexington results.

		Mean	1	2	3	5	6	10	11
L(Hβ)	E34	5.41	5.20	5.56	5.52	5.35	5.41	5.38	5.44
He I	5876	0.13	0.12	0.14	0.12	0.14	0.12	0.12	0.15
He II	4686	.088	.095	.082	.088	.088	.088	.091	.086
C III]	1909+	1.17	1.53	0.78	0.81	1.25	1.41	1.43	1.00
C IV	1549+	1.38	1.20	1.34	1.31	1.54	1.43	1.51	1.32
[O III]	5007+	14.3	16.0	12.7	13.1	15.0	14.5	14.6	14.3
[O III]	52 μ	0.26	0.28	0.26	0.26	0.26	0.27		0.26
[O IV]	26 μ	0.22	0.24	0.21	0.23	0.21	0.21	0.22	0.22
[Ne III]	15.5 μ	1.11	1.14	1.09	1.10	1.10	1.12	1.12	1.10
[Ne III]	3869+	1.44	1.66	1.39	1.27	1.48	1.44	1.45	1.38
Ne IV]	2423+	0.10	0.10	0.11	0.11	0.10	0.08	0.08	0.10
[S III]	9532+	0.52	0.65	0.23	0.37	0.26	0.42	0.76	0.99
[S IV]	10.5 μ	1.43	1.20	1.32	1.57	1.32	1.84	1.38	1.37
L(total)	E34	120	126	109	112	122	124	122	121
T(in)	E4	1.78		1.83	1.79	1.76	1.73	1.76	1.81
T(H+)	E4	1.16		1.11	1.11	1.28	1.14	1.20	1.14
⟨He+⟩/⟨H+⟩		0.91		0.91	0.91	0.91	0.91	0.90	0.90

TABLE 7. Low Ionization PN

Table 4. This model is motivated by conditions in the innermost regions of the Orion nebula. It is plane parallel, the simplest representation of the flow off the face of a molecular cloud.

Table 5. This is the original Meudon PN. These calculations are very sensitive to the Bowen OIII HeII Lyα fluorescence problem since helium is ionized in much of the nebula (see Netzer & Ferland 1984).

Tables 6 and 7. These PNe have the same intermediate temperature central star and chemical composition. The conditions are simpler than the Meudon planetary because the transfer of helium diffuse fields does not dominate the nebula. Table 6 has a low density and resulting high ionization. The high density PN (Table 7) has low ionization. Both have outer radii near the H-ionization front. The models are density bounded and so have a simpler ionization structure than the Meudon PN.

Table 8. This AGN Narrow Line Region model is irradiated by a nonthermal continuum with $f_\nu \propto \nu^{-1.3}$ between 1.36 eV and 50 keV, and no radiation outside this range. This tests high-energy effects, but the results are not sensitive to line transfer since emission lines are well below their thermalization density and Lyα trapping is not important.

Table 9. This table compares 100 km s^{-1} steady-flow shock models. The pre-shock density and magnetic field are 1.0 cm^{-3} and 1 μG, and the pre-shock gas is assumed to be fully ionized. Calculations stopped at 10^3 °K. The Allen (1973) abundances are assumed for the 13 to 15 most abundant elements through iron. Line intensities are given relative to Hβ, and the Hβ flux is given in erg cm^{-2} s^{-1}, emitted into 4π sr. "[FeII] sum" is the sum of lines of the lowest 16 levels of the atom. The line "ΔR" is the depth in cm at which the gas cooled to 10^3 °K.

4. Outstanding issues

Discussion within the meeting highlighted several outstanding issues, which are summarized here.

		Mean	1	2	4	5	6	11
I(Hβ)	E0	1.31	1.33	1.31	1.06	1.37	1.43	1.34
Lyα	1216	34.2	38.3	32.1	37.0	32.4	31.5	34.2
HeI	5876	0.12	0.11	0.13	0.14	0.12	0.13	0.13
HeII	4686	0.24	0.25	0.25		0.25	0.23	0.24
HeII	1640	1.60	1.60	1.74	1.49	1.53	1.56	1.67
CIII]	1909+	2.82	2.90	2.99	2.45	2.87	2.83	2.90
CIV	1549+	3.18	2.70	3.85	2.28	3.69	3.17	3.36
[NII]	6584+	2.33	1.40	3.20	1.21	3.10	2.67	2.40
NIII]	1749+	0.19	0.24	0.24	0.01	0.22	0.22	0.22
NIV]	1487+	0.20	0.20	0.23	0.12	0.22	0.21	0.21
[OI]	6300+	1.61	2.20	1.61	1.41	1.67	1.31	1.46
[OI]	63 μ	1.12	0.25	1.13			1.44	1.64
[OII]	3727+	1.72	1.60	1.44	3.18	1.58	1.30	1.20
OIII]	1663+	0.56	0.35	0.63		0.61	0.57	0.63
[OIII]	5007+	33.1	31.4	34.5	31.1	33.0	32.8	36.0
[OIII]	4363	0.32	0.30	0.34		0.31	0.30	0.33
OIV	1403+	0.36	0.49	0.30		0.36	0.42	0.25
[NeIII]	15.5 μ	1.89	1.50	2.01		1.94	2.05	1.95
[NeIII]	3869+	1.91	1.90	2.51	0.84	2.16	1.72	2.34
[Ne IV]	2423+	0.44	0.52	0.42		0.47	0.41	0.38
[NeV]	3426+	0.52	0.59	0.55		0.53	0.44	0.50
MgII	2798+	1.78	3.50	1.72	1.48	1.23	1.12	1.61
[SiII]	34.8 μ	0.90	1.00	0.96		1.07	0.96	0.52
[SII]	6720+	1.33	2.40	1.01	1.58	0.93	0.99	1.10
[SIII]	9532+	1.88	1.60	2.15	1.73	2.06	1.67	2.08
[SIII]	18.7 μ	0.49	0.36	0.61		0.57	0.52	0.37
[SIV]	10.5 μ	1.05	0.86	1.24	1.23	0.82	0.94	1.22
I(total)	E0	125	131	128	92	128	131	133
T(in)	E4	1.70	1.71	1.70		1.72	1.68	1.68
T(H+)	E4	1.17		1.24	1.12	1.06	1.20	1.23

TABLE 8. NLR Cloud

4.1. *Radiation transport*

Two limiting cases of continuum transport exist: the "outward only" and "on-the-spot" approximations. Most codes use one or the other. Two (Harrington and Rubin) solve the problem correctly. Comparisons suggested that outward only gave better agreement with the exact solutions, especially with the temperature of the illuminated face of the cloud. Outward only was used for most calculations presented here.

4.2. *The BLR*

Models of the Broad Line Region of Active Galactic Nuclei were considered during the workshop, but no consensus was reached in the time available. These are challenging since the gas is dense enough for line transport and thermalization to be important, thus adding an extra layer of complexity on top of those present in conventional nebulae.

4.3. *The atomic database*

The Opacity Project (Seaton 1987) and other large-scale efforts (Verner & Yakovlev 1995) have resulted in a large homogeneous photoionization data base. The corresponding recombination (especially dielectronic) coefficients have been computed for second row elements, but do not yet exist for third row and iron group elements.

		6	7	10	12
I(Hβ)	E–6	6.17	4.24	4.74	2.60
Hα	6563	3.34	3.16	3.18	3.80
H 2-ν		12.8	19.1	16.2	—
He I	5876	0.15	0.14	0.10	0.18
HeII	1640	2.14	0.33		
C II]	2326+	2.54	3.50	2.47	1.97
[C II]	158 μ	0.76	1.23	0.73	1.06
C III]	1909+	6.87	8.83	6.03	8.10
C IV	1549+	8.31	7.83	19.3	18.1
[N II]	6584+	1.48	1.91	1.96	1.43
N III]	1749+	0.75	0.90	0.75	0.87
N IV]	1487+	0.26	0.43	0.23	0.54
N V	1240+	0.02	0.02	0.03	0.12
[O I]	6300+	0.07	0.16	0.17	0.03
[OI]	63.2 μ	0.40	0.24		
[O II]	3727+	6.59	8.69	10.4	10.4
[O III]	5007+	5.88	6.30	3.98	5.54
O III]	1663+	2.06	2.42	1.62	3.11
O IV]	1403+	0.54	1.11	0.98	0.98
[NeII]	12.8 μ	0.40	0.33	0.41	
[NeIII]	3968+	0.63	0.68	0.68	0.56
NeIV]	2423+	0.06	0.27	0.07	0.24
Mg II	2798+	0.25	0.52	0.54	0.06
[Si II]	34.8 μ	4.36	3.95	5.25	4.28
SiIV]	1397+	2.31	2.43		
[S II]	6720+	1.22	1.97	1.51	1.00
[SIII]	9532+	0.61	0.44	0.77	
[ArIII]	7136+	0.22	0.22	0.14	
[CaII]	7292+	0.14	0.14	0.20	0.04
[Fe II]	sum	11.1	2.83	2.72	1.05
ΔR	E16	4.44	4.15	3.41	

TABLE 9. Shocks

Photoionization models considered here do not include iron, which can dominate the thermal equilibrium of lower ionization clouds (Wills, Netzer, & Wills 1985). Work is now underway within the atomic physics community to generate the basic data needed to treat FeII emission properly.

4.4. *Reliability—Photoionization calculations*

This can be judged by examining the scatter among the predictions. Figure 1 shows the relative standard deviation (the standard deviation divided by the mean) for all results presented for the Meudon HII region case (Table 3, the open circles). This is among the simplest cases since He^{++} is not present. The dispersion is plotted against line excitation energy in inverse microns.

Our goal is for the simulations to agree within roughly 10%. This is the realistic limit to the *absolute* accuracy a calculation can hope for since modern calculations of theoretical atomic cross sections and rate coefficients are no more accurate than 10%–15%. The lines with the largest standard deviation are largely those of third row elements, especially [SIII] and [SII]. Currently there is a large dispersion in photoionization cross sections for

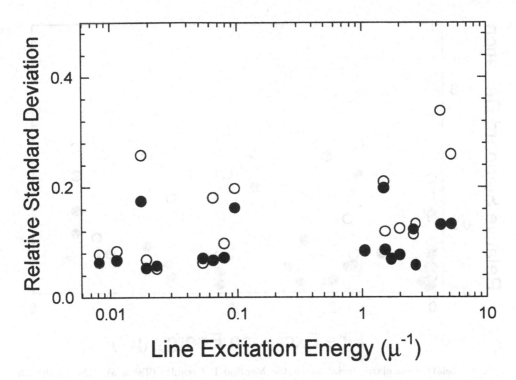

FIGURE 1. Relative standard deviation (standard deviation divided by the mean) for the Meudon HII Region results (Table 3). The results for all calculations are shown as open circles, and the filled circles show those for the subset of the calculations described in the text.

these many-electron systems, and no *total* recombination coefficients now exist (Ali *et al.* 1991).

The Meudon Planetary Nebula (Table 5) presents a more challenging case—a hot star, so that HeII transport problems are present. Results for all calculations are presented as the open circles in Figure 2. The dispersion is ≈30% with some lines larger than 50%, in whole disappointingly large.

The scatter in Figure 2 is most likely the result of different levels of treatment of the He line and continuum transport, a difficult problem (Osterbrock 1989, pp. 23–31). One way to judge the accuracy of the overall transport problem is the luminosity of Hβ. For the Meudon PN the 100% conversion of ionizing photons into hydrogen recombinations (Table 1; Osterbrock 1989) would result in an Hβ luminosity of 2.58×10^{35} erg s^{-1}. The actual Hβ luminosity will be larger, since the Bowen OIII process degrades HeII Lyα into several softer photons. We computed a second relative standard deviation for all lines, including only those calculations with predicted Hβ luminosities close to or larger than the 100% conversion value. These results, the predictions of codes 2, 3, 5, 6, and 11 are shown as filled circles in Figure 2. The mean Hβ intensity for these calculations is $2.70 \pm 0.06 \times 10^{36}$ erg s^{-1}. The scatter for this subset of the calculations is $\approx$ 10%. Clearly the computed spectrum is sensitive to the transport of the ionizing continuum.

We also recomputed a second relative standard deviation for the Meudon HII Region, including only the five calculations listed above, plus code 8. (L(Hβ) and L(total) for code 8 are well within 2 standard deviations of the five previous ones.) The second

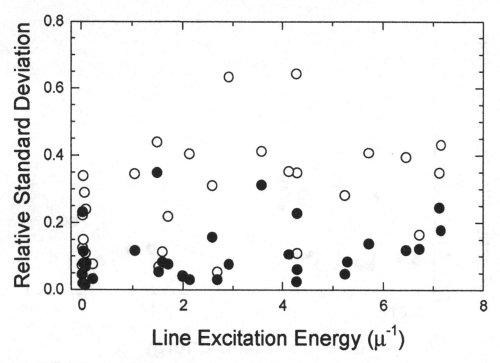

FIGURE 2. Relative standard deviation for the Meudon PN results (Table 5). The results for all calculations are presented as open circles, and those for the calculations which fulfill the Hβ luminosity criterion (see text) by the filled circles.

dispersion is shown on Figure 1 as the filled circles. Clearly, our stated goal of 10% accuracy can be realized.

4.5. *Reliability—Shock calculations*

The most important predictions of the shock models are the intensities of the strong lines relative to Hβ, as these are used to infer shock velocities and elemental abundances. The scatter among the models is largest for lines from the very highest and very lowest ionization states. The former are most sensitive to ionization rates and conditions close to the shock, while the latter are most sensitive to details of the radiative transfer and thermal balance in the recombining gas. The discrepancies between the high ionization line predictions would only lead to ≈ 5 km/s difference between the shock velocities derived. The prediction of Hβ in shocks depends in a complex manner on collisional rates and diffuse radiation transfer, and so is not as simple to predict as in a photoionized nebula.

5. Conclusions

A major goal of this workshop was to establish a set of benchmark nebulae for reference in code development. These are presented in the tables throughout this chapter.

Intercomparisons among the calculations show that an overall spectroscopic accuracy approaching 10% can be realized, but this will require improvements in the atomic data base (especially recombination coefficients of third row elements) and continuum transport. The photoionization/recombination data base now constitutes a major uncertainty for calculations of third row elements. The situation is far better for second row elements, where this 10% accuracy is now realized.

Acknowledgment

The Lexington Meeting on Numerical Simulations of Ionized Nebulae was sponsored by the Graduate School, and the Center for Computational Sciences, of the University of Kentucky. We thank the Director of the CCS, Dr. John Connolly, for his hospitality.

REFERENCES

ALI, B., *et al.* 1991 *Publ. A.S.P.* **103**, 1182

ALLEN, C. W. 1973, *Astrophysical Quantities* (London: Athlone Press)

BINETTE, L., DOPITA, M., & TUOHY, I. R. 1985 *ApJ* **297**, 476

BINETTE, L., WANG, J., VILLAR-MARTIN, M., MARTIN, P. G., & MAGRIS C. G. 1993a *ApJ* **414**, 535

BINETTE, L., WANG, J., ZUO, L., & MAGRIS C. G. 1993b *AJ* **105**, 797

BREGMAN, J. & HARRINGTON, J. 1986 *ApJ* **309**, 833

CLEGG, R. E. S., HARRINGTON, J., BARLOW, M., & WALSH, J. R. 1987 *ApJ* **314**, 551

COX, D. P. 1972 *ApJ* **178**, 143

COX, D. P., & RAYMOND, J. C. 1985 *ApJ* **298**, 651

DRAINE, B. T., & MCKEE, C. F. 1993 *ARAA* **31**, 373

FERLAND, G. J., FABIAN, A. C., & JOHNSTON, R. 1994 *MNRAS* **266**, 399

GARNETT, D., & SHIELDS, G. A. 1987 *ApJ* **317**, 82

GRUENWALD, R. B., & VIEGAS, S. 1992 *ApJS* **78**, 153

HARTIGAN, P., RAYMOND, J., & HARTMANN, L. 1987 *ApJ* **316**, 323

HARRINGTON, J. P. 1968 *ApJ* **152**, 943

HARRINGTON, J. P., SEATON, M. J., ADAMS, S., & LUTZ, J. 1982 *MNRAS* **199**, 517

INNES, D. E. 1992 *A&A* **256**, 660

JONES, A. P., TIELENS, A. G. G. M., HOLLENBACH, D. J., & MCKEE, C. F. 1994 *ApJ* **433**, 797

KALLMAN, T. R., & MCCRAY, R. 1982 *ApJS* **50**, 263

KRAEMER, S. 1986 *ApJ* **307**, 478

NETZER, H. 1993 *ApJ* **411**, 594

NETZER, H., & FERLAND, G. J. 1984 *PASP*

OSTERBROCK, D. E. 1989 *Astrophysics of Gaseous Nebulae and Active Galactic Nuclei* University Science Books; Mill Valley

PÉQUIGNOT, D. 1986 Editor, Workshop on Model Nebulae, Publication de l'Observatoire de Paris

PETITJEAN, P., BOISSON, C., & PÉQUIGNOT, D. 1990 *A&A* **240**, 433

RAYMOND, J. C. 1979 *ApJS* **39**, 1

REES, M., NETZER, H., & FERLAND, G. 1989 *ApJ* **347**, 640

RUBIN, R. H., DUFOUR, R. J., & WALTER, D. K. 1993 *ApJ* **413**, 242

RUBIN, R. H., SIMPSON, J. P., HAAS, M. R., & ERICKSON, E. F. 1991 *ApJ* **374**, 564

SEATON, M. 1987 *J. Phys. B.* **20**, 6363

SHULL, P. 1983 *ApJ* **275**, 611

SIMPSON, J. P., COLGAN, S. W. J., RUBIN, R. H., ERICKSON, E. F., & HAAS, M. R. 1995 *ApJ* submitted

SUTHERLAND, R., & DOPITA, M. 1993 *ApJS* **88**, 253

VERNER, D., & YAKOVLEV, D. 1995 *A&AS* **109**, 125

VIEGAS, S., & CONTINI 1994 *ApJ* **428**, 113

WILLS, B., NETZER, H., & WILLS 1985 *ApJ* **288**, 94

Emission Line Diagnostics

By HAGAI NETZER

School of Physics and Astronomy and the Wise Observatory, Tel Aviv University

This review focus on three major aspects of emission line diagnostics: Line intensities that are used to deduce the physical conditions in the gas, line profiles that are needed to study the gas dynamics, and line variability, that is used to obtain the gas distribution. Applications and examples are given for active galaxies and quasars. The status of research and the outstanding problems in each of these areas are discussed and new observational findings are shown. The more important developments of recent years are due to systematic, combined space and ground-based observations of individual objects (reverberation mapping) as well as studies of large samples of AGNs (e.g. the HST radio-loud sample).

1. Introduction and overview of active galactic nuclei

Observations of emission lines in photoionized nebulae provide important diagnostics of the line emitting gas in three different ways: Line intensities are used to derive the physical conditions in the gas. Density, temperature, optical depth etc. are all related to emission line ratios and absolute fluxes. Line profiles are used to investigate the gas dynamics and the velocity field. Finally, line variability, when correlated with flux variations of the photoionizing continuum, are used to measure the gas distribution and the size of the emission line region.

Active Galactic Nuclei (AGN) are situated in the center of otherwise normal galaxies and show strong emission lines superimposed on strong nonstellar continua. The underlying continuum is observed, in many such objects, at wavelengths short of the Lyman limit. Its integrated ultraviolet luminosity is more than enough to explain the observed line strength and equivalent width assuming a typical "covering factor" (the fraction of the source covered by line emitting gas of moderate to large continuum optical depth) of about 0.1. Two types of line profiles are observed. Narrow lines (300–1000 km s^{-1} FWHM) coming from the Narrow Line Region (NLR) and broad lines (2000–6000 km s^{-1} FWHM) coming from the Broad Line Region (BLR). Some objects (quasars and Seyfert 1 galaxies) show both type of lines. Others (Seyfert 2 galaxies, LINERs) show only narrow lines.

The following review summarizes our understanding of AGNs by way of analyzing their emission line intensities, profiles and variability. I discuss broad-line AGNs and emphasize the more important developments of recent years as well as several outstanding problems.

2. Line intensity and physical conditions

2.1. *Photoionization models*

The line emitting gas in AGNs is photoionized by the central nonstellar continuum. Evidence for this are the correlated line and continuum luminosities, the similarity of the line and continuum light-curves demonstrated in recent reverberation studies, and the overall spectral appearance which is different from the one found in collisionally ionized gases. Photoionization models are therefore a major tool in analyzing the physical conditions in the gas, by way of calculating the various observed line ratios. Such modeling is very successful in the analysis of the low-density gas in HII regions, Planetary Nebulae etc.

(see G. Ferland contribution in this volume). This success is repeated, to some extent, when attempting to model the NLR gas in AGNs. However, modeling the BLR gas is more complicated and no published model can successfully explain the entire observed broad-line spectrum. The reason must be related to the BLR density, thought to be in the range of 10^9 to $10^{12} cm^{-3}$, and the large column density of the BLR clouds. Such densities are intermediate between the low densities in PN and HII regions and the high densities of stellar atmospheres. Calculating the emission line spectrum in low density objects is simple because no line opacity takes place. The very high densities are relatively simple too because non-LTE techniques can be used. This is not the case for the BLR gas which is far from LTE but still of large line and continuum opacity to required special calculational techniques. Evidently, such techniques are far from perfection and the resulting calculations do not match, very well, the observed spectrum. Several attempts to solve this problem, as well as more complete descriptions of the theoretical situation can be found in Davidson & Netzer (1979), Kwan & Krolik (1981), Osterbrock & Mathews (1986), Kwan (1984), Krolik & Kallman (1988), Rees, Netzer & Ferland (1989), Ferland & Persson (1989), Collin-Souffrin *et al.* (1986) and Netzer (1990).

2.2. *Outstanding problems*

The following is a list of several outstanding problems of AGN research related to emission line intensities.

The Lα/Hβ problem (BLR): The hydrogen line spectrum has been a problem since the early discovery, by Baldwin (1977a), of the large deviation of the observed Lα/Hβ ratio from the predicted value. This may be related to either large opacities (e.g. Kwan & Krolik 1981), dust extinction (Netzer & Davidson 1979) or both. A related difficulty was the lack of observed correlation of Lα/Hβ with any other property. This situation has now changed; see §2.3.

Dust and reddening (BLR+NLR): There are clear indications that dust and reddening are affecting the narrow line intensity and profiles (e.g. MacAlpine 1985, Osterbrock & Mathew 1986). There has also been some suggestion of reddening of the broad emission lines (Netzer & Davidson 1979). This is related to the Lα/Hβ problem since the simplest, sometimes the only way for measuring the extinction is by comparing theoretical and observed hydrogen line ratios. The dust location and properties, in AGNs, are still unknown.

The FeII problem (BLR): The intensity of the ultraviolet and optical FeII lines is another example of an unsolved problem, related to line transfer, atomic physics and perhaps metalicity (e.g. Wills, Netzer & Wills, 1985). Such lines are extremely strong in many broad-line AGNs and carry much of the cooling of the BLR gas. Not a single BLR model can fully account for their observed intensity.

The Baldwin relationship (BLR): There is a strong correlation of broad-line equivalent width with continuum luminosity (Baldwin 1977b). This is known to be different for different lines and to depend on the luminosity and perhaps the radio properties. Possible explanations range from continuum variability and changes in ionization parameter to disk orientation and enhanced metalicity.

2.3. *New results*

Recently we (B. Wills, D. Wills, J. Baldwin, G. Ferland, M. Brotherton, Mingsheng Han, I. Browne, H. Netzer) have undertaken a very large spectroscopic study of radio-loud quasars (Wills *et al.* 1993, 1994; Netzer *et al.* 1994, 1995). We are using HST/FOS observations, combined with ground-based data, to study many emission lines in a sample of about 50 intermediate luminosity quasars. The space-borne and ground-based data are

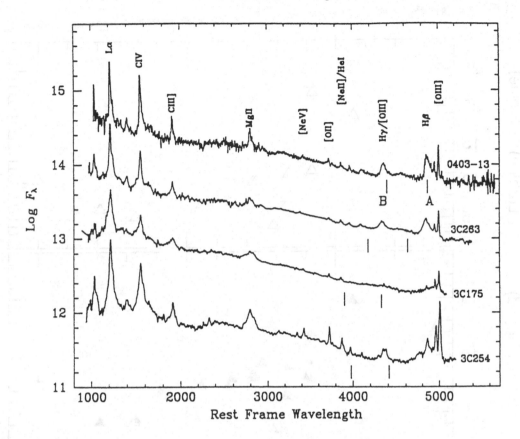

FIGURE 1. Multi-wavelength spectra of 4 radio-loud quasars observed with the HST/FOS and several ground based telescopes. This is part of a large HST sample (e.g. Wills *et al.* 1994) with a coverage from about 1000Å to beyond 5000Å rest wavelength. Observations of Hα are available for some of the objects.

obtained simultaneously and we can measure, for the first time, line fluxes and profiles for all lines from Hα to Lα. The sample was selected according to the radio-lobe flux and is thus free from most known selection effects such as orientation, optical continuum shape and line intensity. Several examples are shown in Fig. 1. For all sources we have a measurement of the radio orientation parameter, R, defined as

$$R = \frac{Radio - core\ flux}{Radio - lobe\ flux} \tag{2.1}$$

This enables us to distinguish core-dominated (large R) from lobe-dominated (small R) sources. (For more explanation and relation to unified schemes of radio-loud AGNs see Browne & Murphy 1987 and Barthal 1989).

An interesting result, based on the study of a sub-sample of 20 such sources with very good Lα and Hβ measurements (Netzer *et al.* 1995), is shown in Fig. 2. The diagram demonstrate the strong correlation of the Lα/Hβ line ratio with the slope of the nonstellar continuum. This, to the best of our knowledge, is the first significant correlation of this line ratio with any other observed property of AGNs. Our study reveals that the correlation is due to the red wings of the lines, i.e. the red wings of the Hβ lines are stronger compared with the red wings of the Lα lines in steeper (i.e. larger

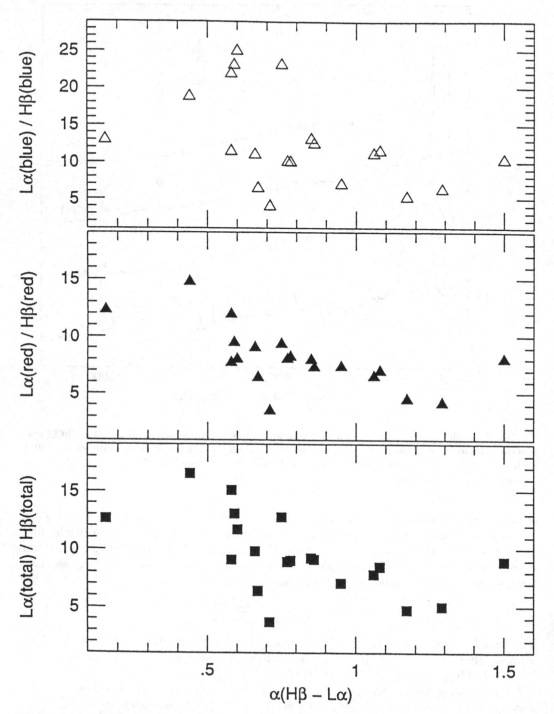

FIGURE 2. Lα/Hβ versus the nonstellar continuum slope, $\alpha(H_\beta - L_\alpha)$, for 20 radio-loud quasars (Netzer *et al.* 1995). Note the strong correlation of the red wings and the integrated line fluxes, and the lack of such a correlation for the blue wings of the lines (top panel).

α in $F_{\nu} \propto \nu^{-\alpha}$) continuum quasars. The correlation is consistent with extinction by galactic type dust with E_{B-V} of 0.1–0.3 mag., assuming the intrinsic $L\alpha/H\beta$ ratio in the sample is about 20.

Line and continuum reddening is not the only explanation for the above correlation and some other effects, to do with the intrinsic continuum shape, and perhaps with the synchrotron component, must be important too. A crucial and important test would be the search for a similar correlation in radio-quiet quasars.

3. Line profiles and gas dynamics

3.1. *Theoretical models*

Line profiles can be used to deduce the gas kinematics and velocity field. This is important for both the BLR and the NLR and has been studied in a large number of papers (see Mathews 1982; Mathews & Capriotti 1985; Osterbrock & Mathews 1986; Espey *et al.* 1989; Kallman *et al.* 1993; and references therein). Most theoretical models assume either outflaw of material, due to acceleration by radiation pressure, or Keplerian virialized motion. The latter can be either a radial infall or "random motion," meaning bound orbits of different inclinations. There were also attempts to consider electron-scattering as an additional broadening mechanism.

3.2. *Outstanding problems*

Some open questions related to the comparison with the observations are listed below.

Infall, outflaw or random motion? (BLR+NLR): Detailed line profile calculations are not very successful in explaining the observed line profiles in some objects. There are cases where different dynamical models fit equally well the observed lines, thus there is no unique solution to the overall velocity field and not even an answer whether it is dominated by infall, outflaw or random motion. Evidently, profile fitting cannot provide a unique description of the velocity field in AGNs.

Different lines have different profiles (BLR+NLR): The various emission lines in the spectrum of many objects differ significantly in their profiles. This must indicate that regions of different dynamics must have different physical conditions. However, in other objects , most profiles are very similar and the physical condition do not change by much from regions of low to high velocities. Models to explain those differences are not yet available.

Different lines have different redshifts (BLR): As demonstrated by Espey *et al.* (1989), Corbin (1990) and others, broad emission line redshifts, as determined from line peaks, can be vastly different in low and high ionization lines. This is not common to all AGNs and seem to depend on luminosity. There is no satisfactory model to explain this phenomenon.

3.3. *New results*

Several recent developments may help to solve these outstanding problems. I shall mention the two I am associated with.

Our new $L\alpha$ and $H\beta$ measurements show that in some objects both lines have similar profiles and redshifts while in others large shifts and big profile differences are present. This complement the results of Espey *et al.* (1989) about redshift differences in lines of different ions and shows that this phenomenon is present also in lines of the same ion. Fig. 3 shows a few examples from Netzer *et al.* (1995). Such profiles, and the resulting $H\beta/L\alpha$ ratio, indicate different physical conditions in different parts of the BLR. The $L\alpha/H\beta$ line ratio is sensitive to optical depth and density and the parts of

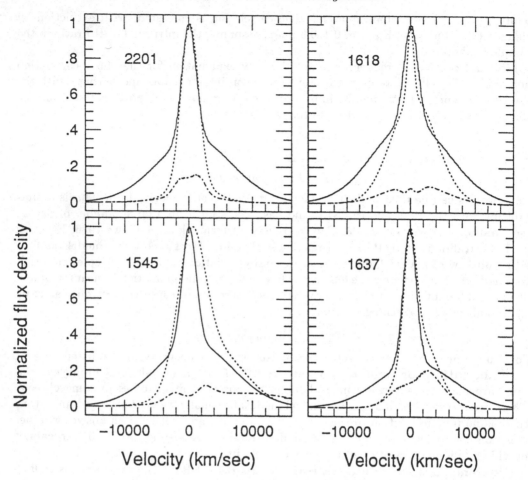

FIGURE 3. Example of normalized, smoothed Lα (solid line) and Hβ (dotted line) profiles, and the Hβ/Lα ratio (dashed-dotted line).

the BLR with different kinematics must also differ in their density and column density. Reddening, if present, can also affects different parts of the BLR in different ways. All this may be related to the BLR orientation to the line of sight if, as suggested by Wills & Browne (1986), the BLR has a flat geometry with cloud motion preferentially in a plane perpendicular to the radio-jet direction. Our work, so far, do not show any clear relation between the measured profiles and R, the orientation index of eqn. 1.

Recently we (Netzer & Laor 1993) have looked into the problem of gas distribution across the entire emission line zone (BLR and NLR) and the lack of strong emission lines with intermediate line width. We noted the similarity between the best estimate of the BLR size, as obtained from reverberation studies,

$$R_{\text{BLR}} \simeq 0.1 L_{46}^{1/2} \, pc, \tag{3.2}$$

and the sublimation distance of dust grains in the strong radiation field,

$$R_{\text{sub}} \simeq 0.2 L_{46}^{1/2} pc. \tag{3.3}$$

We proposed that the so called BLR is the inner part of a continuous dust and gas distribution. This inner part is devoid of dust since grains cannot survive so close to the center. Calculations show that, given an additional assumption of a decreasing ionization parameter with increasing radius, there is a large region just outside of the BLR where line emission is heavily suppressed because most of the ionizing photons are absorbed by the dust. This can explain the overall profile of optical lines (broad bases and strong sharp cores) as well as the weakness of the narrow ultraviolet resonance lines in high luminosity AGNs (the resonance line photons are destroyed by the large number of scatterings in the dusty environment, see Wills *et al.* 1993).

4. Line variability and gas distribution

4.1. *Reverberation mapping*

This area of AGN research has been, by far, the most productive and successful in recent years. Several large campaigns have resulted in detailed, sometimes unexpected results about the BLR size and gas distribution. In my opinion, the most important and successful programs are:

(1) *The AGN watch campaign*. This is the largest collaboration with more than 100 participating observers. Three galaxies have been observed so far, NGC 5548, NGC 3783 and NGC 4151, and several papers have been published (e.g. Clavel *et al.* 1991; Peterson *et al.*, 1991; Maoz *et al.* 1993; Reichert *et al.*, 1994; Stirpe *et al.*, 1994). The most recent study is a very ambitious HST (39 daily observations) IUE and many ground-based telescopes monitoring of NGC 5548 (Korista *et al.*1995).

(2) *The Wise Observatory campaign*. This is the first large campaign of recent years. Optical monitoring of a dozen sources, in 1988, resulted in three positive lag measurements and the first transfer function of the BLR (e.g. Maoz *et al.* 1990; Netzer *et al.* 1990; Maoz *et al.* 1991; Maoz *et al.* 1994).

(3) *The LAG campaign*. This consortium of "Lovers of Active Galaxies" is using several telescopes on La Palma to monitor low luminosity AGNs. It resulted is a few significant measurements of emission line lags (e.g. Wanders *et al.* 1993).

The main goal of these large campaigns is to study the gas distribution across the BLR, using information from line and continuum light-curves (see Peterson 1993 for a review and Krolik *et al.* 1991, for a detailed model). In principle, this is a straightforward problem, involving the recovery of the transfer function and, from it, the gas distribution. In practice, undersampling, noisy data and non-linear response result in great complications and large uncertainties. All methods developed so far have produce ambiguous results and there is yet no model of a single source that satisfactory explains the observed, time-dependent intensity of all emission lines. The theory in this area definitely lags behind the observations, by much more than a few days.

4.2. *Outstanding problems*

The following major problems have been noted:

Different lines show different lags: This has been noted from the very first international campaign and suggests that the ionization of the BLR gas decreases outward. It is not yet clear whether it holds, in a simple way, for all ions. In particular, the strong low-ionization lines of FeII may differ, in this respect, from the similar ionization line of MgII (Maoz *et al.* 1993).

Different "events" show different lags: The first international campaign on NGC 5548 (Clavel *et al.* 1991) clearly shows three continuum "events" followed by three emission-line events of similar duration. As noted by Netzer & Maoz (1990), the lag of the strong emission lines, measured separately for the three events, is very different.

Netzer & Maoz (1990), and Netzer (1993) have interpreted this as a sign of a "thick" BLR geometry.

What line to model? The very different behavior of different lines, and different events, suggest that modeling of the BLR, using the light curves and transfer functions of *individual* lines, are too simplified. Because of the stratification, there is no simple, model independent solution that relates line emissivity to gas distribution. Netzer (1990, 1993) suggested that the *total* line emission is a better indicator of the overall distribution of radiation-bounded clouds.

4.3. *New results*

The various open questions suggest that modeling the BLR gas distribution is better be based on the total line emission, rather than the transfer function of individual lines. The following is an application of this idea to the 1989 data of NGC 5548 (Clavel *et al.*, 1991; Peterson *et al.* 1991).

First, one must obtain the total emission-line light curve, by adding up observed intensities of all emission lines, with allowance for unobserved lines. Next, a model of the gas distribution is constructed such that when convolved with the observed continuum light-curve, the observed total emission light-curve is recovered. The preferred approach is light-curve modeling, i.e. reconstructing the observed line light-curve by repeated guesses on the gas distribution. This is safer and better defined than the inversion techniques and their resulting ill-defined transfer functions. In the case of NGC 5548 shown here I have assumed that the uncertainty in the first period of 50 days is very large, since this part depends on the unknown history of continuum variation.

The geometrical model presented here is a thick spherical shell of variable inner and outer radii. This spherical BLR is made out of numerous small clouds in virial motion. The mass of the clouds is conserved as they move in or out. Some assumptions must be made about the gas pressure, or density, as a function of radius, and the response of the total emission in the lines to the UV continuum variations. Guided by the observed larger lags for lower ionization lines, the ionization parameter must be a decreasing function of the distance to the center. This, plus the assumption of mass conservation, uniquely define the dependence of the covering factor on radius.

As is well known, some emission lines do not respond linearly to the continuum source variations. I find this to be also the case for the total line emission if all three observed events in 1989 are to be explained by the model. This can be caused if the ionizing flux does not vary linearly with the observed UV continuum. The non-linearity assumption, when added to the other model ingredients, results in the theoretical light-curve shown in Fig. 4. Given all the assumptions that enter such models, it is not surprising that transfer functions of individual lines cannot be used to put strong constraints on the gas distribution.

5. Conclusions

Observations of line ratios, line profiles and line variability give a complete description of the physical conditions, the gas dynamics and the gas distribution. Until recently, the main limitation in understanding AGNs has been the lack of high quality data. This has now changed and the many outstanding problems listed in this review are largely due to theoretical difficulties. The hope for the near future is that the new large-scale monitoring efforts, that provide simultaneous information on intensity, profiles and variability, will be used in perfecting such models.

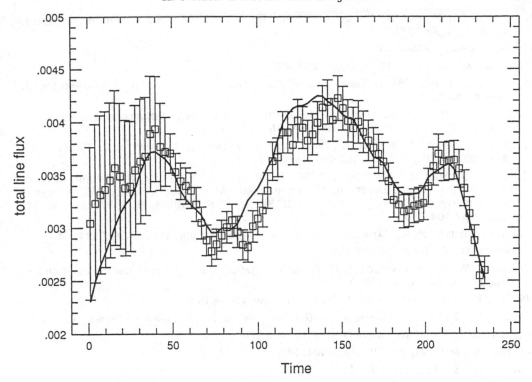

FIGURE 4. Observed (points with error bars) and calculated (smooth curve) total emission line light-curves for NGC 5548.

Acknowledgements

I am greatful to the US–Israel Binational Science Foundation for support of the HST project through grant 8900179. The *Hubble Space Telescope* observations and analysis were supported by grant GO-2578.01-87A from the Space Telescope Science Institute which is operated by AURA, Inc. under NASA contract NAS 5-26555.

REFERENCES

BALDWIN, J. A. 1977a *MNRAS* **178**, 67p

BALDWIN, J. A. 1977b *ApJ* **214**, 679

BARTHAL, P. D. 1989 *ApJ* **336**, 606

BROWNE, I. W. A., & MURPHEY, D. W. 1987 *MNRAS* **226**, 601

CLAVEL, J. *et al.* 1991 *ApJ* **366**, 64

COLLIN-SOUFFRIN, S., DUMONT, A. M., JOLY, M., & PEQUIGNOT, D. 1986 *AA* **166**, 27

CORBIN, M. R. 1990 *ApJ* **357**, 346

DAVIDSON, K. & NETZER, H. 1979 *Rev. Mod. Phys.* **51**, 715

ESPEY, B. R., CARSWELL, R. F., BAILY, J. A., SMITH, M. G., & WARD, M. J. 1989 *ApJ* **342**, 666

FERLAND, G. J., & PERSSON, S. E. 1989 *ApJ* **347**, 656

KALLMAN, T. R., WILKES, B. J., KROLIK, J. H., & GREEN, R. 1993 *ApJ* **308**, 805

KORISTA, K. *et al.* 1995 *ApJ Supp* (in press)

KROLIK, J. H., & KALLMAN, T. R. 1988 *ApJ* **324**, 714

KROLIK, J. H., HORNE, K, KALLMAN, T. R., MALKAN, M. A., EDELSON, R. A., & KRISS, G. A. 1991 *ApJ* **371**, 541

KWAN, J. 1984 *ApJ* **283**, 70

KWAN, J., & KROLIK, J. H. 1981 *ApJ* **250**, 478

MACALPINE, G. M., 1985, in *Astrophysics of Active Galaxies and Quasi-Stellar Objects* (ed. J. Miller), p. 259

MAOZ, D., NETZER, H., LEIBOWITZ, E., BROSCH, N., LAOR, A., MENDELSON, H., BECK, S., MAZEH, T. & ALMOZNINO, E. 1990 *ApJ* **351**, 75

MAOZ, D., NETZER, H., MAZEH, B., BECK, S., ALMOZNINO, E., LEIBOWITZ, E., BROSCH, N., MENDELSON, H., & LAOR, A. 1991 *ApJ* **367**, 493

MAOZ, D., NETZER, H., PETERSON, B. M., BECHTOLD, J., BERTRAM, R., BOCHKAREV, N. G., CARONE, T. E., DIETRICH, M., FILIPPENKO, A. V., KOLLATSCHNY, W., KORISTA, K., SHAPOVALOVA, A. I., SHIELDS, J. C., SMITH, P. S., THIELE, U. & WAGNER, R. M. 1993 *ApJ* **404**, 576

MAOZ, D., SMITH, P. S., JANNUZI, B. T., KASPI, S., & NETZER, H. 1994 *ApJ* **421**, 34

MATHEWS, W. G. 1982 *ApJ* **252**, 39

MATHEWS, W. G., & CAPRIOTTI, E. R. 1985 in *Astrophysics of Active Galaxies and Quasi-Stellar Objects* (ed. J. Miller), p. 185

NETZER, H. 1990 in *Active Galactic Nuclei* Springer-Verlag Pub.

NETZER, H. 1993 in *The Nearest Active Galaxies* (eds. Beckman, Colina, & Netzer), p. 219

NETZER, H. & DAVIDSON, K. 1979 *MNRAS* **187**, 871

NETZER, H., & LAOR, A. 1993 *ApJL* **404**, L51

NETZER, H., & MAOZ, D. 1990 *ApJL* **365**, L5

NETZER, H., MAOZ, D., LAOR, A., MENDELSON, H., BROSCH, N., LEIBOWITZ, E., ALMOZNINO, E. BECK, S., & MAZEH, T. 1990 *ApJ* **353**, 108

NETZER, H., KAZANAS, D., WILLS, B. J., WILLS, D., HAN, MINGSHENG, BROTHERTON, M. S., BALDWIN, J. A., FERLAND, G. J., & BROWNE, I. W. A. 1994 *ApJ* **430**, 131

NETZER, H., BROTHERTON, M. S., WILLS, B. J., MINGSHENG, HAN, WILLS, D., BALDWIN, J. A., FERLAND, G. J., & BROWNE, I. W. A. 1995 *ApJ* (submitted)

OSTERBROCK, D. E., & MATHEWS, W. G. 1986 *Ann. Rev. Ast. Ap.* **24**, 171

PETERSON, B. M. 1993 *PASP* **105**, 247

PETERSON, B. M. *et al.* 1991 *ApJ* **366**, 119

REES, M., NETZER, H., & FERLAND, G. J. 1989 *ApJ* **347**, 640

REICHERT, G. A., *et al.* 1994 *ApJ* **425**, 582

STIRPE, G. M., *et al.* 1994 *ApJ* **425**, 609

WANDERS, I., *et al.* 1993 *A&A* **269**, 39

WILLS, B. J. & BROWNE, I. W. A. 1986 *ApJ* **302**, 56

WILLS, B. J., NETZER, H., & WILLS, D. 1985 *ApJ* **288**, 94

WILLS, B. J., HAN MINGSHENG, NETZER, H., WILLS, D., BALDWIN, J. A., FERLAND, G. J., BROWNE, I. W. A., & BROTHERTON, M. S. 1994 *ApJ* (in press)

WILLS, B. J., NETZER, H., BROTHERTON, M. S., HAN, MINGSHENG, WILLS, D., BALDWIN, J. A., FERLAND, G. J., & BROWNE, I. W. A. 1993 *ApJ* **410**, 534

Ultraviolet Spectroscopy

By REGINALD J. DUFOUR

Department of Space Physics & Astronomy, Rice University, Houston, TX 77251–1892

A review of the field of astronomical ultraviolet spectroscopy with emphasis on *emission lines* in astrophysical plasmas is presented. A brief history of UV spectroscopy instruments is given, followed by a discussion and tabulation of major atlases of UV emission-line objects to date (mid-1994). A discussion of the major diagnostic UV emission lines in the ~ 912–3200Å spectral region that are useful for determining electron densities, temperatures, abundances, and extinction in low- to moderate-density plasmas is given, with examples of applications to selected objects. The review concludes by presenting some recent results from HST, HUT, and IUE on UV emission-line spectroscopy of nebulae and active galaxies.

1. Introduction

The history of ultraviolet (UV) spectroscopy in astronomy spans over three decades now and such observations have led to many discoveries regarding the physical nature of the entire gambit of astronomical objects. Hot astrophysical plasmas have line and continuum emission and absorption processes for which UV spectroscopy can probe the more energetic physical processes that cannot be studied adequately in the optical or infrared. In addition, studies of the UV spectral properties of cooler bodies, such as planetary atmospheres, comets, and interstellar dust provide important information on their physical state and composition.

This article concentrates on reviewing some of the techniques and results from the study of *emission lines* in astronomical UV spectroscopy. Given that the range of astronomical objects from the Earth's geocorona to quasars show UV emission lines and that during the past three decades over two thousand papers have appeared in the literature, including numerous conferences and books, a comprehensive review is unpractical. Therefore, the author will limit this discussion to a review of the various emission-line diagnostics present in UV spectra from approximately the Lyman Limit (912Å) to near the atmospheric cutoff (3200Å). In addition, he will concentrate on recent results during the past few years obtained with the International Ultraviolet Explorer (IUE), the Hubble Space Telescope (HST), and the Hopkins Ultraviolet Telescope (HUT) flown on the 1991 Astro-1 space shuttle mission. Readers who wish to obtain a more comprehensive review of the observations and scientific results from UV spectroscopic studies across the entire repertoire of astronomical objects should began with a study of the book *Exploring the Universe with the IUE Satellite* (Kondo 1987), which contains 36 review articles on UV spectroscopic results for all types of astronomical objects, as well as discussions on the history and future promise of astronomical UV studies.

A good indicator of scientific interest in the field of astronomical UV spectroscopy during the past two decades can be derived from counting the numbers of scientific papers in the *Astrophysical Journal* under the subject keyword "ultraviolet astronomy" which was instigated in 1972. A plot of the numbers during the 1972–1993 period is shown in Figure 1. It shows that during the period 1972 to 1985 that the number of refereed papers on the subject increased from ~25/yr to ~100/yr, then dropped to about ~65/year in the late 1980's, but increased rapidly to the ~100 level again during the early 1990's. I interpret this trend as indicating that the "short-term" scientific

Ap.J. PAPERS ON UV SPECTROSCOPY

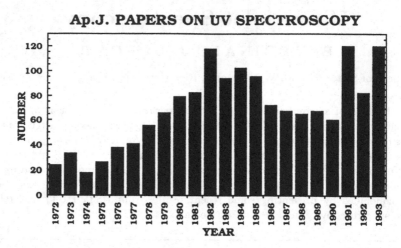

FIGURE 1. Graph of *Astrophysical Journal* papers on "UV Spectroscopy" during 1972–1993.

investigations with the IUE peaked in the mid-1980's and then dropped due to relatively longer time-scale multi-year project research with that satellite, then the rapid rise during the 1991–1993 period was due to the first scientific results from HST and HUT being published. I suspect that during the remaining part of the decade that ApJ publications related to UV spectroscopy will remain at the ~100/yr level due to additional results from IUE-HST-HUT observations and archival research are published, as well as results from EUVE and several UV astronomy space shuttle payloads recently flown or are planned. Not treated in this analysis are papers in the AJ, PASP, or European journals, which collectively add about 50% more to the total of refereed papers per year; or various conferences on spacecraft results or restricted astronomical topics, which probably add another 50% to the numbers of papers per year again.

2. A brief history of UV spectroscopy instruments

Astronomical UV spectroscopic studies began in the early 1960's with rocket flights. The first measurements of the UV continuum spectra of stars were obtained in 1961 with a low resolution objective grating spectrometer launched in an Aerobee rocket (Stecher & Milligan 1962). The first successful UV spectral line studies were made using a similar rocket-borne spectrometer in 1965 (Morton & Spitzer 1966). Subsequent rocket-borne UV spectrometers made the first observations of UV emission lines in hot stars and late-type supergiants. The late-1960's heralded the coming of the series of Orbiting Astronomical Observatories (OAOs), which were satellites specifically directed at obtaining imagery and spectroscopy of astronomical objects in the UV. Four OAO's were built during the late 1960's, and while two (OAO-A and OAO-B) were failures, OAO-2 (a reconstituted clone of OAO-A), was successfully launched in 1968 and was notable in initiating the era of UV spectroscopy from space-borne observatories. OAO-2 carried low-dispersion spectrometers for observations in the 1200–1400Å wavelength range and made hundreds of observations of astronomical sources during its three-year lifetime. A summary of the many scientific results of OAO-2 can be found in NASA SP-310 (Code 1972).

Astronomical UV spectroscopy in decade of the 1970's flourished with a variety of observations made from rockets, balloon-borne telescopes, manned-spacecraft, and unmanned orbiters. The decade also marked the beginning of active European involvement in the field of UV spectroscopy. Notable European successes during this period included a rocket flight obtaining (for then) high resolution UV spectra of the W-R stars γ Vel and ζ Pup over the 900–2300Å region, two balloon-borne high resolution UV spectrometers working in the 2000–3000Å region accessible in the upper atmosphere, and two satellites—TD-1 (ESRO) & ANS (Netherlands), which obtained UV imagery and low resolution spectroscopy of many astronomical sources. The two most notable US instruments for astronomical UV spectroscopy during the decade were the launch of the COPERNICUS (OAO-C) satellite in 1972 and the International Ultraviolet Explorer (IUE) in 1978. The COPERNICUS satellite, developed at Princeton (Rogerson *et al.* 1973), obtained moderate- and high-resolution UV spectra of hundreds of bright stars during its two-year lifetime. Though its single-channel stepping spectrometer limited the practical wavelength coverage in a given observation, the signal-to-noise on the high resolution ($\sim$0.1Å) UV spectra of bright stars obtained by COPERNICUS surpassed anything previously, and it made important contributions to studies of stellar mass-loss and interstellar abundances from observations of various UV resonance absorption and emission lines.

In astronomical UV spectroscopy, the decade of the 1980's clearly belongs to the IUE satellite. Launched on 26 January 1978 into a geosynchronous orbit by a Thor-Delta rocket, the IUE was developed as an international collaborative effort by NASA-SRC-ESA with its sole mission being UV spectroscopy of astronomical bodies over the $\sim$ 1150–3300Å wavelength range (Boggess *et al.* 1978). Composed of a 0.45m Richey-Crétien telescope with two echelle spectrographs and four SEC Vidicon detectors, the IUE has obtained over 100,000 spectra of astronomical objects and continues to operate to date, some 13 years beyond its designed 3-year lifetime ("*...and keeps going...*"—IUE's motto these days). IUE operates in two modes: a high-resolution mode, whereby "short wavelength" (1145–1930Å continuous coverage) and "long wavelength" (1845–1920Å continuous coverage) two dimensional echelle spectra are imaged by one of three (SWP and LWP or LWR, respectively) SEC Vidicon cameras with UV converters; and a low-resolution mode, whereby the cross-disperser grating is removed and a single spectrum is imaged over the 1150–1975Å range with the SWP camera and over the 1910–3300Å region with the LWP/R cameras (Harris & Sonneborn 1987). At high resolution, the point-source resolution is about 0.1–0.2Å (decreasing with longer wavelengths) for the SWP range and about 0.2Å for the LWP/R range; at low resolution, the corresponding resolution varies from 3–7Å with the SWP and $\sim$6Å with the LWP/R cameras. One of the main limitations of the IUE, particularly for emission-line studies, is dynamic range—its A-D converter is only 8-bits (0–255), and the best signal-to-noise (S/N) possible is only 25–30 for multiple exposures and typically only 15–20 on single long exposure spectra (due to a combination of fixed-pattern noise, reseaux marks, and radiation "hits").

The decade of the 1990's heralded the promise of a "new era" in astronomical UV spectroscopy; beginning with the launch of the Hubble Space Telescope (HST) on 24 April 1990. Compared to IUE, HST offered the increased power of a larger aperture (2.4m vs. 0.45m) and higher sensitivity and dynamic range (12-bit vs. 8-bit encoded data) in its spectroscopic instruments for UV spectroscopy. In addition, when the primary mirror aberration problem was discovered, more priority was given to spectroscopy projects than previously anticipated. Thus HST made an immediate impact on the field of astronomical UV spectroscopy, particularly regarding emission line studies. Two of the five original axial science instruments on HST are spectrographs with UV capabilities: FOS (Faint Object Spectrograph) and GHRS (Goddard High Resolution Spectrograph). Both have a variety of gratings for spectroscopy over the 1150–8500Å wavelength range (FOS) and

Instrument	Projected Aperture Size	Resolving Power ($\lambda/\Delta\lambda$)	Time Resolution	Wavelength Range (Å)	Magnitude Limit
FOS	0.085″–3.7″	1,300	30 ms	1150–8500	13.7–20.5–20.2
		250	30 ms	1150–8500	15.3–22.9–21.5
GHRS (Side 1)	0.22″, 1.74″	80,000	50 ms	1150–1700	11–14
		25,000	50 ms	1100–1900	13–16
		2,000	50 ms	1100–1900	17
(Side 2)		80,000	50 ms	1700–3200	11–14
		25,000	50 ms	1150–3200	13–16

TABLE 1. HST UV spectroscopy capabilities

the 1100–3200Å range (GHRS) at a variety of resolutions. Detailed capabilities of these two instruments are listed in Table 1. While the larger aperture of HST is somewhat offset by the small apertures of spectrographs compared to IUE for extended targets, the potential S/N of UV spectroscopy observations with HST exceed that possible with IUE by 1–2 orders of magnitude. In addition, the maximum resolution of GHRS in the far-UV exceeds that of IUE by a factor of about 7 (*cf.* the recently published review by Brandt *et al.* 1994 on instrumental characteristics and a summary of early science results from GHRS).

In December 1990 the Astro-1 space shuttle mission (STS-35) carried into orbit three UV telescopes for imaging (UIT) and spectroscopy (HUT and WUPPIE). The Wisconsin Ultraviolet Photo-Polarimeter Experiment (WUPPIE) is a 0.5m f/10 Cassegrain telescope with a spectropolarimeter for simultaneous observations of the UV energy distribution and polarization over the 1400–3200Å region at 12Å resolution. The Hopkins Ultraviolet Telescope (HUT) is a 0.9m f/2 telescope with a prime-focus spectrograph capable of UV spectroscopy over the 830–1860Å range with a nominal resolution of $\sim$3Å (Davidsen *et al.* 1992). A noteworthy accomplishment during this mission was by HUT obtaining high S/N UV spectra of a variety of faint emission line nebulae and galaxy nuclei in the 830–1150Å region, which previously had not been observed by UV spectrographic instruments. Compared to previous UV spectroscopy instruments, the large aperture, short f-ratio optics, and large slit sizes of HUT enabled high S/N spectra of faint nebulosities to be obtained in relatively short exposures.

While HUT on Astro-1 made spectroscopic observations shortward of the 912Å Lyman limit, generally referred to as the "extreme ultraviolet," the era of EUV astronomy really began with the launch of the Extreme Ultraviolet Explorer satellite (EUVE) in 1992 June. In addition to imagery in the EUV, the satellite is equipped with a spectrometer which covers the 70–760Å region at a resolution of R $\sim$ 250. EUV spectroscopy observations began in mid-1993, after completion of an all-sky survey in four EUV bandpasses (60–180Å, 160–240Å, 345–605Å, and 500–740Å), and the first spectroscopic results are beginning to appear in the literature at the time of writing (Bowyer 1994).

The remainder of the decade looks bright for UV spectroscopy as HST, IUE, and EUVE continue to operate, a second Astro mission is planned for 1995, several UV imagery and spectroscopy instruments are planned for the space shuttle (*e.g.*, ORFEUS, Krämer *et al.* 1990) and rocket flights, and the Far-Ultraviolet Spectroscopic Explorer (now called LYMAN) may be launched into orbit before the decade is over. LYMAN will possibly be the next major orbiting UV spectroscopy mission (though plans for a Russian 1.7m UV spectroscopy telescope were announced in 1990), and is planned to consist

The *IUE Ultraviolet Spectral Atlas* 1983, Wu *et al.*, IUE Newsletter No. 23.

The *IUE Ultraviolet Spectral Atlas, Addendum I* 1991, Wu *et al.*, IUE Newsletter No. 43.

IUE Ultraviolet Spectral Atlas of Selected Astronomical Objects 1992, Wu *et al.*, IUE Reference Publication 1285.

IUE Spectral Atlas of Planetary Nebulae, Central Stars, and Related Objects 1988, Feibelman *et al.*, NASA Reference Publication 1203.

An Ultraviolet Atlas of Quasar and Blazar Spectra 1991, Kinney *et al.*, ApJ Suppl. 75, 645.

An Atlas of Ultraviolet Spectra of Star-Forming Galaxies 1993, Kinney *et al.*, ApJ Suppl. 86, 5.

Ultraviolet Spectra of QSO's, BL Lacertae Objects, and Seyfert Galaxies 1993, Lanzetta, Turnshek & Sandoval, ApJ Suppl. 84, 109.

IUE-ULDA Access Guide No. 1: Dwarf Novae and Nova-like Stars 1989, La Dous, ESA SP 1114.

IUE-ULDA Access Guide No. 2: Comets 1990, Festou, ESA SP 1134.

IUE-ULDA Access Guide No. 3: Normal Galaxies 1992, Longo & Capacioli, ESA SP 1152.

IUE-ULDA Access Guide No. 4: Active Galactic Nuclei 1993, Courvoisier & Paltani, ESA SP 1153A & 1153B.

TABLE 2. Some selected atlases of UV emission-line objects

of a meter-class telescope for dedicated UV *two-dimensional* spectroscopy at moderate resolution in the EUV (100–912Å) and high resolution in the FUV (912–1200Å) (Moos 1990). In addition, in the 1996–1997 period we will have the second HST refurbishment mission, which will install STIS (Space Telescope Imaging Spectrograph), enabling two-dimensional UV-optical spectroscopy to be done with HST.

3. Reference catalogs and atlases

One of the greatest scientific benefits of the long lifetime of the IUE satellite is that it has accumulated over 100,000 UV spectra of astronomical objects covering the 1150–3200Å wavelength range. Such an extensive data archive provides the opportunity to produce extensive compilations of UV spectra of various classes of astronomical objects for the first time. The first efforts to produce atlases of the UV spectra of astronomical objects were made by Benvenuti *et al.* (1982) on supernovae, by Wu *et al.* (1983) and Heck *et al.* (1984) on normal stars, and by Rosa *et al.* (1984) on extragalactic H II regions. During the past several years a number of extensive atlases have appeared, and several that are most relevant to studies of UV emission line spectra are listed in Table 2. A more complete listing of all IUE-related UV spectra atlases to late 1993 is given by Pitts (1993) in IUE Newsletter No. 50.

The four IUE-ULDA "access guides" published by ESA utilize the *Uniform Low Dispersion Archive* of IUE spectra (ULDA) and an access and analysis software package called USSP (Wamsteker *et al.* 1989). The guides are directed towards specific classes of astronomical objects, and contain all of the good UV spectra taken of each class through 1991; while the ULDA/USSP archive and software are intended to provide archive users with a current easily accessible processed dataset and software for analysis of IUE low dispersion spectra. By contrast, the Kinney *et al.* (1991, 1993) atlases present low dispersion spectra of galaxies that are processed with a continuum weighted extraction technique (called *optimal*) that is designed to improve S/N. Since Kinney *et al.* required there be at least three good SWP or LWP/R spectra of each object for inclusion, their atlases contain only a subset of the existing IUE spectra archive on each class of object. They also include references to previous studies of many of the individual objects. This spectral extraction technique results in significant S/N improvements in some spectra over the old "boxcar" processing. Recently, Lanzetta *et al.* (1993) published an atlas of optimal-extracted spectra of QSO's, BL Lac objects, and Seyfert galaxies, which supplements the original Kinney *et al.* 1991 atlas by removing the "3-good spectra" requirement and contained an additional 192 objects compared to the 69 in the original Kinney *et al.* atlas.

All UV spectroscopists will benefit from having a copy of NASA Reference Publication 1285, *IUE Ultraviolet Spectral Atlas of Selected Astronomical Objects* by Wu *et al.* (1992). This atlas includes spectral plots and tabular flux data on the spectra of some 330 astronomical objects ranging from the moon to quasar PKS 2344+092, covering the 1150–3200Å wavelength range, with many of the spectra (stellar and semi-stellar objects) extracted using a gaussian-weighted technique to improve S/N. In addition to the variety of objects with representative spectra, the atlas gives excellent lists of various UV emission and absorption lines in astrophysical spectra, as well as a brief discussion and tabulation of interstellar extinction in the UV. Of great importance and utility to current and future users of low dispersion IUE archive data is the discussion and tabulation of artifacts in IUE SWP and LWP/R spectra.

Figure 2 on the next page shows composite UV spectra of four strong emission line objects taken from the Wu *et al.* (1992) atlas: (a) the symbiotic star RR Telescopii, (b) the high excitation planetary nebula NGC 7027, (c) a bright filament in the Cygnus Loop supernova remnant, and (d) the nucleus of the Seyfert 2 galaxy NGC 1068. In the center of the figure is an identification strip listing some of the stronger UV lines and multiplets seen in all four spectra. While the four objects represent quite different physical conditions of ionization source, temperature, density, and size of the emitting region, it is noteworthy that the stronger UV emission lines are from the same ions.

It is important to note that, with regard to the large IUE UV spectra archive, all of the low dispersion SWP, LWP, and LWR spectra are currently being re-processed with an entirely new software package, called NEWSIPS, which entails a more rigorous geometric registration, image-transfer function, and absolute calibration based on white dwarfs (Nichols *et al.* 1994). In addition, the spectra are extracted using the optimal technique and the data are available in the astronomical FITS standard format. Pending continued NASA funding, this reprocessing will extend into high dispersion spectra as well. Significant S/N improvements are realized in many of the spectra over the old processing and the new extensive (and easily accessible via NASA NDADS) IUE UV spectra archive is expected to be of great scientific value for decades to come.

4. UV emission-line diagnostics

Over the 912–3200Å wavelength range hot astrophysical plasmas of standard (solar) composition emit a wealth of emission lines which can be used to determine electron temperatures (T_e), densities (N_e), and ionic concentrations $(N(X^i))$ in the plasma. These diagnostics can then be used with others in the optical and infrared to develop detailed photo-ionization and/or shock-ionization models of a given object. Some of the "classic" examples in the early literature of these techniques utilizing UV emission-line diagnostics in conjunction with observations at other wavelengths and models are Harrington *et al.*'s

(1982) study of the planetary nebula NGC 7662, Raymond *et al.*'s (1983) study of the Cygnus Loop supernova remnant, Hayes & Nussbaumer's (1986) study of the symbiotic star RR Tel, and Ferland & Osterbrock's (1986) study of Seyfert 2 galaxies. A good summary of some of the more important UV emission-line diagnostics, with plots of several useful line ratios, is the paper by Czyzak, Keyes, & Aller (1986); in addition, the textbooks of Aller (1984) and Osterbrock (1987) contain much information of UV emission-line diagnostics as well. While the early studies were classics in the techniques utilized, newer atomic data has since appeared in the literature for many of the UV-line ions which should be utilized in future studies.

A list of the many UV emission lines in the 1150–3200Å wavelength range found in IUE low dispersion spectra of all types of objects is given in the Wu *et al.* (1992) atlas. The stronger lines seen in the four objects of Figure 2 are: H I Lyα λ1216, N V $\lambda\lambda$1238–1242, Si IV+O IV]+S IV $\lambda\lambda$1394–1406, N IV] $\lambda\lambda$1483–87, C IV $\lambda\lambda$1548–1550, He II λ1640, O III] $\lambda\lambda$1658–67, N III] $\lambda\lambda$1747–1754, Si III] $\lambda\lambda$1883,92, C III] $\lambda\lambda$1907,9, [O III]+C II] $\lambda\lambda$2321–31, [O II] λ2470, Mg II $\lambda\lambda$2796–2803, and O III λ3133. A more detailed list of UV lines seen at high dispersion is found in Penston *et al.* (1983) for the symbiotic star RR Tel (but see also Doschek & Feibelman (1993) and Aufdenberg (1993) for an improved list in the $\sim$1150–2000Å range). For expected emission lines in the EUV-FUV range, a useful reference is Doyle & Keenan (1992), which gives the results of theoretical calculations of line emissivities in the 100–2800Å wavelength range at densities appropriate to stellar sources.

Six physical processes give rise to the formation of emission lines in the ultraviolet: *(a) direct recombination*—electron capture and cascade processes are responsible for the strong He II Baα 1640Å and He II Paschen series seen in high excitation nebulae and the He I $2s^3P - np$ series (as well as Lyα); *(b) electron collisional excitation*—this gives rise to the "forbidden lines" corresponding to magnetic dipole and electric quadrupole transitions such as [Ne III] λ1815, [O III] $\lambda\lambda$2321+31, [O II] λ2470, and certain resonance transitions such as N V $\lambda\lambda$1238+42, C IV $\lambda\lambda$1548+50, and Mg II $\lambda\lambda$2796+2803; *(c) intercombination*—resulting in permitted (electric dipole) and forbidden (electric quadrupole or magnetic dipole) multiplets closely space in wavelength, such as N IV] $\lambda\lambda$1483–87, O III] $\lambda\lambda$1658–1667, N III] $\lambda\lambda$1747–54, Si III] $\lambda\lambda$1883–92, C III] $\lambda\lambda$1907+9, and C II] $\lambda\lambda$2325–2329 (lines that are among the strongest in UV spectra); *(d) dielectronic recombination*—important for many far-UV line multiplets such as C II λ1335, C III $\lambda\lambda$977+1176, N III λ991, N IV λ923, as well as many of the extreme UV lines of ions of the CNO elements, it can be important for many of the lines of CNO ions in the UV (*cf.* Nussbaumer & Storey 1984 for a thorough discussion and tabulation of the various lines); *(e) Bowen fluorescence*—gives rise to exceptionally strong lines of O III (compared to simple recombination) at 2837Å, 3023Å, 3045Å, and particularly 3133Å, in the near-UV spectra of high excitation planetary nebulae and symbiotic stars; and *(f) charge exchange*—important for determining the ionization equilibrium near ionization fronts, it also can affect the low-level populations of metastable ions like [O II] and [S II]. Williams (1994) gives an interesting chart of various emission lines from these processes for ions of astrophysical interest in this volume.

In nebular plasmas, the level populations of a given ion are determined by the equilibrium between radiative decay and collisions to/from a given excited level i and other levels j. The statistical equilibrium equations (Osterbrock 1989, p. 57) take the form:

$$\sum_{j\neq i} n_j N_e q_{ji} + \sum_{j>i} n_j A_{ji} = \sum_{j\neq i} n_i N_e q_{ij} + \sum_{j<i} n_i A_{ij}, \qquad (1)$$

with the left hand side giving transitions populating level i and the right hand side giving

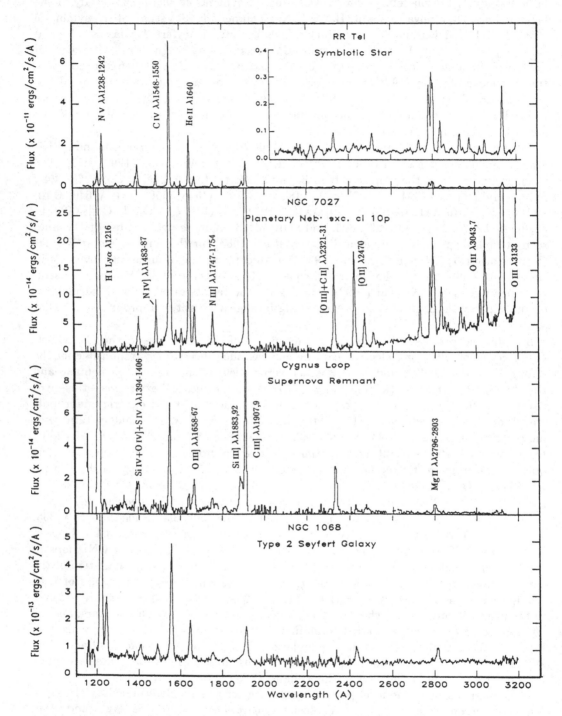

FIGURE 2. Example IUE low dispersion spectra of selected emission-line objects adopted from the atlas of Wu *et al.* (1992)

transitions that depopulate level i. A_{ij} are the radiative transition probabilities for $i \to j$ (to all lower levels) and q_{ij} are the collisional excitation/deexcitation rates for level i. For $i > j$, the collisional deexcitation rate is:

$$q_{ij} = \frac{8.63 \times 10^{-6}}{\sqrt{T_e}} \frac{\Omega(j,i)}{\omega_j}, \tag{2}$$

and for $j < i$, the collisional excitation rate can be written:

$$q_{ji} = \frac{\omega_j}{\omega_i} q_{ij} e^{-E_{ij}/kT_e}, \tag{3}$$

where Ω is the relevant collision strength (which have only a weak dependence on T_e), the ω's are the relevant statistical weights, E_{ij} is the energy of level i above j (normally the ground level), and T_e is the electron temperature. The emissivity of the $i \to j$ transitions can then be written as

$$4\pi I_{ij} = A_{ij} N(X^i) n_i E_{ij}, \tag{4}$$

where $N(X^i)$ is the ionic abundance, n_i is the fractional level population, and E_{ij}. Therefore, the emissivity (or observed intensity) ratio of two excited levels physically depend on how the two level populations are influenced by collisional processes that populate and depopulate the two levels. When collisional deexcitation rates for the two levels are very different the line ratio is density sensitive; when collisional excitation rates for the two levels are very different the line ratio is temperature sensitive.

4.1. Density diagnostics

Emission line ratios of two metastable transitions from upper levels of similar excitation energy to a given lower level are useful diagnostics of electron density when the lifetimes of the upper levels are significantly different and the plasma density is near the "critical density" of the longer lived level. For a given level i, the critical density is:

$$N_c(i) = \sum_{j<i} A_{ij} / \sum_{j\neq i} q_{ij}, \tag{5}$$

where A_{ij} is the radiative transition probabilities for $i \to j$ (to all lower levels) and q_{ij} are the collisional excitation/deexcitation rates for level i.

Since collisional excitations have an energy threshold, but collisional deexcitations do not, the collisional deexcitation-to-radiative decay rate for an excited level is relatively insensitive to T_e (eqn. 5). Therefore, radiative decay from multiplet excited levels of a given ion with similar excitation energies to a common lower level are predominantly sensitive to level lifetimes (given comparable collisional deexcitation rates), and thus are good diagnostics of electron density over a limited range of density near the critical densities of the longer-lived upper level. For $N_e << N_c$ for any two multiplet levels, their intensity ratio is effectively equivalent to the ratio of collision strengths (low density limit condition). By contrast, when $N_e >> N_c$, the intensity ratio is fixed by the ratio of statistical weights and transition probabilities (the high density limit condition, where LTE applies rigorously).

At optical wavelengths, the $^2D-^4S$ ground transition doublet for np^3 ions of [O II] and [S II] (and to a lesser extent [N I], [Cl III], [Ar IV], and [K V]) have been used as density indicators in nebulae. In the UV, the [Ne IV] $\lambda2422/\lambda2424$ doublet can be used as a density diagnostic for high excitation objects like planetary nebulae and supernova remnants. However, the UV is special in that intercombination lines of C II], C III], N III], O IV], and Si III] exist and provide density diagnostics for nebulae, particularly at higher densities ($> 10^4$ cm^{-3}) appropriate for symbiotic stars, proto-PN, and HH-objects.

Of particular utility are the $^3P_{1,2}-^1S_0$ doublets of C III] and Si III] which are in the $\lambda \sim 1900$ spectral region. The doublets consist of a highly forbidden magnetic quadrupole transition ($nsnp$ $^3P_2-$ $^1S_0$ C III] 1906.8Å & Si III] 1882.7Å) and an intercombination, or "semi-forbidden" electric dipole transition ($nsnp$ $^3P_1 - ns^2$ $^1S_0$ C III] 1908.6Å & Si III] 1892.0Å). As the density increases, the shortward forbidden line is more collisionally quenched and the ratios of $\lambda1907/\lambda1909$ and $\lambda1883/\lambda1892$ decrease from a low-density limit ratio of ~1.5 (set by the ratio of collision strengths) to a high-density limit ratio of "near zero" (set by the ratio of statistical weights and transition probabilities. Plots of these two ratios are given in Figure 3, taken from the recent study by Keenan, Feibelman, & Berrington (1992) which uses more modern atomic data than previous investigations. Note that the C III] doublet is a good density indicator over the $3 \times 10^3 \leq N_e \leq 2 \times 10^5$ cm^{-3} range, while the Si III] is a good indicator for higher densities, $3 \times 10^4 \leq N_e \leq 3 \times 10^{10}$ cm^{-3}. The ratios are very insensitive to T_e, as can be seen in Fig. 3, where the four lines represent the ratio for $T_e = 5000$ K (*short-dashed line*), 10,000 K (*solid line*), 15,000 K (*long dashed line*), and 20,000 K (*dashed-dotted line*). Another doublet in the UV useful for density diagnostics in highly excited plasmas is N IV] $\lambda1486.5/\lambda1483.3$, which increases by over four orders of magnitude in the range $10^6 \leq N_e \leq 10^{11}$ cm^{-3} (Czyzak *et al.* 1986).

The N III] $2s^22p^2P - 2s2p^{2} {}^4P$ intercombination multiplet near $\lambda \sim 1750$Å (1746.8Å, 1748.7Å, 1749.7Å, 1754.0Å, and 1752.2Å lines) provides several density-sensitive line ratio diagnostics (Czyzak *et al.* 1986; Keenan *et al.* 1994) which can be exploited with high resolution UV spectra. Czyzak *et al.* presented plots of three N III] density diagnostic line ratios: N III] $\lambda1752.2/\lambda1749.7$ is useful over both a low density range ($\sim 100 - 3000$ cm^{-3}) and over a high density range ($\sim 10^8 - 10^{10}$ cm^{-3}); similarly, the N III] $\lambda1754.0/\lambda1749.7$ ratio is a good diagnostic over two density ranges, for $N_e < 3000$ cm^{-3} and for $10^9 \leq N_e \leq 3 \times 10^{11}$ cm^{-3}; thirdly, the N III] $\lambda1748.7/\lambda1752.2$ ratio is useful in the high density range $3 \times 10^8 \leq N_e \leq 3 \times 10^{11}$ cm^{-3}. Keenan *et al.* (1994) noted four useful density diagnostic line ratios of N III]: $\lambda1754.0/\lambda1749.7$, $\lambda1752.2/\lambda1749.7$, $\lambda1748.6/\lambda1749.7$, and $\lambda1746.8/\lambda1749.7$, and discussed the effects of using modern atomic data (collision strengths and oscillator strengths) for N III on the line ratio diagnostics. They presented plots of two ratios, $\lambda1754.0/\lambda1749.7$ and $\lambda1752.2/\lambda1749.7$, computed with the modern atomic data and compared to the older results of Czyzak *et al.* (1986). These two plots are reproduced in Figure 4. They note that the newer atomic data results in difference as much as 24% in the line ratios for lower densities.

The intercombination multiplets of C II] near $\lambda \sim 2325$Å (2323.5Å, 2324.7Å, 2325.4Å, 2326.9Å, and 2328.1Å) and O IV] near $\lambda \sim 1400$Å (1397.2Å, 1399.8Å, 1401.2Å, 1404.8Å, and 1407.4Å) are similar to that of N III] and provide comparable sets of density diagnostics. However, both the C II] $\lambda\lambda2325$ and O IV] $\lambda\lambda1402$ multiplets are blended with lines from other ions ([O III] $\lambda2321.0$ & N II] $\lambda2325.9$ in the first case; and S IV] $\lambda1398.1,\lambda1404.8,\lambda1406.0$ & Si IV $\lambda1402.8$ in the second) which make measurements difficult even at high resolution. Figure 5 shows two plots of the regions of the O IV] and C II] lines from IUE high dispersion spectra of the symbiotic star RR Tel, which illustrates the resolution obtainable with IUE for a stellar object (extended nebulae have significantly degraded resolution with the large aperture). Observations of these multiplets in emission-line stars and nebulae with GHRS on the Hubble Space Telescope offer the promise of much improved wavelength resolution and photometric accuracy, as is illustrated at this meeting in the poster paper of Walter *et al.* (1994) for a GHRS spectrum of the C II] multiplet in the Orion Nebula.

Use of the UV C II], N III], and O IV] multiplets for density diagnostics is facilitated my the availability of modern collision strengths calculated over a wide range of temperatures

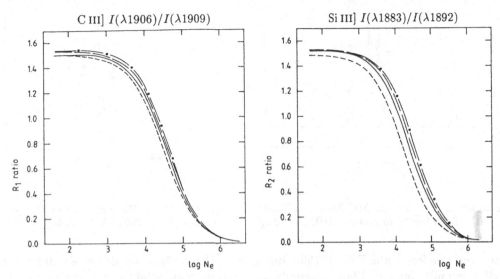

FIGURE 3. Plots of the N_e diagnostic line ratios of C III]$\lambda1907/\lambda1909$ (left) and Si III]$\lambda1883/\lambda1892$ (right) as a function of density N_e from Keenan *et al.* (1992).

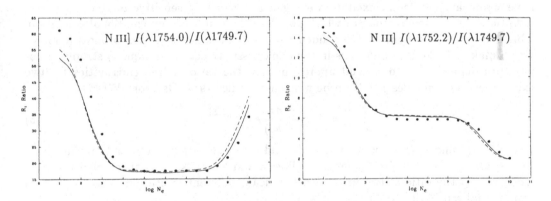

FIGURE 4. Plots of two N_e diagnostic line ratios of N III]: $\lambda1754.0/\lambda1749.7$ (*left*) and $\lambda1752.2/\lambda1749.7$ (*right*), taken from Keenan *et al.* (1994). Lines represent results for three T_e (5,000 K, 10,000 K, and 20,000 K) and the dots represent previous results based on older atomic data.

by Blum & Pradhan (1992) using an 8-state R-matrix method developed for the Opacity Project). However, for N III, Stafford, Bell & Hibbert (1994) derived collision strengths using an 11-state R-matrix code which differ from those of Blum and Pradhan by as much as 25% for $T_e = 10,000$ K. Therefore, observationalists who derive joy at being able to measure a line ratio to better than 10% with modern instruments need to be mellowed by the fact that the atomic data used to derive densities, or temperatures and abundances, are still uncertain at the 10–30% level. The review by Pradhan at this conference summarizes the best modern atomic data to date for use in emission-line diagnostics. Of particular note and usefulness is the recent compilation of collision strengths (and their weak, but important dependence on T_e) in Volume 57 of *Atomic*

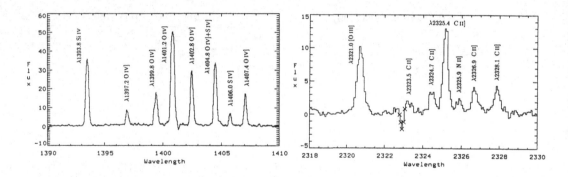

FIGURE 5. Plots of the O IV] 1402Å multiplet (*left*) and the C II] 2325Å multiplet (*right*) in RR Tel taken from archival IUE high dispersion spectra (SWP 29535 & LWR 2021).

Data and Nuclear Data Tables (1994 July issue). Hopefully, a future issue will have a similar compilation of oscillator strengths and transition probabilities for astrophysically interesting ions.

4.2. *Temperature diagnostics*

The ratio of two emission lines from an ion produced by radiative decay from levels which have significantly different excitation energies are primarily sensitive to the energy distribution of the free electron gas which populate the two levels by collisional excitation. This is due to the exponential dependence of the collisional excitation process (eqn. 3), which makes the level populations in the line emissivity equation (eqn. 4) strongly temperature dependent. To a good approximation, the ratio of line emissivities for two excited levels of energies $j > i$ can be written as (Aller 1984; Osterbrock 1989):

$$\frac{I_i}{I_j} = K \frac{(1 + ax)}{(1 + bx)} e^{\frac{-(E_i - E_j)}{kT_e}}, \tag{6}$$

where K, a, and b are constants that depend on collision strengths and transition A-values, and $x = 10^{-2} N_e / \sqrt{T_e}$. For many lines a and b are $<< 1$, so at low densities (i.e. when $x < 1$) the ratio of two lines with significant different excitation energies have an exponential sensitivity to T_e.

Over the optical spectral region accessible to ground-based spectroscopy, "auroral-to-nebular" forbidden line ratios ($^1S_0 - ^1D_2 / ^1D_2 - ^3P_{1,2}$) of LS-coupled ground state configurations of np^2 ions such as [N II] $\lambda 5755 / \lambda\lambda 6548 + 83$ and [O III] $\lambda 4363 / \lambda\lambda 4959 + 5007$ have been used to derive T_e in nebulae, emission-line stars, and galaxies. Correspondingly, "transauroral" transitions from the $^1S_0$ level to the ^3P ground multiplet (the strongest transition being $^1S_0 - ^3P_1$) have lines in the UV spectral region which can be used with the optical nebular transitions for temperature diagnostics. In addition, np^4 ions such as O I, Ne III, and Ar III have similar LS-coupled configurations with UV transauroral lines. A list of the UV transauroral lines and the corresponding nebular transition lines for eight ions prominent in nebular spectra is given in Table 3 (wavelengths taken from Osterbrock 1989).

In addition to the ground-state configuration transitions, intercombination transitions from $^5S_2 \rightarrow ^3P_2$ and from $^5S_1 \rightarrow ^3P_1$ produce UV lines that are observed in many nebular objects—particularly O III] 1660.8–1666.2Å & N II] 2139.0–2142.8. These intercombination lines can be compared to the usually strong lines of [O III] and [N II] to

Ion	Transauroral $(^3P_1 - {}^1S_0)$	Nebular $(^3P_2 - {}^1D_2)$	$R^*(\frac{Neb}{TrA})$
[N II]	3062.8	6583.4	1082
[O III]	2321.0	5006.9	762
[Ne V]	1575.2	3425.9	507
[S III]	3721.7	9530.9	753
[Ar V]	2691.1	7005.7	29.5
[O I]	2972.3	6300.3	680
[Ne III]	1814.6	3868.8	610
[Ar III]	3109.2	7135.8	64.4

*Line ratio computed for $N_e = 10^3$ cm^{-3} and $T_e = 10^4$ K

TABLE 3. UV/optical—transauroral/nebular line ratios for T_e determinations

derive T_e and are better high temperature diagnostics than the auroral and transauroral lines of [O III] and [N II]. Czyzak *et al.* (1986) and Keenan & Aggarwal (1987) present diagrams of the [O III]/O III] UV line ratio ($\lambda2322$)/($\lambda\lambda1660+ 1666$) versus T_e and of the O III]/[O III] UV-optical line ratio ($\lambda\lambda1660+1666$)/ ($\lambda\lambda4959+5007$) versus the optical auroral-to-nebular [O III] ($\lambda4363$)/($\lambda\lambda4959+5007$) ratio. The Keenan & Aggarwal plots are based on more modern atomic data and are presented in Figure 6. The UV lines of O III] and N II] provide unambiguous determinations of T_e for $N_e < 10^5$cm^{-3} and, when combined with T_e results from the optical auroral-to-nebular transitions, can provide unique information on the extent of temperature fluctuations in gaseous nebulae. However, as noted by Nussbaumer & Storey (1984), if dielectronic recombination contributes to the ^5S levels in an object, then one will obtain elevated temperatures from the UV lines (and abundances).

Dielectronic recombination lines of several ions of the CNO element group exist in the UV and can be compared with collisionally excited lines for T_e determinations. The emissivity of a dielectronic line can be written as:

$$4\pi I_D(\lambda) = N_e N(X^{i+1})\alpha_D(N_e, T_e)E_\lambda, \tag{7}$$

where $\alpha_D(N_e, T_e)$ is the effective dielectronic recombination coefficient for the $X^{i+1}+e \rightarrow X^i + h\nu$ recombination process. The dielectronic recombination coefficient in eqn. (7) is relatively insensitive to density but has a temperature dependence which can be characterized roughly as $\sim T^{-3/2}exp[-E/kT]$. The ratio of a collisionally excited line of an ion of state X^{i+1} to a dielectronic recombination line from state X^i has a strong dependence on T_e (with $exp[-\Delta E/kT]$ being the energy difference between the two transition states above ground). Nussbaumer & Storey (1984) discuss this method in detail as applied to UV spectral lines, in addition to providing information on the relevant dielectronic recombination coefficients for various lines of C I, C II, C III, N I, N II, N III, N IV, O I, O II, O III, O IV, and O V. Some of the more commonly observable UV line ratios available for such purpose are (C III $\lambda1908$)/(C II $\lambda1335$), (C IV $\lambda1549$)/(C III $\lambda2297$ or $\lambda977$), (N IV $\lambda1486$)/(N III $\lambda2064$ or $\lambda991$), (N V $\lambda1240$)/(N IV $\lambda1719$), and (O IV $\lambda1402$)/(O III $\lambda835$). Unfortunately O I and O II do not have any significant dielectronic recombination lines in the UV to use with the UV O III] and [O II] lines. In addition, the UV dielectronic lines are usually weak in nebular spectra and their past use for T_e diagnostics (and abundances) have been limited to bright symbiotic stars (*cf.* Nussbaumer & Stencel 1987; Nussbaumer *et al.* 1988, and references therein).

$R_1 = I(\lambda 2322)/I(\lambda\lambda 1660 + 1666)$

$R_2 = I(\lambda\lambda 1660 + 1666)/I(\lambda\lambda 4959 + 5007)$

$R_3 = I(\lambda 4363)/I(\lambda\lambda 4959 + 5007)$

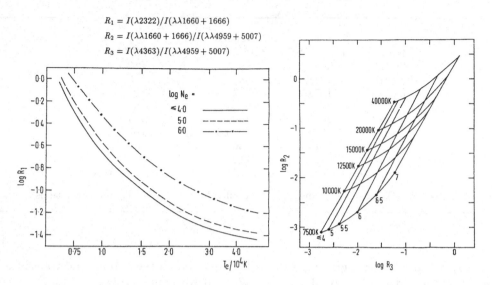

FIGURE 6. UV-intercombination/optical-forbidden line ratios of O III vs. T_e (*left*) and [O III] (auroral/nebular) (*right*) from Keenan & Aggarwal (1987).

4.3. *Abundances*

In addition to being probes of physical conditions in various ionization zones of astrophysical plasmas, UV emission lines provide fundamental data on abundances of the elements in the plasmas. Since abundances derived from emission lines is the topic of the review by M. Peimbert at this conference, I will limit my discussion to the impact that UV lines have in the field. Compared to ionic emission lines in optical spectra, the UV is unique in that it contains lines from abundant ions that have no optical counterparts: $C^{+,+2,+3}$, $N^{+2,+3,+4}$, $O^{+3,+4,+5}$, Ne^{+3}, $S^{+3,+4}$, $Mg^{+,+4}$, $Si^{+,+2,+3}$, $Al^{+,+2}$, and $Fe^{+3,+5}$. Included among these are the astrophysically important elements C, N, O, Mg, Si, and Fe for which determining abundances of is fundamental to understanding stellar nucleosynthesis processes and galactic chemical evolution. During the decade of the 1980's UV spectra of numerous emission-line objects were made with the IUE satellite which enabled the first comprehensive determination of abundances of C, N, Mg, and Si in a wide variety of objects ranging from comets to AGN (*cf.* the various reviews in Kondo 1987).

Traditionally two approaches have been applied to calculating abundances in nebular plasmas: (a) direct calculation of ionic abundance ratios from line strengths and applying empirical ionization corrections to get net elemental abundances, and (b) varying abundances (and other input parameters) in a photo-or shock-ionization model until a match to the observed spectrum is found. Aspects of the model approach is discussed in the review by G. Ferland at this conference, so I will limit my discussion to aspects of direct calculations of nebular abundances from UV emission lines. The vast majority of UV emission lines observed arise from collisional excitation or recombination (normal and dielectronic) for which their emissivities (eqns. 4 & 6) are functions of T_e, N_e, and ionic number density $N(X^i)$. The ratio of number densities of two ions follow from the ratio of two emission lines as:

$$\frac{N(X_k)}{N(X_l)} = \frac{I_{obs}(\lambda_k)}{I_{obs}(\lambda_l)} \frac{j(\lambda_l)}{j(\lambda_k)} = \frac{I_{obs}(\lambda_k)}{I_{obs}(\lambda_l)} \times f(\text{atomic data}, N_e^k, N_e^l, T_e^k, T_e^l), \qquad (8)$$

where the I_{obs}'s are the observed line intensities and the j's are the line emissivities. Note that on the right side of eqn. (8) "f" represents the functional dependence of the emissivities to atomic data (collision strengths, transition probabilities, wavelengths, and various constants) and the electron temperatures and densities *appropriate to the zones in a nebula where each of the two ions exist.*

It has been common practice to relate ionic abundances to hydrogen and calculate $N(X)/N(H^+)$ from UV-optical-IR line intensities relative to Hβ. For such purposes, Aller (1984) provided coefficients for deriving ionic abundances of 13 ions relative to H^+ that have prominent collisionally excited lines in the UV. His formulae are based on the assumption of an isothermal nebula of sufficiently low density where collisional deexcitation is negligible ($N_e < 10^4 cm^{-3}$) and are based on the compilation of atomic data by Mendoza (1983). They are applicable for most photoionized nebulae and emission-line galaxies of low to moderate density. However, care should be exercised in their use since newer atomic data are available for many of the ions and at high densities dielectronic recombination and radiative transfer problems for some of the UV resonance lines (*e.g.* C IV $\lambda1549$ and N V $\lambda1240$) become important.

Another procedure to perform calculations of ionic abundances, as well as T_e and N_e diagnostics from UV-optical-IR lines is to use a code that solves the statistical balance equation (eqn. 1) to get level populations and line emissivities for a specified (T_e, N_e), then use eqn. 8 with the observed line intensities to derive ionic abundance ratios. For most ions of astrophysical interest, a 5-level approximation is adequate and such a code has been developed and exported by De Robertis, Dufour, & Hunt (1987). During the past year, R. A. Shaw at ST ScI and the author have been upgrading this code with modern atomic data and including additional ions of particular importance in the UV (C II, C III, N III, O IV, and Si III). The code has been integrated into STSDAS/IRAF and is available with the current release of STSDAS (Shaw & Dufour 1993, 1994). While the original FIVEL code of De Robertis *et al.* was an interactive FORTRAN code, the new NEBULAR package has both interactive and automated options, where one can either input line ratios to derive T_e or N_e or ionic abundance ratios; or input a table of *observed* UV-optical line strengths and the program will perform reddening corrections using the Galactic extinction law of Seaton (1979), derive T_e and N_e from numerous ionic line ratios, construct a "diagnostic diagram" of $logN_e$ vs. $logT_e$, decide whether the data permits a 3-, 2-, or 1-zone (T_e, N_e) model (with the zones based on ionization potentials of various ions), then derive ionic abundances for most ions with UV or optical emission lines. Information about obtaining and using the program can be obtained via email to Dick Shaw at ST ScI (shaw@stsci.edu). We are continuously in the process of upgrading the program to include more ions as new atomic data becomes available, and plan in the future to add alternative reddening laws (*e.g.* LMC & SMC laws), include recombination lines of He I and He II, and improve the graphical and user-friendly aspects of the program.

While ionic abundance ratios are derivable directly from line intensity ratios, deriving total elemental abundance for some elements require corrections for ionic states without observable lines. Such corrections involve "empirical ICF's" (ionization correction factors) based on directly observed ionic ratios such as O^+/O^{+2} or ionization fractions computed by models. Optical spectra alone provide only a few ionic ratios for such purpose (*e.g.* O^+/O^{+2}, S^+/S^{+2}, Ar^{+2}/O^{+3}), while UV spectra provide a wealth of additional ionic ratios, particularly for high ionization objects. A good example is nitrogen, for which optical spectra contain only [N II] $\lambda\lambda6548,83$—nominally indicative of only a small fraction of the total N abundance. By contrast, UV spectra contain prominent lines of N III], N IV], and N V which represent a large fraction of the existing states of N in most nebular plasmas. Carbon is an even more extreme example, for which optical spectra provide only the C II λ4267 recombination line in any strength, while UV lines of C II], C III], and C IV exist which nominally represent a large fraction of the existing gaseous-phase carbon. In addition, C abundances derived from the (generally weak) C II λ4267 line for planetary nebulae have generally been much higher than from the UV lines (*cf.* Barker 1991 and references therein). An excellent discussion of the C abundance problem in PN has been presented by Rola & Stasińska (1994), who conclude that much of the discrepancy between the optical C II and UV C III] results is due to observational errors and selection effects with the weak optical C II line. Silicon is another element where deep UV spectra provide direct determination of the abundances of a majority of its existing ionic states and for which optical spectra have no significant lines.

4.4. *Extinction*

Corrections for interstellar extinction is a "necessary evil" in studies of UV spectra, particularly when comparisons have to be made to optical and infrared spectra. Fortunately the UV contains recombination lines of He I and He II which are relatively insensitive to T_e and N_e and facilitate both the determination of the magnitude and wavelength dependence of extinction for many types of objects. The stronger UV He I lines are of the triplet series, $2s^3S \rightarrow np^3P°$, at 2663Å, 2696Å, 2733Å, 2763Å, 2829Å, 2945Å, and 3188Å, which can be compared to optical triplet and singlet lines such as those at 5876Å, 4471Å, 4026Å, 4921Å, and 6678Å. High excitation objects have strong UV lines of He II such as the Bα 1640Å line and Paschen series lines at 2253, 2306, 2511, 2733, and 3203 Å, which can be compared to He II Pα 4686Å and other optical He II lines. Seaton (1979) performed such a study of the heavily reddened planetary nebula NGC 7027 and developed an analytical "mean Galactic extinction curve" which has been widely used in nebular UV spectroscopy of Galactic objects. In addition to the recombination lines, collisionally excited metastable multiplet UV lines with optical counterparts can be used for extinction determinations (*e.g.* [O II], [O III], [Ne III], [Ar IV], and [Ne V]; Aller 1984), though the theoretical UV-optical line ratios are normally very sensitive to T_e.

Whenever practical, UV spectroscopy studies should evaluate both the magnitude and wavelength dependence of extinction from UV-optical emission-line comparisons. It is well established that the UV extinction along high-density lines of sight in the Galaxy (*e.g.* Mathis 1989), such as in the Orion Nebula, show significant deviations from the standard Galactic law of Seaton (1979) or Cardelli *et al.* (1989). In addition, the UV extinction in the LMC (Clayton & Martin 1985; Fitzpatrick 1985) and SMC (*e.g.* Nandy *et al.* 1982) shows variations, particularly in the relative weakness of the 2175Å bump and a steeper rise in the far UV for galaxies of lower metallicity than the Milky Way. Variations in extinction among Local Group spirals such as M31 and M33 has also been found (Hutchings *et al.* 1992) based on IUE spectroscopic studies.

5. Some recent UV emission-line studies with HST, HUT, & IUE

Since 1990 several UV spectroscopy studies of emission-line nebulae and galaxies have appeared in the literature based on observations with HST FOS & GHRS and the HUT spectrometer flown during the Astro-1 shuttle mission. In the remainder of this paper I will describe several of the excellent results obtained with these new-generation UV spectroscopy instruments.

5.1. *H*II *regions*

UV spectroscopy of Galactic and extragalactic H II regions provide the best means of determining the abundance of carbon in the ISM of galaxies via the C III] $\lambda1909$ and C II] $\lambda2325$ emission lines. The IUE has been used extensively for such studies (*cf.* Dufour 1987), with recent analyses of CNO abundances in the Orion Nebula (Walter, Dufour, & Hester 1992) and the Lagoon Nebula (Peimbert, Torres-Peimbert, & Dufour 1993) reported. In addition, Rubin, Dufour, & Walter (1993) reported the first analysis of Si-C abundances in an H II region (Orion) using high dispersion IUE observations. These and previous studies indicate that C/O is approximately solar in M42 and M8, but that temperature fluctuations in an H II region need to be evaluated carefully for accurate determination of C abundances from the UV lines. In addition, Rubin *et al.* 1993 found the gaseous-phase Si (from the UV Si III] lines) abundance to be significantly depleted (Si/H $\sim$ 1/8 solar) in Orion. Studies of extragalactic H II regions with IUE generally indicated that C/O decreases with O/H in the ISM of metal-poor systems, but that C/N$\sim$constant (Dufour 1988).

Observations of the UV emission-line spectra of H II regions with the IUE are limited in S/N and resolution due to the limited dynamic range of its vidicon detectors and spectral impurity by the extended objects filling the large aperture. Observations of several extragalactic H II regions with the HST FOS have been recently conducted during Cycles 2 & 3 with excellent results. A comparison of HST FOS and IUE low dispersion UV spectra of SBS 0335-052 and 30 Doradus in the LMC has been presented by Dufour *et al.* (1993), which illustrate vividly how the higher dynamic range and better resolution of FOS permits ten-fold improvements in S/N of various weak emission lines compared to what is possible with IUE. Poster papers of FOS spectra of the Orion Nebula (Rubin *et al.* 1994) and the Lagoon Nebula (Cox 1994) are presented at this conference. In addition, a poster comparing HST GHRS spectra of the C II] $\lambda\lambda2325$ multiplet in Orion with IUE high dispersion data is presented by Walter *et al.* (1994).

An illustration of the excellent S/N and resolution afforded by HST FOS UV spectra of H II regions, Figure 7 presents spectra of the Orion Nebula and SMC N88A (an H II region in the Small Magellanic Cloud) over the 1150–3200Å wavelength range. The two spectra illustrate the various emission lines which can be seen in a moderate excitation metal-rich and dusty Galactic H II region (Orion) compared to an high excitation metal-poor extragalactic H II region (N88A). Despite the spectra being rebinned for display and smoothed by a 3-point gaussian, the high resolution of the "low dispersion" FOS spectra is evident, with the multiplets of O III] $\lambda\lambda1663$, Si III] $\lambda\lambda1888$, and C II] $\lambda\lambda2325$ being partially or cleanly resolved. This higher resolution available on FOS spectra of H II regions is very important for detection and measurement of emission lines above the usually strong UV continuum in H II regions (due to scattering by dust or inclusion of hot stars in the aperture). Another significant advantage of FOS spectra is that longer wavelength optical-nearIR spectra can be obtained with the same aperture as the UV spectra, thus enabling accurate comparisons of spectral lines over a broad wavelength range. One of the historical problems with IUE spectra of extended sources has been the "tie-in" between UV spectral measurements with optical spectra due to the usually different aperture sizes and orientations used. This problem is most evident in carbon abundances derived (from C III] $\lambda\lambda1909$) for the most metal-poor galaxy known, IZw18 (Dufour, Garnett, & Shields 1988; Dufour & Hester 1990), where imagery data is necessary to accurately tie-in the IUE UV and optical line measurements.

The impact that the high quality UV spectra of H II regions now possible with HST FOS will have on the problem of CNO (and Si) abundances and chemical evolution is

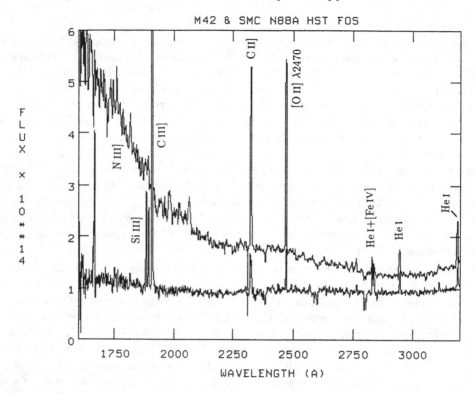

FIGURE 7. HST FOS spectra of two H II regions in the 1600–3300Å region (gratings G190H & G270H): M42 (Orion Nebula, *top*) and SMC N88A (a compact high excitation H II region in the Small Magellanic Cloud, *bottom*).

evident from the study of C/O abundance variations in seven extragalactic H II regions by Garnett *et al.* (1994). HST FOS spectra were obtained with the G190H grating, which enabled measurement of both the C III] 1909Å and O III] 1663Å lines and direct determination of the C^{+2}/O^{+2} in a manner that is insensitive to T_e, N_e, and extinction corrections. Their results for C/O and C/N, the latter based on optical measurements of [N II] 6583Å and nebular model ICF's, are shown in Figure 8. They found that C/O increases with O/H, consistent with a power law having an index of 0.43 ± 0.09 over the range of $-4.7 \le \log(O/H) \le -3.6$. In addition, unlike previous studies, they found that C/N increases with O/H in irregular galaxies, but is lower for solar neighborhood stars and H II regions. This suggests that the most metal-poor galaxies support the idea that their ISM enrichment is only from massive stars, while that of more metal-rich systems, like the Galaxy, have higher C/O due to a delayed enrichment from intermediate mass stars. They interpret the C/N results as indicating that the bulk of the nitrogen production is decoupled from the synthesis of carbon in the Galaxy. The Garnett *et al.* spectra also provide more accurate determination of Si/C and Si/O in metal-poor H II regions than previously possible (from the cleanly resolved Si III] and C III] lines, and detection of O III] in the UV), and the preliminary analysis (Dufour *et al.* 1994) suggest that the gaseous-phase Si/O ratio in the H II regions is generally lower than solar by a factor of ~2–3.

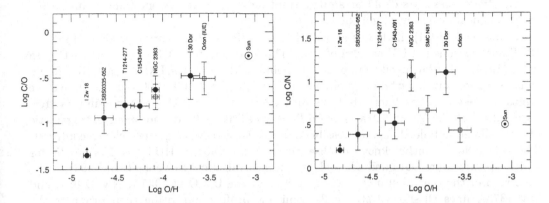

FIGURE 8. The variation of C/O (*left*) and C/N (*right*) for extragalactic H II regions, based on HST FOS observations by Garnett *et al.* (1994).

5.2. *Planetary Nebulae*

During it history, the IUE satellite has made extensive observations of the UV spectra of planetary nebulae (PNe), with ~2800 spectra currently existing in the archives. Since numerous Galactic PNe are bright and have relatively low extinction, the IUE has enjoyed excellent success in obtaining PN spectra (as well as of their central stars) at both low and high dispersion. UV emission-line spectra of PNe span a large range in their ionization characteristics, as is evident from the *IUE Spectral Atlas of Planetary Nebulae, Central Stars, and Related Objects* by Feibelman *et al.* (1988). A review of their UV spectral characteristics and early scientific results of IUE studies of PNe has been given by Köppen & Aller (1987). UV spectroscopy of PNe and of their central stars permit: (a) additional T_e and N_e diagnostics to be made from UV-line multiplets, particularly for the high ionization inner zones of the shells; (b) improved abundance determinations of certain elements, most importantly of C, N, Si, and Mg which have few or no optical lines; (c) evaluation of dielectronic recombination effects for lines of C III, C IV, N IV], N V, O III], O IV], etc.; (d) evaluate internal extinction and optical depth effects in resonance lines; (e) permit determination of effective temperatures and mass-loss rates for their central stars; and (f) provide additional constraints on physical parameters necessary for realistic theoretical models of PNe to be developed.

The most extensive study of PNe abundances from modern UV-optical spectra is by Perinotto (1991). He compiled data on 209 Galactic PNe and studied correlations between He, C, N, O, and Ne, and compared the results to the expectations of modern stellar evolution and nucleosynthesis theory for intermediate-mass stars. Perinotto concluded that the enrichment of N in type I PNe (He- & N-rich PNe arising from the more massive progenitors) is due mostly to the ON cycle, while that for type II–III PNe is due to the CN cycle.

The IUE has also had success in studying the spectra of extragalactic PNe, particularly in the Magellanic Clouds (*cf.* the review by Dufour 1990 and spectra of 5 SMC and 6 LMC PNe in the Feibelman *et al.* 1988 atlas), where it is possible to assess the effects of C-N enrichment in PN from low metallicity progenitors. Many PN in the Clouds were found to have significant enrichments of C and N compared to that in H II regions (Aller *et al.* 1987), suggesting that the shell compositions reflect the products of the third dredge-up phase of intermediate-mass star AGB evolution, where primary carbon from the core is ejected into the RG envelope prior to PN formation. If so, then PNe from intermediate-mass stars could be an important factor in C and N-enrichment during the early chemical evolution of galaxies.

To date (mid-1994), little has appeared in the refereed literature regarding HST and HUT observations the UV spectra of PNe. The first reported FOS study was by Dopita *et al.* (1994) of the low-excitation planetary nebula SMP 85 in the LMC; which they found to be a dense, young, carbon-rich object—being only about 500–1000 years in time since ionization began. It is likely that the future will see HST FOS, and possibly GHRS, observation of the UV spectra of more PNe, most likely faint and small extragalactic objects, Galactic Halo PNe, and possibly a few Galactic PNe of special scientific interest. The only PNe the author knows of observations were made by HUT was NGC 1535, for which spectra were obtained on and off the central star (Bowers *et al.* 1994). The spectrum of the central star is interesting in that the UV O vi 1035Å, N v 1240Å, and O v 1375Å lines all showed strong P-Cygni line profiles indicating their origin in the immediate vicinity of the star (all were weak or absent in the spectrum off the star).

5.3. *Supernova remnants*

The UV spectra of supernova remnants (SNRs) are rich in emission lines due to the high temperatures and ionization levels associated with the shocked gas. Studies of the many UV emission-line diagnostics nominally present provide important measurements of the shock and post-shock conditions in the gas. SNRs have been extensively observed with IUE, with the Cygnus Loop and Crab Nebula being the favorite Galactic SNRs (due to low extinction and large angular size), followed by SNRs in the LMC (*cf.* the review by Blair & Panagia 1987). In particular, the Cygnus Loop has been the target of extensive studies by the IUE (spectra of several positions) coupled with multi-wavelength spectral and imagery data and spatial model analyses (*e.g.* Raymond *et al.* 1988; Hester, Raymond, & Blair 1994) to assess in detail the complex nature of the shocked ISM structures visible. The Far-UV (FUV) region also contains several high-excitation UV lines important for SNR shock diagnostics; among the most important is O vi at 1032+38Å, for which Raymond *et al.* (1983) noted is the dominant line energy loss mechanism in the post-shock flow. FUV spectrophotometric observations of the Cygnus Loop have been made with the VOYAGER 1 & 2 spacecraft (Shemansky, Sandel, & Broadfoot 1979; Vancura *et al.* 1993), with the second study detecting C iii 977Å, O vi 1035Å, and Lyα 1215Å.

SNRs were high-priority targets for HUT during the Astro-1 mission and observations for three have been reported in the literature: the Cygnus Loop (Blair *et al.* 1991; Long *et al.* 1992), the Crab Nebula (Blair *et al.* 1992), and N49 in the LMC (Vancura *et al.* 1992b). The spectra obtained for two filaments in the Cygnus Loop are shown in Figure 9 (kindly provided to the author by W. P. Blair). Comparison of the two spectra is very interesting in that the radiative filament shows emission lines from several ionic states of C, N, and O; while the spectrum of the nonradiative filament is dominated by O vi λλ1032+1038, which is partially resolved, as well as other high ionization species due to the higher shock velocity and relative greater incompleteness compared to the radiative filament. Blair *et al.* (1991) modeled the radiative shock spectrum and concluded that it is dominated by a fast $v \sim 165$ km s^{-1} incomplete shock, but slower shocks may contribute to the observed spectrum. The nonradiative filament spectrum was discussed and modeled by Long *et al.* (1992), who concluded that much of the FUV-UV spectrum is consistent with a shock of velocity 175–185 km s^{-1} propagating into a low density

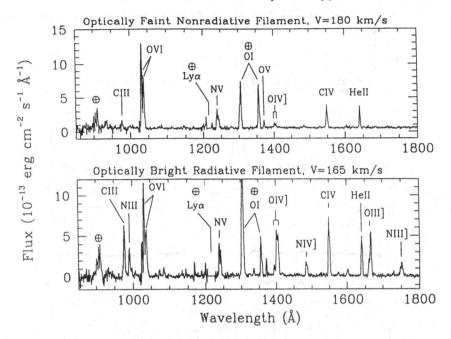

FIGURE 9. HUT FUV-UV spectra of two filaments in the Cygnus Loop SNR. The top spectrum (Blair *et al.* 1991) is of a high velocity nonradiative filament with a very incomplete recombination zone, while the bottom spectrum (Long *et al.* 1992) is of a slower radiative filament which is more complete.

medium, but that lower shock velocities indicated by optical Hα profiles suggest that the shock is rapidly decelerating as it encounters denser material.

The HUT spectrum of the LMC SNR N49 (Vancura *et al.* 1992b) showed emission lines of O VI λλ1035, O IV] λλ1402, C IV λλ1549, and He II λ1640 from the remnant. They interpreted the HUT spectral results based on a model for N49 previously developed from a multiwavelength (IUE UV spectra, ground-based imagery and spectroscopy, and Einstein X-ray imagery) study (Vancura *et al.* 1992a). The most useful diagnostic from the HUT spectrum was O VI λλ1035Å, which they note as originating in faint low density shocks with $v \sim$ 190–270 km s^{-1} tranversing clouds with $n \sim$ 20–40 km s^{-1}, rather than in the main $v \sim$ 730 km s^{-1} blast wave of the remnant. They presented a shock model with a distribution of shock velocities to match well the entire FUV-UV-optical spectral line strengths observed. By contrast, the Crab Nebula spectrum showed few FUV emission lines, but showed a heavily reddened synchrotron continuum down to $\lambda \sim$ 1000Å with only two emission lines, C IV λλ1549 and He II λ1640, detected. The C IV line showed a double-peaked profile, representative of the two sides of an expanding shell of velocity $v_{exp} \sim \pm 1100$ km s^{-1}; while the He II line showed only one blueshifted component representative of the near side of the shell.

5.4. *Emission-line galaxies*

During its 15+ years of operation, the IUE has obtained over 5000 UV spectra of AGN and starburst galaxies. A summary of most of the principal studies during the first half of this period can be found in the reviews of "Starburst Galaxies," "Active Galactic Nuclei," "Blazars," and "Quasars" in *Exploring the Universe with the IUE Satellite* (Kondo 1987). In recent years much attention has been given to using the large IUE UV spectra database for the construction of atlases of the various types (*cf.* Table 2),

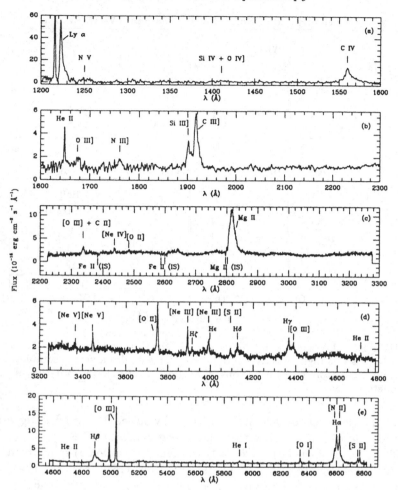

FIGURE 10. HST FOS spectra of the weak Seyfert 1 nucleus of NGC 1566 from Kriss *et al.* (1991) illustrating complete UV-optical spectral coverage capabilities of FOS over 1150–6800Å.

UV broad-line region variability studies (*e.g.* Clavel *et al.* 1991; Koratkar & Gaskell 1991 for NGC 5548; Zheng, Fang, & Binette 1992 for Fairall 9; Reichert *et al.* 1994 for NGC 3783), studies of the Baldwin Effect in Quasars (Kinney, Rivolo, & Koratkar 1990), and development of a new classification scheme for QSO's based on combined UV-optical spectral characteristics (Francis *et al.* 1992). In addition, numerous UV spectral studies of individual objects made with the IUE continue to appear in the literature. Among the more innovative use of IUE large aperture spectra has been to utilize the fact that the $10'' \times 20''$ large aperture *images* extended objects in UV emission lines and the 2D line features in low dispersion spectra can be deconvolved to spatially map the emission structure. This technique has been applied to resolve emission knots in the Seyfert galaxy NGC 1068 (Bruhweiler & Altner 1988) and study the spatial distribution of emission in the core of the LINER galaxy NGC 3998 (*e.g.* Reichert *et al.* 1992). It is expected that the NEWSIPS-processed low dispersion IUE archive with its improved geometric processing will enable improved UV emission-line maps of extended objects to be made in the future.

Even with the aberration problem, HST FOS and GHRS UV spectra of the central regions of various emission-line galaxies have been successful. Among the first objects to be studied by FOS included the Quasar UM 675 (Beaver *et al.* 1991), 3C 273 (Bahcall *et al.* 1991), the nucleus of NGC 1068 (Caganoff *et al.* 1991), and the weak Seyfert galaxy NGC 1566 (Kriss *et al.* 1991). More recently, as HST comes into more "routine" operation, multi-object UV spectroscopy studies of AGN have appeared in the literature (*e.g.* Laor *et al.* 1994). In addition to having high sensitivity and large dynamic range, FOS has the capability of taking spectra of sub-arcsecond regions in and near AGN covering a wide UV-optical wavelength range. An example of such high quality spectra for NGC 1566 from Kriss *et al.* is shown in Figure 10. Such spectra enable UV-optical emission-line diagnostics, including velocity profiles, to be made of the stellar cores of AGN without significant contamination by emission from adjacent sources. An even more difficult problem (due to the relatively much more luminous nuclear region) is to study the spectra of regions near the central source in AGN, one of the first results from GHRS was obtaining a high resolution UV spectrum, covering the 1150–1450Å & 1490–1750Å wavelength range, of a starburst knot near the nucleus of NGC 1068 (Hutchings *et al.* 1991). The spectrum indicated that the knot consisted of thousands of OB stars which were formed only $\sim$ 3 million years ago. Now with the correction afforded to FOS and GHRS by COSTAR, such studies are now even more practical for AGN and other types of emission-line nuclei of galaxies.

Among the most successful of the HUT observations during the Astro-1 mission were FUV-UV spectra of the Seyfert galaxies NGC 4151 (Kriss *et al.* 1992a) and NGC 1068 (Kriss *et al.* 1992b). The NGC 1068 FUV-UV spectrum is reproduced in Figure 11. Kriss *et al.* noted that the FUV lines of C III 977Å, N III 991Å, and O VI 1035Å were found to be stronger than expected. T_e's derived from the ratios of the FUV lines of C III and N III to the UV intercombination lines of C III] 1909Å and N III] 1750Å, were 26,700 K and 24,000 K, respectively. These high values and the very strong O VI line led them to conclude that shock heating is an important mechanism in the nucleus of NGC 1068. By contrast, the FUV lines of C III and N III appear in *absorption* (blueshifted by about 200–1330 km s^{-1}) in the nucleus of NGC 4151 and the O VI lines have both emission and blueshifted absorption components. Other emission lines, such as C IV 1549Å, had narrow emission, blueshifted narrow absorption, and broad emission wings. They concluded that the spectral features were consistent with high density ($n \sim 10^{9.5}$ cm^{-3}) outflowing material from the broad-line region.

6. CONCLUDING REMARKS

The author wishes to express his gratitude in being invited to present a review paper on "UV Spectroscopy" a conference honoring two pioneers in the field of emission-lines of astrophysical plasmas—Don Osterbrock and Mike Seaton. To Don I owe the gratitude of a son, since he fathered me into the field of nebular spectroscopy as his 13th (lucky!) Ph.D. student at Wisconsin and for continual motivation to continue my studies in the two decades since. To Mike I owe almost an equivalent gratitude for fathering many of the "young turks" of modern atomic physics who are now providing me and other spectroscopists modern atomic data for use in our analyses.

The present and the future is bright for us in the field of UV spectroscopy. The IUE continues to acquire *scientifically important* UV spectra of all kinds of astronomical objects and the astronomical community needs to keep the pressure on NASA to continue support of its operation. HST is "fixed" and continues the promise of acquiring UV spectra of exceptional quality of many faint objects not previously reachable with IUE in

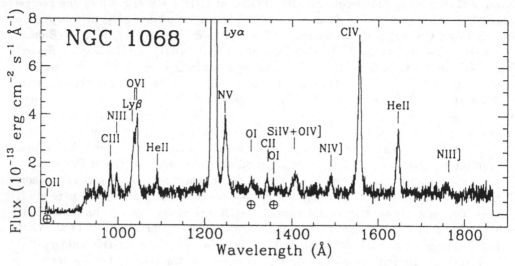

FIGURE 11. HUT FUV-UV spectrum of the nucleus of the Seyfert 1 galaxy NGC 1068 from Kriss *et al.* (1992b). Note the strong O VI 1035Å line, which Kriss *et al.* argues as being indicative of shock ionization in the nucleus.

the coming decade (as well as offering the promise of being able to obtain 2D spectra later in the decade). HUT has obtained high quality spectra in the FUV during the Astro-1 mission and we hope for even more success for it during the 1995 Astro-2 mission. EUVE is now in the middle of its lifespan with exciting results for the EUV region only now coming out in press. The promise of Lyman for enabling us to have an extended future of being able to do EUV-FUV spectroscopy by an orbiting spacecraft, potentially as versatile as IUE, is important to all of us and we should continue to urge its strong support by NASA.

Finally, I thank my many colleagues for many useful conversations and collaborative studies over the past two decades. I also wish to thank NASA/IUE for a decade of support for my IUE-related research via grant NAG5-262 (which is ending after over a decade) and to AURA/ST ScI for current support of collaborative UV-spectroscopic investigations with HST through grants GO.4267-91B, GO.4382-92A, GO.4383-92A, and GO.5474-93A, as well as NASA grant NCC2-5008.

REFERENCES

ALLER, L. H. 1984 *Physics of Thermal Gaseous Nebulae*, Dordrecht: Reidel

ALLER, L. H. *et al.* 1987 *ApJ* **320**, 159

AUFDENBERG, J. P. 1993 *ApJS* **87**, 337

BAHCALL, J. N. *et al.* 1991 *ApJL* **377**, L5

BARKER, T. 1991 *ApJ* **371**, 217

BEAVER, E. A. *et al.* 1991 *ApJL* **377**, L1

BENVENUTI, P. *et al.* 1982 NASA SP-1046

BLAIR, W. P. & PANAGIA, N. 1987 in *Exploring the Universe with the IUE Satellite* (ed. Y. Kondo) Dordrecht: Reidel, p. 549

BLAIR, W. P. *et al.* 1991 *ApJL* **379**, L33

BLAIR, W. P. *et al.* 1992 *ApJ* **399**, 611

BLUM, R. D. & PRADHAN, A. K. 1992 *ApJS* **80**, 425

BOGGESS, A. *et al.* 1978 *Nature* **275**, 372

BOWYER, S. 1994 *Science* **263**, 55

BOWERS, C. W. *et al.* 1994 in preparation.

BRANDT, J. C. *et al.* 1994 *PASP* **106**, 890

BRUHWEILER, F. C. & ALTNER, B. 1988 in *A Decade of UV Astronomy with the IUE Satellite*, Vol. 2, (ed. E. J. Rolfe) Noordwijk: ESTC, 319

CAGANOFF, S. *et al.* 1991 *ApJL* **377**, L9

CARDELLI, J. A., CLAYTON, G. C. & MATHIS, J. S. 1989 *ApJ* **345**, 245

CLAVEL, J. *et al.* 1991 *ApJ* **366**, 64

CLAYTON, G. & MARTIN, P. 1985 *ApJ* **288**, 558

CODE, A. D. (ed.) 1972 *The Scientific Results from the Orbiting Astronomical Observatory (OAO-2)*, NASA SP-310

COX, N. 1994 poster paper, this conference

CZYZAK, S. J., KEYES, C. D. & ALLER, L. H. 1986 *ApJS* **61**, 159

DAVIDSEN, A. F. *et al.* 1992 *ApJ* **392**, 264

DEROBERTIS, M. M., DUFOUR, R. J. & HUNT, R. W. 1987 *JRASC* **81**, 195

DOPITA, M. A. *et al.* 1994 *ApJ* **426**, 150

DOSCHEK, G. A. & FEIBELMAN, W. A. 1993 *ApJS* **87**, 331

DOYLE, J. G. & KEENAN, F. P. 1992 *AAp* **264**, 173

DUFOUR, R. J. 1986 *PASP* **98**, 1025

DUFOUR, R. J. 1990 in *Evolution in Astrophysics*, (ed. E. J. Rolfe), ESA SP-310, p. 117

DUFOUR, R. J. 1991 *PASP* **103**, 857

DUFOUR, R. J., GARNETT, D. R. & SHIELDS, G. A. 1988 *ApJ* **332**, 752

DUFOUR, R. J. & HESTER, J. J. 1990 *ApJ* **350**, 149

DUFOUR, R. J. *et al.* 1993 *RMAA* **27**, 115

DUFOUR, R. J. *et al.* 1994 in preparation

EDELSON, R. *et al.* 1992 *ApJS* **83**, 1

FEIBLEMAN, W. A. 1994 *PASP* **106**, 756

FERLAND, G. J. & OSTERBROCK, D. E. 1986 *ApJ* **300**, 658

FITZPATRICK, E. L. 1985 *ApJ* **299**, 219

FRANCIS, P. J. *et al.* 1992 *ApJ* **398**, 476

GARNETT, D. R. *et al.* 1994 *ApJ* in press

HARRINGTON, J. P., SEATON, M. J., ADAMS, S. & LUTZ, J. H. 1982 *MNRAS* **199**, 517

HARRIS, A. W. & SONNEBORN, G. 1987 in *Exploring the Universe with the IUE Satellite* (ed. Y. Kondo) Dordrecht: Reidel, p. 729

HAYES, M. A. & NUSSBAUMER, H. 1986 *A&Ap* **161**, 287

HESTER, J. J., RAYMOND, J. C. & BLAIR, W. P. 1994 *ApJ* **420**, 721

HECK, A., EGRET, D., JASCHEK, M. & JASCHEK, C. 1984 IUE Low-Resolution Spectra Atlas: Part 1. Normal Stars, ESA SP-1052.

HUTCHINGS, J. B. *et al.* 1991 *ApJL* **377**, L25

HUTCHINGS, J. B. *et al.* 1992 *ApJL* **400**, L35

KEENAN, F. P. & AGGARWAL, K. M. 1987 *ApJ* **319**, 403

KEENAN, F. P., FEIBELMAN, W. A. & BERRINGTON, K. A. 1992 *ApJ* **389**, 443

KEENAN, F. P. *et al.* 1994 *ApJ* **423**, 882

KINNEY, A. L., BOHLIN, R. C., BLADES, J. C. & YORK, D. G. 1991 *ApJS* **75**, 645

KINNEY, A. L. *et al.* 1993 *ApJS* **86**, 5

KINNEY, A. L., RIVOLO, A. R. & KORATKAR, A. P. 1990 *ApJ* **357**, 338

KONDO, Y. (ed.) 1989 *Exploring the Universe with the IUE Satellite*, Dordrecht: Reidel

KÖPPEN, J. & ALLER, L. H. 1987 in *Exploring the Universe with the IUE Satellite* (ed. Y. Kondo) Dordrecht: Reidel, p. 589

KORATKAR, A. P. & GASKELL C. M. 1991 *ApJ* **375**, 85

KRÄMER, G. *et al.* 1990 in *Observatories in Earth Orbit and Beyond* (ed. Y. Kondo) Dordrecht: Kluwer, p. 177.

KRISS, G. A. *et al.* 1991 *ApJL* **377**, L13

KRISS, G. A. *et al.* 1992a *ApJ* **392**, 485

KRISS, G. A., DAVIDSEN, A. F., BLAIR, W. P., FERGUSON, H. C. & LONG, K. S. 1992b *ApJL* **394**, L37

LANZETTA, K. M., TURNSHEK, D. A. & SANDOVAL, J. 1993 *ApJS* **84**, 109

LAOR, A. *et al.* 1994 *ApJ* **420**, 110

LONG, K. S. *et al.* 1992 *ApJ* **400**, 214

MATHIS, J. S. 1989 *AnRevAAp* **28**, 37

MENDOZA, C. 1983 in *IAU Symp. No. 103: Planetary Nebulae* (ed. D. R. Flower) Dordrecht: Reidel, p. 143

MOOS, W. 1990 in *Observatories in Earth Orbit and Beyond* (ed. Y. Kondo) Dordrecht: Kluwer, p. 171

MORTON, D. C. & SPITZER, L. 1966 *ApJ* **144**, 1

NANDY, K. *et al.* 1982 *MNRAS* **201**, 1P

NICHOLS, J. S., GARHART, M. P., DE LA PEÑA, M. D. & LEVAY K. L. 1994 IUE Newsletter No. 53, 1

NUSSBAUMER, H. 1986 *AAp* **155**, 205

NUSSBAUMER, H., SCHILD, H., SCHMID, H. M. & VOGEL, M. 1988 *AAp* **198**, 179

NUSSBAUMER, H. & STENCEL, R. E. 1987 in Exploring the Universe with the IUE Satellite (ed. Y. Kondo) Dordrecht: Reidel, p. 203

NUSSBAUMER, H. & STOREY, P. J. 1984 *AApSup* **56**, 293

OSTERBROCK, D. E. 1989 *Astrophysics of Gaseous Nebulae and AGN*, Mill Valley: University Science Books

PEIMBERT, M., TORRES-PEIMBERT, S. & DUFOUR, R. J. 1993 *ApJ* **418**, 760

PENSTON, M. V. *et al.* 1983 *MNRAS* **202**, 833

PERINOTTO, M. 1991 *ApJS* **76**, 687

PITTS, P. S. 1993 IUE Newsletter No. 50, p. 21

RAYMOND, J. C., BLAIR, W. P., FESEN, R. A. & GULL, T. R. 1983 *ApJ* **275**, 636

RAYMOND, J. C. *et al.* 1988 *ApJ* **324**, 869

REICHERT, G. A. *et al.* 1992 *ApJ* **387**, 536

REICHERT, G. A. *et al.* 1994 *ApJ* **425**, 582

ROGERSON, J. B. *et al.* 1973 *ApJL* **181**, L97

ROLA, C. & STASIŃSKA, G. 1994 *AAp* **282**, 199

ROSA, M., JOUBERT, M. & BENVENUTI, P. 1984 *AApS* **57**, 351

RUBIN, R. H., DUFOUR, R. J. & WALTER, D. K. 1993 *ApJ* **413**, 242

RUBIN, R. H. *et al.* 1994 poster paper, this conference

SEATON, M. J. 1979 *MNRAS* **187**, 73P

SHAW, R. A. & DUFOUR, R. J. 1993 *ASP Conf. Ser.* Vol. 61, p. 327

SHAW, R. A. & DUFOUR, R. J. 1994 in preparation (to be submitted for publication in *PASP*)

SHEMANSKY, D. E., SANDEL, B. R. & BROADFOOT, A. L. 1979 *ApJ* **231**, 35

STAFFORD, R. P., BELL, K. L. & HIBBERT, A. 1994 *MNRAS* **266**, 715

STECHER, T. P. & MILLIGAN, J. E. 1962 *ApJ* **136**, 1

VANCURA, O., BLAIR, W. P., LONG, K. S. & RAYMOND, J. C. 1992a *ApJ* **394**, 158

VANCURA, O. *et al.* 1992b *ApJ* **401**, 220

VANCURA, O., BLAIR, W. P., LONG, K. S., RAYMOND, J. C. & HOLBERG, J. B. 1993 *ApJ* **417**, 663

WALTER, D. K. *et al.* 1994 poster paper, this conference

WALTER, D. K., DUFOUR, R. J. & HESTER, J. J. 1992 *ApJ* **397**, 196

WAMSTEKER, W. *et al.* 1989 *AApS* **79**, 1

WILLIAMS, R. E. 1994 poster paper, this conference.

WU, C.-C. *et al.* 1983 IUE Newsletter No. 22

WU, C.-C. *et al.* 1992 IUE Ultraviolet Spectral Atlas of Selected Astronomical Objects, NASA RP-1285

ZHENG, W., FANG, L.-Z. & BINETTE, L. 1992 *ApJ* **392**, 74

Infrared Emission Lines as Probes of Gaseous Nebulae

By HARRIET L. DINERSTEIN

Astronomy Department, University of Texas at Austin, Austin, TX 78712

Infrared emission lines are just coming into their own as diagnostic tools for the study of gaseous nebulae. Various infrared line ratios provide the means to determine physical parameters such as extinction, density, temperature, and ionic and elemental abundances. In certain regimes, infrared lines provide essentially the *only* opportunity for determining these parameters, or even for detecting the gas. Examples include regions of high extinction, low temperature, and predominantly neutral material. I review the general properties of infrared emission lines and their characteristics as nebular diagnostics, and cite a number of illustrative applications. The latter are drawn from a wide range of fields, including star formation and H II regions, planetary nebulae, nova and supernova remnants, shocked and photodissociated gas, and AGN and starburst galaxies.

1. Introduction

The phrase "infrared emission lines," like so many other topics covered at this meeting, is extremely broad. Therefore I will begin by defining and limiting the scope of this review. I will discuss only lines that arise from gas-phase atoms, ions, and molecules, and will not include spectral features produced by interstellar dust. In this review, "infrared" will mean the spectral region 1–200 μm, which corresponds to certain types of astronomical detectors; shorter wavelengths will be called "far red" (with apologies to Don Osterbrock), and longer wavelengths, "submillimeter." Another boundary condition is that I will restrict myself to low densities, $n \leq 10^8$ cm^{-3}. Apart from these constraints, I will attempt to be as general as possible, though with no pretentions to completeness. This review is organized by physical properties and methods of analysis, rather than by class of astronomical object. In selecting applications I have tried to sample diverse areas of astronomy, in order to illustrate the breadth and power of the various methods for measuring and constraining astrophysically interesting parameters.

One might begin by asking "why infrared?" Perhaps it is not quite as necessary to raise this question as it was a decade or two ago, when infrared instrumentation was relatively primitive, and one needed to justify the extra effort required to make observations in the infrared region. However, it remains true that infrared emission lines have a number of special characteristics that translate into distinctive advantages, and a few limitations. Some special qualities of infrared emission lines include: (1) access to regions with high visual extinctions; (2) avoidance of the temperature sensitivity that is essentially intrinsic to most optical and ultraviolet emission (hereafter, optical/uv) lines; and (3) and the ability to sample species that lack observable lines in the optical/uv, either because of the vagaries of atomic structure or because the gas is too cool to emit in lines from high energy levels.

Most of the observed infrared lines (other than recombination lines of H and He) connect fine-structure levels within ground state p^n configurations of abundant ions (where n is an integer between 1 and 5). A good synopsis of the physics of fine-structure lines can be found in Simpson (1975), an influential paper in the 1970's, a period of rapid advances in infrared spectroscopy. Other discussions of these lines can be found in Aller

(1984, ch. 5), Watson (1985), and Osterbrock (1989, chs. 3 and 5). What follows here are just a few general points.

The upper levels of the fine-structure lines are populated by collisions, usually with electrons. For upper levels with small excitation energies χ, the populations are substantial. Whereas the excitation energies of the optical/uv lines are larger than the typical energy of the free electrons, for the infrared lines $\chi = h\nu \ll kT$ and the exponential term is nearly equal to unity. The ratio of populations of an excited level (1) to the ground level (0) in a two-level atom is

$$\frac{n_1}{n_0} = \frac{n_e q_{01}}{A_{10}} \left[1 + \left(1 + \frac{n_e q_{10}}{A_{10}} \right)^{-1} \right] ; \quad q_{01} \propto \Omega(T) T^{-0.5} e^{-\chi/kT} \approx \Omega(T) T^{-0.5} \quad (1.1)$$

(after Osterbrock 1989, eq. 3.25), where A_{10} is the radiative transition probability, q_{01} and q_{10} are the collisional excitation and de-excitation probabilities respectively, and $\Omega(T)$ is the "collision strength," which depends only weakly on temperature. For lines in which both levels belong to the same electronic configuration, downward radiative transitions are forbidden by electric dipole selection rules. Combined with their low frequencies, this yields low transition probabilities, $A_{10} \leq 10^{-2}$ s^{-1}, for the fine-structure lines. However, the large populations of these levels tend to compensate for the low A values, making the net volume emissivities of the infrared lines (in energy units) comparable to those of the optical/uv lines, which arise from much smaller populations in higher-energy levels.

From equation 1.1 one can see that the small excitation energies of the infrared lines insure a weak dependence of the emergent line strengths on electron temperature T_e compared to the optical/uv lines. This has both advantages and disadvantages: it is helpful for deducing ionic abundances, since one need not know the exact value of T_e; however, such lines therefore give no information about the temperature. The low transition probabilities, combined with high collisional de-excitation rates, lead to small critical densities, where, for level i,

$$n_{\rm crit}(i) = \sum_{j<i} A_{ij} / \sum_{j\neq 1} q_{ij} \quad (1.2)$$

(Osterbrock, eq. 3.31). As a result, the infrared lines are sensitive to density and clumping (*i.e.*, Rubin 1989). They are good density indicators for low-density regions, but "saturate" to asymptotic values at relatively low densities, whereupon they become good mass tracers.

The history of the observations of the infrared emission lines more or less begins in the 1970's, when these lines were first observed in bright, compact H II regions and planetary nebulae. By 1982, the list of detected lines ran to fifteen (*e.g.*, Dinerstein 1983). In the 1980's infrared lines began to be observed from a more diverse group of sources, such as novae, supernovae and supernova remnants, and photodissociation regions. Within the last few years, there has been a rapid growth in infrared spectroscopy of extragalactic sources, including active galactic nuclei and starburst galaxies. The fact that it is no longer feasible or sensible to present a comprehensive list of detected infrared emission lines in a review such as this one is a measure of the maturity of the field and the diversity of the astronomical sources and regions that benefit from being studied with infrared line spectroscopy.

2. Gas diagnostics

The first step in the analysis of emission line strengths from nebulae is generally the derivation of physical parameters that characterize the local conditions in the gas, primarily density and temperature (see the synopsis of the standard or "conventional" analysis of H II regions, in Dinerstein 1990). This is as true for infrared lines as for optical/uv lines, and the specific methods used with infrared lines are essentially analogs of techniques developed for optical/uv lines.

2.1. *Extinction (A_V)*

Determination of the extinction towards the line-emitting material is a prerequisite to the derivation of other properties, since further analysis requires knowledge of the true line intensities, corrected for extinction. The usual approach is to measure the intensity ratios among lines whose intrinsic relative strengths are considered known. Discrepancies are ascribed to extinction, assumed to follow a particular law $A(\lambda)$, and a best fit is found for the overall scaling factor, usually characterized by A_V or sometimes A_K (*i.e.*, Mathis 1990). The most commonly used set of line ratios is the recombination spectrum of H I; this is called the *hydrogen decrement* method. In the optical, one uses the Balmer lines (*e.g.*, Hβ/Hα, Hγ/Hα), whose relative intensities in the absence of extinction are known to high accuracy and have little dependence on temperature or density (Osterbrock 1989, ch. 4.2). The main transitions used in the infrared are the Paschen and Brackett series, which extend out to 4.05 μm (Br α). These lines are used to determine extinctions to H II regions in our own Galaxy, and in other galaxies (*e.g.*, Lester *et al.* 1990). A couple of caveats should be kept in mind, however. First, where the geometry is complex and the observations integrate over material with a range of A_V values, the effective extinction law may have a form quite different from the standard one, and may have to be determined empirically by sampling lines at various wavelengths and interpolating (*i.e.*, Lester, Dinerstein, & Rank 1979). Second, there are situations, such as high densities and high optical depths, where the intrinsic recombination decrement deviates strongly from the classic "Case B" values. Examples of the latter range from the "line excess" problem in young stellar objects (Thompson 1982; Natta *et al.* 1988), to broad lines in active galactic nuclei (Kawara, Nishida, & Phillips 1989; Goodrich 1990).

When H I recombination lines are not available, one can use pairs or sets of lines with different wavelengths that arise from the same upper level of any species. An early example of this method was a comparison the strengths of [S II] lines near 1 μm with others near 4000 Å (Fig. 1) to determine the extinction towards the metal-rich knots in the Cas A supernova remnant (Searle 1971); these knots are composed entirely of heavy elements and contain no H or He at all. To determine extinctions towards regions where hydrogen is molecular, it is common to use H_2 quadrupole lines from common upper levels; applications include the shocked gas in Orion (*e.g.*, Geballe *et al.* 1986) and the evolved bipolar nebula GL 618 (Latter *et al.* 1992). The rich energy level structure of Fe II (Nussbaumer & Storey 1988) provides a multitude of line ratios to use in partially ionized regions such supernova remnants (see Fig. 1).

2.2. *Electron density (n_e)*

The honorees of this conference were among the pioneers in using optical line intensity ratios to deduce densities in ionized nebulae (Seaton & Osterbrock 1957; updated in Fig. 5 of Osterbrock 1988). Intensity ratios of pairs of infrared lines can similarly be used to infer gas densities. A comprehensive discussion of infrared density indicators is given by Rubin (1989). The critical densities of the infrared lines fall mostly in the range 10^2 cm^{-3} < n_{crit} < 10^6 cm^{-3}, with a general tendency of the longer-wavelength lines to have lower values than shorter-wavelength lines because of their lower A values (equation 1.2). For example, $\log n_{crit}$([O III] 51.8, 88.4 μm) = (3.5, 2.7), while $\log n_{crit}$([N II] 122, 204 μm) = (2.4, 1.6). On the other hand, the critical densities of [S III] 18.7, 33.5 μm are higher

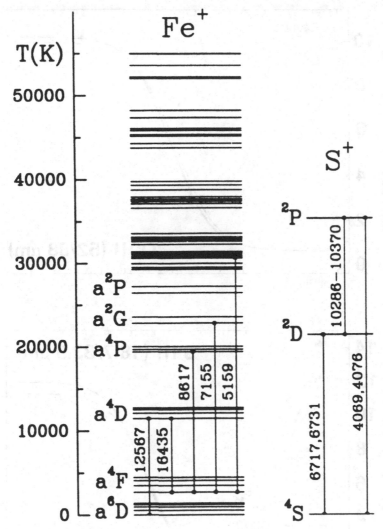

FIGURE 1. Energy levels of Fe II & S II. (From Oliva, Moorwood, & Danziger 1991)

than those of some optical lines (see Fig. 2). These three pairs of lines have been used extensively to study obscured Galactic H II regions (*e.g.*, Moorwood *et al.* 1980; Colgan *et al.* 1993). Ratios of infrared to optical lines can also be used as density indicators (*e.g.*, Keenan & Conlon 1993).

In partially ionized regions where the gas is heated by shocks and/or by photoelectrons donated by species such as C^+, $Si+$, and Fe^+, local gas densities can be derived from ratios of [Fe II] lines. This species provides a multitude of strong emission lines in the infrared, particularly between 1.25 μm and 1.9 μm, from which density-sensitive ratios can be formed (Nussbaumer & Storey 1988). Most of these [Fe II] ratios are useful density indicators at relatively high densities, $10^4 < n < 10^{5.5}$ cm^{-3} (*e.g.*, Oliva, Moorwood, & Danziger 1989).

Studies that have measured densities from more than one pair of infrared lines often find that different indicators yield different nominal values of n_e. This should not come

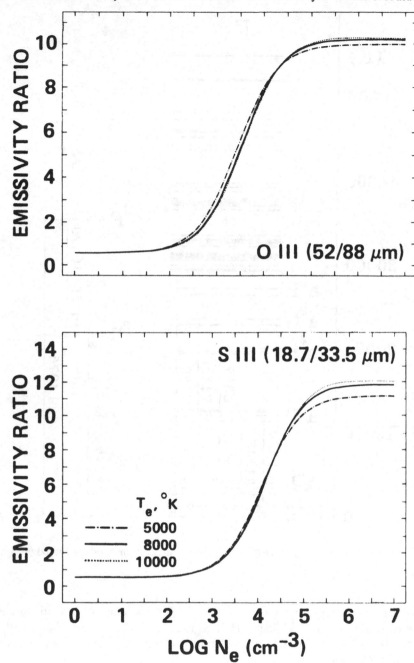

FIGURE 2. [O III], [S III] lines as n_e diagnostics. (From Rubin 1989)

as a surprise. It behooves us to bear in mind the complexity of real nebulae and the limitations of diagnostics:

"It must always be remembered that mean temperatures and densities derived from these diagnostic ratios are complicated averages over the structure of the nebula, weighted by the emission processes themselves. Thus the mean temperature is weighted toward high temperature, where the collisionally excited emission lines are all strongest, and the mean density is weighted toward the critical densities, where the emission is most efficient."

(Osterbrock 1988, p. 419). For ratios involving infrared lines only, temperature effects are minimal, but density is an important consideration. Nebulae are almost inevitably inhomogeneous, probably both on large and small scales, and diagnostics from lines with different critical densities will be biased accordingly. The effects of such biases can propagate further in the analysis. If the "wrong" (single-value) density is inferred, then the wrong emissivity will be used in the subsequent calculations, and an incorrect abundance will be derived. Rubin (1989) examined the effects of such "density bias" on ionic ratios, for several representative cases of inhomogeneous nebulae. He found that the errors in the derived abundance vary widely for different sets of lines, ranging from negligible (*e.g.*, < 10% for N^{++}/O^{++}) to significant (> 50% for several other ratios under reasonable conditions); his analysis also included many optical/uv line diagnostics.

2.3. *Electron Temperature* (T_e)

Electron temperatures (T_e) of ionized nebulae can be determined from intensity ratios of collisionally excited lines from levels of widely separated energies. The pre-eminent optical line indicator for T_e is the ratio of [O III] 4363 Å to one or both of the stronger [O III] lines at 4959, 5007 Å (*e.g.*, Osterbrock 1989, Fig. 5.1; Dinerstein 1990). As pointed out in Section 1, infrared fine-structure lines are relatively insensitive to T_e. Thus, by themselves they are not useful for measuring T_e in typical hot, ionized nebulae; however, one can construct T_e diagnostics by comparing infrared lines from low-lying levels with optical/uv lines from higher-excitation levels. This can provide a means to derive T_e values for regions that lack optical/uv T_e diagnostics. For example, [Ne V] 24.3 μm can be combined with [Ne V] 3426Å, and [O IV} 25.9 μm similarly with far-uv lines, to infer T_e for the high-ionization zones of planetary nebulae (Shure *et al.* 1983a, b; Rowlands *et al.* 1989). In regions that do not produce optical/uv emission lines, infrared line ratios can be used as temperature indicators in particular regimes, *e.g.*, [O I] 145/63 μm at $T \leq 500$ K. However, because infrared lines are strongly affected by collisions at relatively low densities, their dependences on n and T are often difficult to separate. It may be necessary to measure at least two line ratios from one ion in order to derive meaningful constraints on T (Keenan, Burke, & Aggarwal 1991). In some cases it is necessary to combine lines from more than one species and assume a value for their relative abundances. For example, in photodissociation regions, [O I] 63, 145 μm and [C II] 157 μm can be used together to constrain n and T, at least in certain regimes (Fig. 3, from Watson 1985; also Tielens & Hollenbach 1985).

There is no reason to expect that real nebulae are isothermal. As was the case for density, observations will include emission from material at a range of temperatures, particularly if made with large apertures. The spread in temperatures is likely to be modest ($\leq 30\%$) compared to density inhomogeneities, which could cover several orders of magnitude. However, for optical/uv lines, with their exponential emissivity dependence on T_e, even modest T_e variations produce enormous changes in line strengths and inferred parameters. A formalism for characterizing the temperature distribution in non-isothermal nebulae was introduced by Peimbert (1967), who defined a mean temperature T_o and an rms "temperature fluctuation parameter" t_2. These parameters can be related to, and therefore measured by, intensity ratios of lines with different temperature dependences (*i.e.*, Peimbert & Costero 1979). If the spread in T_e is large, the use of a single T_e value (determined from the standard diagnostics) to determine ionic abundances can

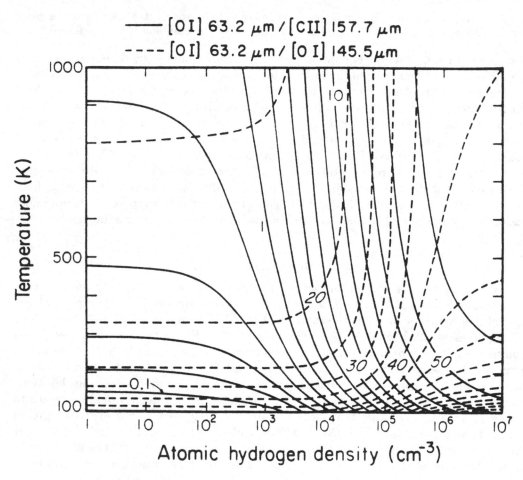

FIGURE 3. Mapping of the infrared [C II],[O I] line ratios to n and T. (From Watson 1985).

yield inaccurate results. A spread of just a few percent can lead to abundance errors of a factor of two or worse, for very temperature-sensitive lines!

The rich infrared-optical [O III] spectrum offers a particularly attractive opportunity to constrain and characterize the temperature properties of nebulae. The ground 3P term gives rise to the strong far-infrared lines at 52 and 88 μm that have already been discussed as density indicators, whereas the higher levels produce the traditional optical [O III] temperature diagnostic (Fig. 4). By combining observations of a single far-infrared line (*e.g.*, 88 μm) with two optical lines, Dinerstein (1983) showed that it was possible to solve simultaneously for n_e and T_e from a single ion. In Dinerstein, Lester, & Werner (1985, hereafter DLW), both far-infrared [O III] lines were compared with 5007 Å; the mapping of this ratio in the (n_e, T_e) plane is shown in Fig. 5. The latter technique has the advantage of providing a determination of T_e that is independent of the 4363/5007 Å line ratio, and is less weighted or "biased" by small amounts of high-temperature gas than the optical ratio. Indeed, if one can measure *four* [O III] lines in a particular nebula and therefore construct *three* diagnostic line ratios, it should be possible to extract a measure of the inhomogeneity of the gas in addition to "mean" values of n_e and T_e. As

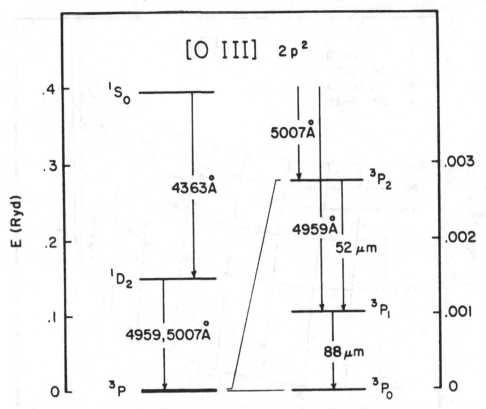

FIGURE 4. Energy level diagram for [O III]. (From Dinerstein, Lester, & Werner 1985.)

shown in DLW (Appendix), using Peimbert's (1967) formalism,

$$T_e(5007\text{ Å}/52\ \mu\text{m}) \approx T_o \quad \text{and} \quad T_e(4363\text{ Å}/5007\text{ Å}) \approx T_o(1+3t^2) \tag{2.3}$$

making it (in principle) straightforward to solve for values of both T_o and t^2. The accuracy of this method, however, depends critically on one's ability to calibrate the absolute far-infrared fluxes correctly, since ratios are taken relative to optical line fluxes measured with completely different instrumentation and flux standards. DLW concluded that there was evidence for significant non-zero values of t^2 in four observed planetary nebulae, with an average value of about .06. Compensation for this degree of inhomogeneity would elevate the O/H abundance by nearly a factor of two. These values of t^2 are only slightly higher than recent values derived for several galactic H II regions from various optical indicators (Peimbert 1993 and references therein). However, new and better-calibrated measurements of the far-infrared lines in NGC 6543, formerly found by DLW to have $t^2 > 0.1$, suggest that the actual value is smaller (Dinerstein *et al.* 1995). The promise of this method, and implications for the nebular oxygen abundances, await a larger set of accurately calibrated far-infrared line fluxes for optically well-studied nebulae, which are likely to become available in the next few years, from airborne studies and from new infrared space observatories.

3. Ionization and excitation mechanisms

Further analysis of nebular emission lines can be categorized as falling in two areas: (1) determination of additional properties *intrinsic to the nebular gas*, such as elemental

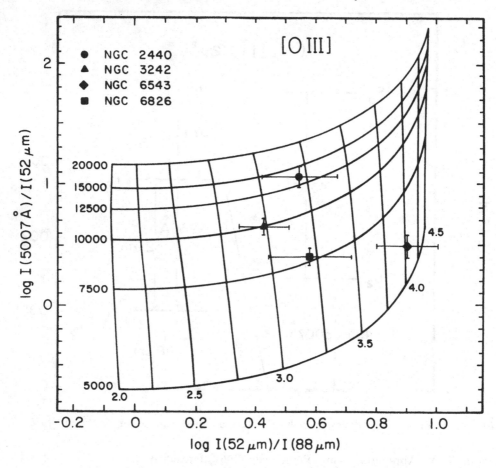

FIGURE 5. Infrared/optical [O III] diagnostics. (From Dinerstein, Lester, & Werner 1985.)

abundances, geometry, or total mass; or (2) derivation of properties of the *agents or mechanisms responsible* for ionizing and heating the gas, such as central star temperatures or shock velocities. To address the type of question, one must first decide what mechanism is responsible for the line emission; this determines the particular parameters o be evaluated. Knowledge of the excitation mechanism is often crucial to deriving meaningful results on quantitative properties of the gas as well.

3.1. *Photoionization: Characterizing the UV radiation field*

For photoionized nebulae, estimates of the surface temperatures of the central stars have long been made using observations of optical/uv emission lines from ions with different ionization windows. (By "window" I mean the range between the ionization potential of the next lower ionization state $\chi(X^{n-1})$, which is the minimum energy needed to create X^n, and $\chi(X^n)$, the energy it takes to ionize and thus destroy X^n.) In principle, the stellar temperatures can be calibrated empirically, from nebulae with central stars of known surface temperatures, but in practice this is rarely done. Instead, nebular ionization models are usually used at least to extend or interpolate such calibrations, and are often used as stand-alone calibrators. For example, mid-infrared spectra of obscured H II regions sample two strong ionic lines from species with very different lower bounds in χ: [Ne II] 12.8 μm (21.6–40.9 eV) and [S IV] 10.5 μm (34.8–40.7 eV). In H II regions ionized by normal thermal stellar spectra, the observed ratio of these two lines gives a first-order indication of the temperature of whether the central star (*e.g.*, Rank *et al.* 1978).

When airborne or space infrared observations are available, a larger variety of ions may be sampled, and the application of detailed models is more fruitful. Sequences of model nebulae can be constructed specifically to fit observations of the infrared lines for a particular nebula (*e.g.*, Simpson & Rubin 1984). While this may seem unnecessarily involved, nebular models developed for optical/uv observations have shown that the ionization balance in a nebula is significantly affected by nebular properties. One of the most critical is the local ratio of photons to atoms, or "ionization parameter;" changes in this parameter can mimic or mask changes in the stellar temperature (*e.g.*, Evans & Dopita 1985). Ionization ratios formed from infrared lines are affected by the ionization parameter, the density structure and inhomogeneities, and internal non-gray opacity due to dust (Herter, Helfer, & Pipher 1983) or atoms (*e.g.*, Rubin 1985). Geometry is another potentially important factor. Relaxing the assumption of spherical symmetry, for example in "blister" H II regions located on the edge of a dense cloud, changes the predicted ionization structure (Rubin *et al.* 1991). Thus, there is not always a unique correspondence between a particular observed line ratio and stellar temperature; determination of the latter will always require additional assumptions or constraints on the nebular structure.

A similar moral appears in the story of the He I 2.06 μm/H I(Br γ) 2.16 μm line ratio, which is relatively easy to observe in moderately obscured H II regions, and is therefore attractive as a possible indicator of ionizing star properties. It is well known that the ionic abundance ratio He^+/H^+ varies with the surface temperature of the ionizing star (*e.g.*, Osterbrock 1989, Fig. 2.5). However, this does not translate in a simple way into the variation of this particular line ratio, because the energy level structure of He I makes it prone to strong radiative-transfer effects (see Osterbrock 1989, Fig. 4.4). Indeed, the strength of the 2.06 μm line is sensitive to nebular geometry, density, and velocity structure as well as to optical depth (Shields 1993). Other infrared He I lines do not suffer from as many problems, but are intrinsically weaker and hence more difficult to observe.

Geometry and density effects can only do so much; the presence of highly ionized species in a photoionized nebula requires a relatively hot source for the radiation field. Highly ionized species are often seen in the optical/uv spectra of planetary nebulae with particularly hot central stars; such nebulae are good places to look for infrared emission from ions of very high ionization potential, sometimes called "coronal lines." Examples include [Ne VI] 7.65 μm (Pottasch *et al.* 1985), [Si VI] 1.96 μm, and [Si VII] 2.48 μm (Ashley & Hyland 1988).

High ionization species are also found in the ejecta of recent classical nova events. Indeed, infrared coronal lines were first identified in a nova (Grasdalen & Joyce 1976). Since then it has been found that many, if not all, novae show high-ionization infrared lines such as [Si VI], [Si VII], [Al XIII], [Ne VI], etc., starting about 10 to 300 days after visual maximum (see reviews by Gehrz 1988; Dinerstein & Benjamin 1993). The record-setting infrared line is [Si IX] 3.92, requiring 401 eV to produce (Greenhouse *et al.* 1990). The origin of the coronal ions has been controversial. Greenhouse *et al.* argue that they arise from gas in collisional ionization equilibrium at coronal temperatures. However, the observed near-infrared line-to-continuum ratios in several recent novae (Benjamin & Dinerstein 1990) support the view that the gas is thermally cool, and the coronal ions are produced through photoiniozation by the hard radiation field of the hot white dwarf (Williams 1991).

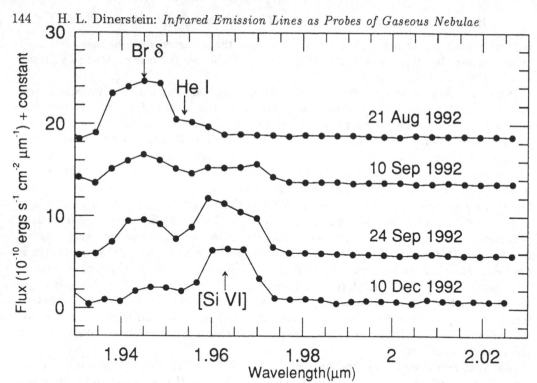

FIGURE 6. Evolution of the spectrum of N Cyg 1992 near 2.0 μm, during the transition to the "coronal phase." Note the declining strength of H I Br δ 1.94 μm and dramatically increasing strength of [Si VI] 1.96 μm on a timescale of a few weeks. (From Dinerstein & Benjamin 1993)

Novae also provide a fascinating opportunity to watch a nebular spectrum evolve, in response to the changing surface temperature of the white dwarf and declining density of the ejecta. During the first year or two after visual maximum, both of these effects drive the ionization balance towards higher ionization states. Figure 6 shows a particularly dramatic example of spectral evolution: Nova Cyg 1992 in the K-band during its transition to the "coronal" phase. If conditions in the ejecta can be properly modelled, the changing spectrum can be used to infer the spectral evolution of the white dwarf. The composition of the ejecta, which are enriched in metals due to the thermonuclear runaway and in some cases may also include material originating from the interior of the white dwarf, is also of interest (Section IV).

The presence of lines from species with a wide range in ionization potential is often taken as an indicator of ionization by a power-law continuum, for example from an active galactic nucleus (AGN). Several groups have predicted intensity ratios for infrared emission lines, for various models of AGN's, starburst galaxies, and other related objects (Voit 1992; Spinoglio & Malkan 1992). The idea is to construct diagnostic diagrams, by plotting pairs of line ratios against each other, in which these different types of regions can be easily separated and distinguished. The technique is similar to indicators based on optical and far red lines developed by Baldwin *et al.* (1981) and Osterbrock, Tran, & Veilleux (1992) respectively. It has also been suggested that the intensities of some infrared coronal lines can be enhanced by laser effects in AGNs (Greenhouse *et al.* 1993).

3.2. *Photodissocation regions*

The collisionally excited emission lines in ionized nebulae are populated by interactions with free electrons that are produced by the photoionization of atoms, primarily hydrogen. Photons with energies slightly less than 13.6 eV penetrate somewhat farther, and dominate the energetics of the region just beyond the edge of the Strömgren sphere. The last decade has seen a dramatic growth of interest in such transition regions between ionized nebulae and adjacent molecular clouds, where the gas is warm but mostly neutral. In these regions, the electrons are photoelectrons from grains, or from gaseous species with ionization potentials lower than that of H, of which the most abundant is neutral atomic carbon. Models of these so-called "photodissociation regions" or PDRs (Tielens & Hollenbach 1985, hereafter TH) predict that their temperatures are of order $\approx$ 100–1000K, and that they cool primarily through infrared and submillimeter spectral lines. There are also significant H_2 populations in excited vibration-rotation levels in these regions.

The literature contains two types of theoretical studies of PDR's: those that focus on the details of the near-infrared H_2 quadrupole emission spectrum, and those that emphasize the energetics, chemistry, and longer-wavelength lines. The first type of model treats the populations of each excited H_2 vibrational-rotational level individually, in order to predict accurate intensities of observable emission lines (Black & van Dishoeck 1987; Sternberg & Dalgarno 1989). The two main mechanisms for populating these levels are collisions (as in a thermalized, shock-heated gas), and fluorescence or "uv-pumping." The latter operates by the absorption of uv photons (with 912 Å $< \lambda <$ 1100 Å) which excite the molecules to higher electronic states, followed by a radiative cascade through the levels that give rise to the infrared lines. This process was discussed 20 years ago (*i.e.*, Black & Dalgarno 1976), but not actually observed in the interstellar medium until 10 years later (Hayashi *et al.* 1985; Sellgren 1986). Both the absolute efficiency (line brightness per unit mass) and the relative line intensities of fluorescent H_2 emission are dramatically different from those of a thermalized, collisionally-excited spectrum. In particular, only a fluorescent spectrum will show significant emission in lines from high vibrational levels, $v \geq 3$. Two examples of "pure" or "radiative" fluorescent regions are the reflection nebula NGC 2023 (Gatley *et al.* 1987) and the planetary nebula Hubble 12 (Dinerstein *et al.* 1988; Ramsay *et al.* 1993). However, if the illuminated material is sufficiently dense, the spectrum will be modified by collisional effects, which tend to thermalize the lowest-lying energy levels ($v \leq 2$) but not the higher vibrational states. This is called a "collisional fluorescent" spectrum by Sternberg & Dalgarno (1989). Another complicating factor is that gas illuminated by X-rays (a harder radiation field than that from ordinary hot stars) may also produce line ratios that are intermediate between the "pure" collisional and fluorescent cases (see article by A. Dalgarno in this volume).

Several techniques have been used to distinguish fluorescent excitation of the H_2 lines from emission produced entirely by collisions in shock-heated gas. One method that has been popular because of its simplicity and ease of measurement is to simply take the ratio of the $v =$ 2–1 S(1) line at 2.247 μm to the $v =$ 1–0 S(1) line at 2.121 μm. These lines are sufficiently close in wavelength that they can be observed simultaneously with some instruments. Under typical collisional excitation conditions, this ratio is about 1/10; for fluorescence, it is about 1/2. However, this diagnostic is fallible; in particular, at densities approaching 10^4 cm^{-3} the $v = 2$ level may be depopulated by collisions, even if the original "pump" was uv radiation. A more sophisticated approach that requires additional data is to plot the data in a level-population vs. excitation energy diagram (*e.g.*, Hasegawa *et al.* 1987). In such a diagram, collisionally populated levels will fall along a straight line defined by a single excitation temperature, or a smooth curve for a mixture of thermalized components with a range in temperature (a more realistic case for shocked gas). For a fluorescently excited gas, the vibration or rotation temperatures calculated from different pairs or sets of lines will have different slopes. The difference is illustrated in Fig. 7. A more efficient approach for simply detecting fluorescent emission (as opposed to studying its spectrum) is to observe lines arising from very high v levels, that are absent from a thermal spectrum (*e.g.*, Luhman *et al.* 1994). Most of the strongest high-v lines, such as the $v = 6$–4 lines, fall in the J and H bands.

The second class of PDR models trace the chemistry, heating, and cooling of uv-irradiated molecular clouds, from the ionization front to deep into the cloud. The path length is specified either by N(H) or by A_V; the latter is closely related to the increasing extinction seen by the uv photons. Temperatures in these regions range from as high as 1000K to as low as 50K, depending on the physical conditions, and the strongest emission lines are the far-infrared fine-structure lines of [O I] and [C II]. These are the main cooling lines for atomic gas over a very wide range of physical conditions, approximately $20K < T < 2000K$ and $2 < \log n < 6$ (see Fig. 10 of TH). As discussed above, the intensity ratios among these lines are diagnostics of temperature and density (*e.g.*, Fig. 3). The two key parameters of PDR models are the intensity of the uv radiation field, expressed in multiples of a "standard" interstellar intensity, denoted χ or $G = I_{UV}/I_{UV}(IS)$, and the gas density n. The model yields the run of temperature for given values of G and n. PDR models have been computed for conditions ranging from dense, clumpy media (Burton, Hollenbach, & Tielens 1990) to low-density extended gas in galaxies (Hollenbach, Takahashi, & Tielens 1991). The predictions of how the line intensity ratios map into parameters G and n have been applied to situations as diverse as galactic nuclei (Wolfire, Tielens, & Hollenbach 1990; Carral *et al.* 1994) and planetary nebulae (Dinerstein *et al.* 1995b).

3.3. *Shocks*

The *ultimate* energy source for the collisionally excited emission lines in ionized nebulae and photodissociation regions is the radiation field, although collisions are the *proximate* cause of the line emission. An alternate heating mechanism is the mechanical action of interstellar shocks (supersonic compressions). Conditions in the "post-shock" region, the material that has already passed through the shock, depend on the shock velocity v_s and the "pre-shock density" n_o (the initial density of the gas before it encounters the shock). These are the two primary parameters that are usually extracted from attempts to fit observations of shock-produced emission lines. The ionization, chemical, and thermal structures of the shock and post-shock zone are also affected by the magnetic field strength. Shocks are classified as "J-type" or "C-type" depending on whether there is a discontinuity or "jump" (J) in the gas properties at the interface, or a gradual or "continuous" (C) transition between the pre- and post-conditions (Draine 1980; McKee, Chernoff, & Hollenbach 1984). Recent reviews of the theory of interstellar shocks have been presented by Shull & Draine (1987) and Draine & McKee (1993).

Relatively slow shocks ($v_s \leq 50$ km s^{-1}) moving into molecular clouds cool primarily through infrared and submillimeter lines of neutral and molecular species, such as [O I] 63, 145 μm, [C I] 370, 610 μm, and, at $v_s \geq 10$ km s^{-1}, pure-rotation transitions of H_2 (Draine, Roberge, & Dalgarno 1983). The relative strengths of the different lines vary with v_s, n_o, and the shock type and structure. Given a set of models, the shock parameters can be constrained by plotting the loci of observed line ratios in (v_s, n_o) phase space (*e.g.*, Fig. 6 of McKee, Chernoff, & Hollenbach 1984). Faster shocks dissociate the H_2 molecules, produce a high-temperature zone that illuminates the pre-shock gas with uv radiation (the radiative "precursor"), and ionize some species (Hollenbach & McKee 1989). In such cases the H_2 molecules may re-form on the grain surfaces after the shock passes, and be released into the gas phase. When a number of different lines are

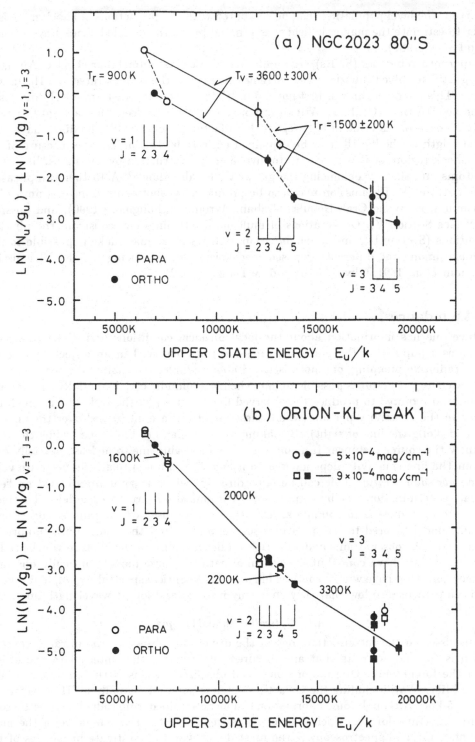

FIGURE 7. Population/excitation-energy diagrams for fluorescent H_2 emission (left), and for shock-excited emission (right). Note the disparate values of T_{vib} and T_{rot} for fluorescence, and the smooth slope of the thermalized gas. (From Hasegawa *et al.* 1987).

observed, the level populations can be compared with the predictions of shock models in order to estimate the magnetic field strength and probe the detailed shock physics (*e.g.*, Brand *et al.* 1988).

Supernova remnants (SNRs) are an obvious site of fast interstellar shocks. An interesting, still unsolved, puzzle is the extreme strength of the near-infrared [Fe II] lines in SNRs. This phenomenon was first noticed in IC 443, where the 1.64 μm [Fe II] line vastly outshines H I Br γ (Graham, Wright, & Longmore 1987). Indeed, the value of this ratio in SNRs exceeds its value in H II regions by a factor of about 500! The dilemma is: can the strength of the [Fe II] lines be explained entirely by the ionization structure of the post-shock region, or it is necessary to invoke grain destruction by the shock, liberating Fe atoms and thereby enhancing the gas-phase Fe abundance? A further complication is that strong [Fe II] emission may also be produced by photoionization by a power-law continuum, as in the Crab Nebula (Graham, Wright, & Longmore 1990), and possibly AGN (see Section V). Observations of multiple [Fe II] lines can constrain the physical conditions (Section II), and in combination with H_2 lines, may make it possible to distinguish among the different proposed mechanisms for enhancing the [Fe II] emission (Graham *et al.* 1991; Oliva, Moorwood, & Danziger 1991).

4. Abundances

Once one has information about the local physical conditions and understands the mechanism populating the particular emission lines observed in an object (*e.g.*, collisions, radiative pumping, or—not discussed above—direct recombination or recombination cascades), it becomes possible to infer chemical abundances. Essentially, the amount of material required to produce the observed lines is given by the ratio of the line luminosity to the emissivity per ion (where these quantities are integrated over the angular size and along the line of sight). By taking ratios of pairs of lines, one avoids the need to know the source distance. In many cases, ratios enable one to at least partially cancel out the line emissivity dependences on n and T, yielding a relatively accurate value of the *abundance ratio*. This can be done directly (which is sometimes called an "empirical" determination), or by comparison with ionization structure models. The most elaborate procedure is to compile an extensive set of observed line fluxes, construct a specific model tailored to the particular source, and "tune" the model until the predictions match the observations in detail. Below I discuss some of the areas in which studies of infrared lines have contributed novel information. These include optically obscured sources, ions that have with no observable lines in the optical spectral region, and regions with temperatures so low that they emit only infrared and longer wavelength lines.

4.1. *Optically "invisible" gas*

As discussed in the Introduction, one of the drivers for observing the infrared spectral region is to study objects that are obscured by dust, whether internal to the object or in the foreground. One major domain of obscured regions is the entire inner disk of the Milky Way, including the Galactic Center region and virtually all H II regions at $R_{gal} \leq 5$ kpc. Although some information can be determined from radio line spectroscopy (including values for T_e, which is inversely correlated with the abundance of the main coolants), infrared spectroscopy is the most direct way to investigate properties of the ionizing stars and to assess global trends such as abundance gradients. Mid-infrared observations ($8 \leq \lambda \leq 25$ μm) are useful for studying S, Ar, and Ne, since more than one ion of each element is represented in this spectral region (*e.g.*, Herter, Helfer, & Pipher 1983). Some of these lines, however, are inaccessible from the ground and require airborne platforms such as the KAO (Kuiper Airborne Observatory) or satellites (*e.g.*, IRAS; Simpson & Rubin 1990). It can be useful to observe some of these infrared lines even in optically *unobscured* sources such as planetary nebulae, where they enable one to measure additional ions of particular elements (Dinerstein 1980; Pottasch *et al.* 1986). At longer wavelengths there is the useful line trio of [O III] 52, 88 μm, and [N III] 57 μm. We have already discussed some uses of the first pair (Section II). Addition of the 57 μm line provides a measurement of N^{++}/O^{++} which is relatively independent of local conditions, because the critical densities of all three lines are similar (Moorwood *et al.* 1980; Dinerstein 1986). Observations of these lines in H II regions ranging from the Galactic Center to outside the solar circle indicate elevated values of N/O in the inner Galaxy, although the behavior is not necessarily a simple linear gradient (Lester *et al.* 1987).

Another type of "invisible" region is gas that is too cool to excite strong line emission in optical/uv transitions, for example gas with extremely high abundances of coolants such as O, N, and S. Such material can be found in gas processed by explosive nucleosynthesis, in supernova remnants and the ejecta of classical novae. When the densities become low enough that the ground state fine-structure lines are not collisionally suppressed, the gas cools extremely efficiently through these lines and the temperature plummets from of order 10,000 K to below 1000 K, causing the optical/uv lines to disappear. Such an "infrared catastrophe" has been predicted to occur in evolving nova ejecta, shifting the main radiative cooling from the optical/uv into the far-infrared lines (Itoh 1981a; Ferland *et al.* 1984). However, such infrared lines have not yet been directly seen. IRAS saw broad-band infrared emission from several old novae (Dinerstein 1986b), but it has not yet been possible to demonstrate whether these fluxes are due to lines or to thermal emission from dust because of their faintness. Likewise, pure-metal inclusions in old supernova remnants should emit strongly in infrared lines from their constituent elements (Itoh 1981b, c; Dopita, Binette, & Tuohy 1984). Although there is some tantalizing evidence for lines from Cas A (Dinerstein *et al.* 1987), the presence of a strong thermal continuum from heated dust makes the lines harder to detect (Dwek *et al.* 1987).

Significant overabundances of particular species can produce strong infrared emission lines in novae and supernovae even while the ejecta are relatively hot. For example, the extremely strong 12.8 μm [Ne II] line seen in QU Vul led to the designation of a new class of novae, the "neon novae" (Gehrz, Grasdalen, & Hackwell 1985). Overabundances of Ne, Si, Mg, and other nearby elements are taken as evidence that this class of novae represents thermonuclear explosions on the surfaces of massive, O-Ne-Mg white dwarfs rather than C-O white dwarfs. An advantage of infrared lines is that they can be seen even if dust condenses in the ejecta, creating internal extinction that can hamper observations of ultraviolet emission lines. The first direct demonstration that Type Ia supernovae produce large amounts of iron was the observation of the 1.64 μm [Fe II] emission line in SN 1983n in M83 (Graham *et al.* 1986). Some further discussion of abundances in supernovae follows in Section 3.3.

4.2. *Ionic abundance ratios*

One nagging issue in the use of infrared lines to measure abundance ratios has been the discrepancy between the derived values and values determined from other methods. (I have previously called this the "zero point" issue.) An acute example has been the difference between N/O ratios determined from the far-infrared [O III], [N III] lines and values determined *for the same objects* from optical lines of [O II] and [N II]. As of several years ago, the infrared determinations of N/O for optically visible H II regions like Orion and M17 were yielding values two to three times higher than the optically determined values (Simpson *et al.* 1986; Lester *et al.* 1987).

Various possibilities were proposed as being responsible for this uncomfortable situation. First, does the problem lie in translating from line intensities to ionic abundances? The infrared lines are relatively insensitive to temperature, and their density dependence is well compensated for by deriving the density from the far-infrared [O III] lines; furthermore, the density "bias" introduced by clumping is minimal for this particular line set (Rubin 1989). On the other hand, the optical lines *are* temperature-sensitive, and furthermore the [O II] lines may be enhanced by dielectric recombination, causing the [N II]/[O II] line ratio to yield too low a value (Rubin 1986). Another possibility lies in the translation from the ionic ratio N^{++}/O^{++} to the elemental ratio, N/O. Rubin *et al.* (1988) advocated application of a correction factor for the relative volumes of N^{++} and O^{++}, which begin to decouple at stellar temperatures cooler than about 37,000 K. However, their correction factors are based on the fractional volume of these ionization zones integrated over the entire nebula, and therefore substantially overestimate the effect for H II regions that are much larger than the instrument beam used to make the observations, which has been the case for most of the far-infrared observations to date. Furthermore, most of the luminous H II regions studied in the galactic gradient surveys are probably ionized by stars hotter than 37,000 K, except for those close to the Galactic Center.

Finally, the discrepancy might be attributed to the use of incorrect atomic constants, especially collision strengths. Accurate atomic parameters for the far-infrared [O III] have been available for a decade (*e.g.*, Aggarwal 1983), but the value of the [57 μm] collision strength, which is crucial for the ionic ratio, has been uncertain. During the late 1980's, an older value (Nussbaumer & Storey 1979) was replaced by a higher, unpublished value (Butler & Storey 1986). However, the latter has now been superceded by a still larger value, about twice the original one (Blum & Pradhan 1992). This new value lowers the Orion N^{++}/O^{++} ratio to 0.27, which is not far from the recent optical redetermination from N^+/O^+ of 0.23 (Baldwin *et al.* 1991). Furthermore, when the new collision strength is applied to planetary nebulae with hot central stars, there is no longer any evidence for a systematic difference between N/O determined from the infrared and the optical lines (Dinerstein *et al.* 1995a, b).

4.3. *Altered & evolving abundances*

Infrared lines can provide evidence for the destruction of dust grains and the subsequent liberation of atoms, enhancing the gas-phase abundances of refractory elements relative to their usual depleted values in the interstellar medium. Such an effect may be responsible for the observed strength of near-infrared [Fe II] lines in supernova remnants, as discussed in Section 3.3. A similar effect is seen in the 35 μm [Si II] line in the Galactic Center region (Herter *et al.* 1989) and in shocked material in Orion (Haas, Hollenbach, & Erickson 1991).

A more exotic situation occurs in supernovae, where actual elemental abundances can change with time due to radioactive decay. In particular, the decay chain of ^{56}Ni $\rightarrow$ ^{56}Co $\rightarrow$ ^{56}Fe can be tracked through infrared lines. The first indication that this sequence might be observable was the identification of a feature at 10.5 μm in SN 1987A as being due to [Co II] rather than [S IV] (Rank *et al.* 1988). At shorter wavelengths, spectral monitoring of SN 1987A revealed a dramatic effect: the decay of a [Co II] line and simultaneous growth of an adjacent [Fe II] feature at 1.54 μm (see Fig. 8, from Varani *et al.* 1990). Nickel was also seen in SN 1987A via [Ni II] lines at 6.63 μm (Wooden *et al.* 1993) and elsewhere. The spectral region 15–30 μm contains lines of all three elements,

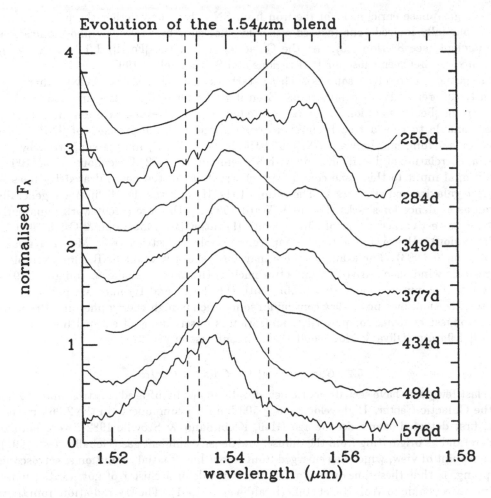

FIGURE 8. Evolution of the spectrum of SN 1987A. The vertical dashed lines indicate, from left to right, the wavelengths of [Fe II] 1.533 μm, [Fe I] 1.535 μm, and [Co II] 1.547 μm. Note the dramatic shift of the line peak as the cobalt decays into iron. (From Varansi *et al.* 1990)

Ni, Co, and Fe, and many of these were seen in SN 1987A (Moseley *et al.* 1989). If the physics of the expanding supernova envelope is sufficiently well understood, or the local conditions can be strongly constrained by diagnostic line ratios, these line strengths can be used to deduce the manufactured masses of these elements. A less dramatic but still important probe of supernova physics is the use of the strong near-infrared He I lines at 1.08, 2.06 μm, which may be the most sensitive indicators of trace amounts of He in a supernova, as, for example, in SN 1993J in M81 (Swartz *et al.* 1993).

5. Some current issues: A sampler

5.1. *The origin of strong [Fe II] lines*

One of the issues currently inspiring much discussion and gnashing of teeth is the origin of the very strong [Fe II] lines in some sources, particularly compared with their H I or H$_2$ emission. We have already touched on this question in the context of supernova remnants, where it is unclear whether the line strengths require actual grain destruction and/or gas-phase enrichment of the iron from SN debris, or whether the presence of a deep, partially ionized zone (caused either by shocks or by power-law photoionization) is sufficient (see Section 3.3). At the Galactic Center, the [Fe II]/H I line ratios are intermediate between those for H II regions and SNRs (DePoy 1992).

In spatially unresolved sources such as starbursts in galaxies and AGN's, there is a slightly different "[Fe II] conundrum." Even if we do not fully understand the cause of the strong [Fe II] emission in SNRs, a number of investigators have assumed that it is reasonable to consider such emission to be a tracer of the presence of SNRs. The case for a strong link between SNRs and the infrared [Fe II] lines is supported by the spatial correlation of [Fe II] emission with SN remnants in M82 (Greenhouse *et al.* 1991). Additional input to this issue comes from observations of the near-infrared H_2 lines in Seyfert galaxies and starbursts. The ratios of the H_2 and the [Fe II] lines are generally taken as evidence for shocks (Forbes & Ward 1993), with more recent work apparently putting to rest earlier claims of "fluorescent" H_2 ratios (see Moorwood & Oliva 1990 for a discussion). However, the issue is far from completely settled (*e.g.*, Mouri, Kawara, & Taniguchi 1993). For example, there may be a shock without SNRs, for example a large-scale wind shock driven by an active nucleus (Kawara, Nishida, & Taniguchi 1988; Lester & Thompson 1990). Or possibly both the [Fe II] and H_2 lines are produced by photoionization, by a power-law continuum source rich in high-energy photons. Probably the strongest evidence for power-law ionization is observations of coronal lines such as [Si VI] 1.96 μm (Oliva & Moorwood 1990; Greenhouse *et al.* 1993).

5.2. *Curiosities at the Galactic Center*

The last few years have seen dramatic new developments in infrared spectroscopic studies of the Galactic Center. High velocity ($v \geq 500$ km s^{-1}) components in the 2.06 μm He I line, first detected over a decade ago (Hall, Kleinmann, & Scoville 1982), were hard to understand as originating from the diffuse, low-density ambient gas (Geballe *et al.* 1991). A new point of view, which has emerged from recent high spatial resolution spectroscopic mapping, is that these lines arise in the stellar winds of a cluster of hot, exotic, mass-losing objects akin to Wolf-Rayet stars (Krabbe *et al.* 1991). The UV radiation from these stars ionizes the adjacent material, but the stellar winds also drive fast shocks into the adjacent material, producing a hot, wind-swept "cavity" (Lutz, Krabbe, & Genzel 1993). The conditions in this cavity produce a variety of [Fe III] lines, whose ratios constrain the temperature and density of the gas (see their Fig. 6). The extent to which this cluster of hot stars is a paradigm for the central regions of other galaxies is unclear, but it does provide a sobering reminder that the Galactic Center is a special environment which may host many different kinds of unusual phenomena (and not "merely" black holes!).

5.3. *Infrared spectroscopy of AGN*

We have already mentioned infrared spectroscopy of AGN in the context of line intensity ratios of [Fe II], H I, and H_2. There are other interesting applications of infrared spectroscopy to AGN. One is to look at infrared H I lines in Seyfert 2 galaxies, where all the optical lines are relatively narrow, and search for a reddened broad-line component, or "buried" Seyfert 1 (*e.g.*, Hines 1991). In the current paradigm of "unified" theories for AGN, such objects are viewed nearly "edge-on," so that the narrow-line gas (located farther from the central engine) is seen directly, whereas the inner, broad-line region is hidden by a dusty torus (*e.g.*, Antonucci 1993). In the near-infrared the extinction is lower, so the broad-line region is more easily seen and therefore the broad wings become more apparent in the line profiles as one goes to longer and longer wavelengths. A second use of infrared spectroscopy is to observe the familiar "optical" lines, such as [O III] 5007 Å and Hα, in high-redshift objects (Hill, Thompson, & Elston 1993).

6. Concluding remarks

This review has not, of course, covered every conceivable application of infrared spectroscopy. One major area I have omitted is the use of line velocities and profiles to study the kinematics and dynamics of astronomical sources. This type of study is not as unique to infrared lines as are some of the other topics discussed above, except in studying regions of heavy obscuration, such as the nuclei of starburst galaxies such as M 82 (*e.g.*, Achtermann & Lacy 1995). So far there has been only a limited amount of high-resolution infrared spectroscopy of emission lines, primarily due to sensitivity limitations. We can anticipate explosive growth of this area in the future, as infrared astronomers become less "signal-starved" with the advent of improved instruments and facilities. On the instrumentation side, individual detectors have already achieved high quantum efficiencies, but for spectroscopic applications, there is a tremendous advantage to using large detector arrays to obtain simultaneous coverage of a large number of spectral resolution elements.

The next decade will also see a major step forward in the light-gathering power of telescopes. On the ground, there will be several 8–10 m class telescopes in operation. Some of them more infrared-optimized than others, but all will undoubtedly be used for at least some infrared spectroscopy: Keck I & II and the Subaru Telescope, on Mauna Kea; Gemini North & South, in Hawaii & Chile; the VLT, in Chile, and the Hobby-Eberly Telescope in Texas. For spectral regions that are blocked from the ground, ESA's Infrared Space Observatory (ISO), scheduled to be launched in late 1995, will be the first space telescope to do extensive infrared spectroscopy. The premier facility for continuing infrared spectroscopic work at these wavelengths will be the Stratospheric Facility for Infrared Astronomy (SOFIA), a 3 m airborne telescope (Erickson & Davidson 1995). SOFIA's ten-fold increase in collecting area and smaller beam size, compared with the KAO, will give it the sensitivity to do infrared spectroscopy on relatively faint and distant sources, and its operational mode will encourage frequent upgrading of auxiliary instrumentation such as infrared spectrometers. At this writing, it appears likely that SOFIA will be in operation by around the year 2000. It is to be hoped that other, more advanced infrared space observatories, such as SIRTF, will follow soon after.

The utility of infrared emission line observations will also be improved by advances on the theoretical side. One major uncertainty in the interpretation of infrared lines has been the values of some of the atomic constants. Fortunately, this situation is now beginning to be remedied (*e.g.*, Blum & Pradhan 1992). Improved models of the chemical, ionization, and thermal structure of photoionized regions, photodissociation regions, and interstellar shocks may help resolve some of the present anomalies and discrepancies that have arisen from studies of infrared emission lines, and permit stronger conclusions to be drawn on the roles of shocks and photoionization in distant galaxies.

Infrared spectroscopy is rapidly becoming a tool that is used by a large number of astronomers, rather than being the domain of a specialized few. This is likely to become even more true in the future, when facility infrared spectrometers on large telescopes are readily available to guest observers without specialized training. Thus, it is useful for all of us to be aware of the potential of infrared emission lines for nebular astrophysics. The techniques reviewed in this paper are merely the beginning of the harvest to be reaped from infrared line spectroscopy in the future.

I am grateful to the organizers of this conference for giving me the opportunity to present this review and to participate in this very special occasion, honoring two scientists who have played such a major role in advancing our understanding of the properties and physics of nebulae. I also wish to acknowledge research support from NSF grant AST 91-15101 and HST grant GO 3880.01-91A during the preparation of this review.

REFERENCES

ACHTERMANN, J. M., & LACY, J. H. 1995 *ApJ* **439**, 163

AGGARWAL, K. M. 1983 *ApJS* **52**, 387

ALLER, L. H. 1984 *Physics of Thermal Gaseous Nebulae* (Dordrecht: Reidel)

ANTONUCCI, R. 1993 *ARAA* **31**, 473

ASHLEY, M. C. B., & HYLAND, A. R. 1988 *ApJ* **331**, 532

BALDWIN, J. A., FERLANDS, G. F., MARTIN, P. G., CORBIN, M. R., COTA, S. A., PETERSON, B. M., & SLETTEBAK, A. 1991 *ApJ* **374**, 580

BALDWIN, J. A., PHILLIPS, M. M., & TERLEVICH, R. 1981 *PASP* **93**, 5

BENJAMIN, R. A., & DINERSTEIN, H. L. 1990 *AJ* **100**, 1588

BLACK, J. H., & DALGARNO, A. 1976 *ApJ* **203**, 132

BLACK, J. H., & VAN DISHOECK, E. F. 1987 *ApJ* **322**, 412

BLUM, R. D., & PRADHAN, A. K. 1992 *ApJS* **80**, 425

BRAND, P. W. J. L., MOORHOUSE, A., BURTON, M. G., GEBALLE, T. R., BIRD, M., & WADE, R. 1988 *ApJ* **334**, L103

BURTON, M. G., HOLLENBACH, D. J., & TIELENS, A. G. G. M. 1990 *ApJ* **365**, 620

BUTLER, K., & STOREY, P. J. 1986 private communciation

CARRAL, P., HOLLENBACH, D. J., LORD, S. D., COLGAN, S. W. J., HAAS, M. R., RUBIN, R. H., & ERICKSON, E. F. 1994 *ApJ* **423**, 223

COLGAN, S. W. J., HAAS, M. R., ERICKSON, E. F., RUBIN, R. H., SIMPSON, J. P., & RUSSELL, R. W. 1993 *ApJ* **413**, 237

DEPOY, D. L. 1992 *ApJ* **398**, 512

DINERSTEIN, H. L. 1980 *ApJ* **237**, 486

DINERSTEIN, H. L. 1983 in *Planetary Nebulae, IAU Symposium 103* (ed. D. R. Flower) Dordrecht: Reidel, 79

DINERSTEIN, H. L. 1986a *PASP* **98**, 979

DINERSTEIN, H. L. 1986b *AJ* **92**, 1381

DINERSTEIN, H. L. 1990 in *The Interstellar Medium in Galaxies* (eds. H. A. Thronson, Jr. & J. M. Shull) Dordrecht: Kluwer, 257

DINERSTEIN, H. L., & BENJAMIN, R. A. 1993 *RevMexA&A* **27**, 33

DINERSTEIN, H. L., HAAS, M. R., ERICKSON, E. F., & WERNER, M. W. 1995a in the *Airborne Astronomy Symposium on the Galactic Ecosystem: From Gas to Stars to Dust* (eds. M. R. Haas, J. A. Davidson, & E. F. Erickson) ASP Conference Series, vol. 73, in press

DINERSTEIN, H. L., HAAS, M. R., ERICKSON, E. F., & WERNER, M. W. 1995b in preparation

DINERSTEIN, H. L., LESTER, D. F., CARR, J. S., & HARVEY, P. M. 1988 *ApJ* **327**, L27

DINERSTEIN, H. L., LESTER, D. F., RANK, D. M., WERNER, M. W., & WOODEN, D. H. 1987 *ApJ* **312**, 314

DOPITA, M. A., BINETTE, L., & TUOHY, I. R. 1984 *ApJ* **282**, 142

DRAINE, B. T. 1980 *ApJ* **241**, 1021

DRAINE, B. T., & MCKEE, C. F. 1993 *ARAA* **31**, 373

DRAINE, B. T., ROBERGE, W. G., & DALGARNO, A. 1983 *ApJ* **264**, 485

DWEK, E., DINERSTEIN, H. L., GILLETT, F. C., HAUSER, M. G., & RICE, W. L. 1987 *ApJ* **315**, 571

ERICKSON, E. F., & DAVIDSON, J. A. 1995 in the *Airborne Astronomy Symposium on the Galactic Ecosystem: From Gas to Stars to Dust* (eds. M. R. Haas, J. A. Davidson, & E. F. Erickson) ASP Conference Series, vol. 73, in press

EVANS, I. N. & DOPITA, M. A. 1985 *ApJS* **58**, 125

FERLAND, G. J., WILLIAMS, R. E., LAMBERT, D. L., SHIELDS, G. A., SLOVAK, M., GOND-HALEKAR, P. M., & TRURAN, J. W. 1984 *ApJ* **281**, 194

FORBES, D. A., & WARD, M. J. 1993 *ApJ* **416**, 150

GATLEY, I., *et al.* 1987 *ApJ* **318**, L73

GEBALLE, T. R., KRISCIUNAS, K., BAILEY, J. A., & WADE, R. 1991 *ApJ* **370**, L73

GEBALLE, T. R., PERSSON, S. E., SIMON, T., LONSDALE, C. J., & MCGREGOR, P. J. 1986 *ApJ* **302**, 693

GEHRZ, R. D. 1988 *ARAA* **26**, 377

GEHRZ, R. D., GRASDALEN, G. L., & HACKWELL, J. A. 1985 *ApJ* **298**, L47; **306**, L163

GOODRICH, R. W. 1990 *ApJ* **355**, 88

GRAHAM, J. R., MEIKLE, W. P. S., ALLEN, D. A., LONGMORE, A. J., & WILLIAMS, P. M. 1986 *MNRAS* **218**, 93

GRAHAM, J. R., WRIGHT, G. S., & LONGMORE, A. J. 1987 *ApJ* **313**, 847

GRAHAM, J. R., WRIGHT, G. S., & LONGMORE, A. J. 1990 *ApJ* **352**, 172

GRAHAM, J. R., WRIGHT, G. S., HESTER, J. J., & LONGMORE, A. J. 1991 *AJ* **101**, 175

GRASDALEN, G. L., & JOYCE, R. R. 1976 *Nature* **259**, 187

GREENHOUSE, M. A., FELDMAN, U., SMITH, H. A., KLAPISCH, M., BHATIA, A. K., & BAR-SHALOM, A. 1993 *ApJS* **88**, 23

GREENHOUSE, M. A., GRASDALEN, G. L., WOODWARD, C. E., BENSON, J., GEHRZ, R. D., ROSENTHAL, E., & SKRUTSKIE, M. F. 1990 *ApJ* **352** 307

GREENHOUSE, M. A., WOODWARD, C. E., THRONSON, H. A., RUDY, R. J., ROSSANO, G. S., ERWIN, P., & PUETTER, R. C. 1991 *ApJ* **383**, 164

HAAS, M. R., HOLLENBACH, D., & ERICKSON, E. F. 1991 *ApJ* **374**, 555

HASEGAWA, T., GATLEY, I., GARDEN, R. P., BRAND, P. W. L., OHISHI, M., HAYASHI, M., & KAIFU, N. 1987 *ApJ* **318**, L77

HAYASHI, M., HASEGAWA, T., GATLEY, I., GARDEN, R., & KAIFU, N. 1985 *MNRAS* **215**, 31P

HERTER, T., HELFER, H. L., & PIPHER, J. L. 1983 *A&AS* **51**, 195

HERTER, T., GULL, G. E., MEGEATH, S. T., ROWLANDS, N., & HOUCK, J. R. 1989 *ApJ* **343**, 696

HILL, G. J., THOMPSON, T. L., & ELSTON, R. 1993 *ApJ* **414**, L1

HINES, D. C. 1991 *ApJ* **374**, L9

HOLLENBACH, D. J., & MCKEE, C. F. 1989 *ApJ* **342**, 306

HOLLENBACH, D. J., TAKAHASHI, T., & TIELENS, A. G. G. M. 1991 *ApJ* **377**, 192

ITOH, H. 1981a *PASJ* **33**, 743

ITOH, H. 1981b *PASJ* **33**, 1

ITOH, H. 1981c *PASJ* **33**, 521

KAWARA, K., NISHIDA, M., & PHILLIPS, M. M. 1989 *ApJ* **337**, 230

KAWARA, K., NISHIDA, M., & TANIGUCHI, Y. 1988 *ApJ* **328**, L41

KEENAN, F. P. & CONLON, E. S. 1993 *ApJ* **410**, 426

KEENAN, F. P., BURKE, V. M., & AGGARWAL, K. M. 1991 *ApJ* **371**, 636

KRABBE, A., GENZEL, R., DRAPATZ, S., & ROTACIUC, V. 1991 *ApJ* **382**, L19

LATTER, W. B., MALONEY, P. R., KELLY, D. M., BLACK, J. H., RIEKE, G. H., & RIEKE, M. J. 1992 *ApJ* **389**, 347

LESTER, D. F., CARR, J. S., JOY, M., & GAFFNEY, N. 1990 *ApJ* **352**, 544

LESTER, D. F., DINERSTEIN, H. L., & RANK, D. M. 1979 *ApJ* **229**, 981

LESTER, D. F., DINERSTEIN, H. L., WERNER, M. W., WATSON, D. M., GENZEL, R., & STOREY, J. W. V. 1987 *ApJ* **320**, 578

LESTER, D. F., & THOMPSON, K. L. 1990 *ApJ* **348**, L49

LUHMAN, M. L., JAFFE, D. T., KELLER, K. D., & PAK, S. 1994 *ApJ* **436**, L185

LUTZ, D., KRABBE, A., & GENZEL, R. 1993 *ApJ* **418**, 244

MATHIS, J. S. 1990 *ARAA* **28**, 37

MCKEE, C. F., CHERNOFF, D. F., & HOLLENBACH, D. J. 1984 in *Galactic & Extragalactic Infrared Spectroscopy* (eds. M. F. Kessler & J. P. Phillips) Dordrecht: Reidel, 103

MOORWOOD, A. F. M., BALUTEAU, J.-P., ANDEREGG, M., CORON, N., BIRAUD, Y., & FITTON, B. 1980 *ApJ* **238**, 565

MOORWOOD, A. F. M., & OLIVA, E. 1990 *A&A* **239**, 78

MOURI, H., KAWARA, K., & TANIGUCHI, Y. 1993 *ApJ* **406**, 52

MOSELEY, S. H., DWEK, E., GLACCUM, W., GRAHAM, J. R., LOEWENSTEIN, R. F., & SILVERBERG, R. F. 1989 *ApJ* **347**, 1119

NATTA, A., GIOVANARDI, C., PALLA, F., & EVANS, N. J. II 1988 *ApJ* **327**, 817

NUSSBAUMER, H., & STOREY, P. J. 1979 *A&A* **71**, L5

NUSSBAUMER, H., & STOREY, P. J. 1988 *A&A* **193**, 327

OLIVA, E., & MOORWOOD, A. F. M. 1990 *ApJ* **348**, L5

OLIVA, E., MOORWOOD, A. F. M., & DANZIGER, I. J. 1989 *A&A* **214**, 307

OLIVA, E., MOORWOOD, A. F. M., & DANZIGER, I. J. 1991 *A&A* **240**, 453

OSTERBROCK, D. E. 1988 *PASP* **100**, 412

OSTERBROCK, D. E. 1989 *Astrophysics of Gaseous Nebulae and Active Galactic Nuclei* (Mill Valley, CA: University Science Books)

OSTERBROCK, D. E., TRAN, H. D., & VEILLEUX, S. 1992 *ApJ* **389**, 196

PEIMBERT, M. 1967 *ApJ* **150**, 825

PEIMBERT, M. 1993 *RevMexA&A* **27**, 9

PEIMBERT, M. & COSTERO, R. 1969 *Bol. Obs. Tonantzintla y Tacubaya* **5**, 3

POTTASCH, S. R., PREITE-MARTINEZ, A., OLNON, F. M., RAIMOND, E., BEINTEMA, D. A., & HABING, H. J. 1985 *A&A* **143**, L11

POTTASCH, S. R., PREITE-MARTINEZ, A., OLNON, F. M., MO, J.-E., & KINGMA, S. 1986 *A&A* **161**, 363

RAMSAY, S. K., CHRYSOSTOMOU, A., GEBALLE, T. R., BRAND, P. W. J. L., & MOUNTAIN, M. 1993 *MNRAS* **263**, 695

RANK, D. M., DINERSTEIN, H. L., LESTER, D. F., BREGMAN, J. D., AITKEN, D. K., & JONES, B. 1978 *MNRAS* **185**, 179

RANK, D. M., PINTO, P. A., WOOSLEY, S. E., BREGMAN, J. D., WITTEBORN, F. C., AXELROD, T. S., & COHEN, M. 1988 *Nature* **331**, 505

ROWLANDS, N., HOUCK, J. R., HERTER, T., GULL, G. E., & SKRUTSKIE, M. F. 1989 *ApJ* **341**, 901

RUBIN, R. 1989 *ApJS* **69**, 897

RUBIN, R. 1985 *ApJS* **57**, 349

RUBIN, R. H., SIMPSON, J. P., ERICKSON, E. F., & HAAS, M. R. 1988 *ApJ* **337**, 377

RUBIN, R., SIMPSON, J. P., HAAS, M. R., & ERICKSON, E. F. 1991 *ApJ* **374**, 564

SEARLE, L. 1971 *ApJ* **165**, 259

SEATON, M. J. & OSTERBROCK, D. E. 1957 *ApJ* **125**, 66

SELLGREN, K. 1986 *ApJ* **305**, 399

SHIELDS, J. C. 1993 *ApJ* **419**, 181

SHULL, J. M., & DRAINE, B. T. 1987 in *Interstellar Processes* (eds. D. J. Hollenbach & H. A. Thronson) Dordrecht: Reidel, 283

SHURE, M. A., HERTER, T., HOUCK, J. R., BRIOTTA, D. A., JR., FORREST, W. J., GULL, G. E., & McCARTHY, J. F. 1983a *ApJ* **270**, 645

SHURE, M. A., HERTER, T., & HOUCK, J. R. 1983b *ApJ* **274**, 646

SIMPSON, J. P. 1975 *A&A* **39**, 43

SIMPSON, J. P. 1984 *ApJ* **281**, 184

SIMPSON, J. P., RUBIN, R. H., ERICKSON, E. F., & HAAS, M. R. 1986 *ApJ* **311**, 895

SIMPSON, J. P., & RUBIN, R. 1990 *ApJ* **354**, 165

SPINOGLIO, L., & MALKAN, M. A. 1992 *ApJ* **399**, 504

STERNBERG, A., & DALGARNO, A. 1989 *ApJ* **338**, 197

SWARTZ, D. A., CLOCCHIATTI, A., BENJAMIN, R., LESTER, D. F., & WHEELER, J. C. 1993 *Nature* **365**, 232

THOMPSON, R. I. 1982 *ApJ* **257**, 171

TIELENS, A. G. G. M. & HOLLENBACH, D. 1985 *ApJ* **291**, 722 (TH)

VARANI, G.-F., MEIKLE, W. P. S., SPYROMILIO, J., & ALLEN, D. A. 1990 *MNRAS* **245**, 570

VOIT, G. M. 1992 *ApJ* **399**, 495

WATSON, D. M. 1985 *Phys. Scripta* **T11**, 33

WILLIAMS, R. E. 1991 in *IAU Colloquium 122, Physics of Classical Novae* (eds. A. Cassatella & R. Viotti) Berlin: Springer, 215

WOLFIRE, M. G., TIELENS, A. G. G. M., & HOLLENBACH, D. J. 1990 *ApJ* **358**, 116

WOODEN, D. H., RANK, D. M., BREGMAN, J. D., WITTEBORN, F. C., TIELENS, A. G. G. M., COHEN, M., PINTO, P. A., & AXELROD, T. S. 1993 *ApJS* **88**, 477

Molecular Emission Line Diagnostics in Astrophysical Environments

By A. DALGARNO

Harvard-Smithsonian Center for Astrophysics, 60 Garden Street, Cambridge, MA 02138 USA

A brief selection is presented of the ways in which molecular emission lines have provided unique information on astrophysical environments.

---

1. Introduction

Molecules have been detected in a broad range of astronomical objects—the atmospheres of stars, diffuse, translucent and dense interstellar clouds, photon-dominated regions (PDRs), circumstellar shells, HII regions, planetary nebulae, stellar outflows, stellar winds, jets, Herbig-Haro objects, novae, supernova remnants and the ejecta of Supernova 1987a. Their presence is a controlling force in the determination of the thermal balance, the ionization structure, the dynamics and the evolution of the entities in which they reside. Molecules provide unique diagnostic probes of the physical nature of their environment, yielding information on the densities, the temperatures, the magnetic fields, the velocities, the isotopic composition, the radiation fields, the masses and the ages.

In dense molecular clouds where star formation takes place, molecules have been detected in remarkable diversity. Over twenty of them have been detected in external galaxies. Many additional species have been discovered in circumstellar shells.

I will not attempt to survey the myriad ways in which molecules serve as diagnostic probes. I will instead make an arbitrary personal selection, beginning not with emission lines but with the absorption lines of CN which provided the first measurement of the temperature of the cosmic blackbody background radiation field.

2. Absorption by CN

Absorption by the CN molecule has been measured towards many stars. Several lines have been observed showing that CN is present in its low-lying rotational levels with rotational quantum numbers $j = 0$, 1 and 2. The measured absorption strengths yield the relative populations N_j of the individual rotational levels integrated along the line of sight to the parent star. From the population ratio, an excitation temperature can be determined

$$\frac{N_j}{N_o} = (2j + 1)exp(-\Delta E_j / RT_{exc})$$

where ΔE_j is the energy of excitation above the $j = 0$ level. The latest data lead to a value of T_{exc} of 2.742 ± 0.02K for $j = 1$ and of 2.679 ± 0.06K for $j = 2$ (Roth, Meyer & Hawkins 1993). The direct measurements of the temperature of the cosmic background radiation of the COBE mission gives a value of $T = 2.76 \pm 0.01$K. It is clear that the CN rotational populations are providing a measure of the background temperature at the wavelengths of 2.64mm and 1.32mm corresponding to the $j = 1 - 0$ and $j = 2 - 0$ transitions. The CN excitation temperatures can be corrected empirically for the contribution to the level populations by electron impact excitation and the inferred blackbody temperature is $2.79 \pm 002K$ (Roth *et al.* 1993). The COBE measurements are restricted to the Galaxy. At least in principle, measurements of CN, and other molecular species, could be carried out for other galaxies.

3. Emission from LiH

De Bernardis *et al.* (1993), Maoli, Melchiorri & Tosti (1994) and Signore *et al.* (1994) have discussed the possibility that the Thompson scattering of the background radiation by the primordial molecule LiH led to a temperature fluctuation in the blackbody spectrum that might be detectable. In the early Universe, nucleosynthesis created a small amount of ^{7}Li, some of which was converted into ^{7}LiH. Lepp & Shull (1984) noted that formation of LiH would occur by radiative association of neutral lithium and neutral hydrogen

$$Li + H \rightarrow LiH + h\nu.$$

Dalgarno & Lepp (1985) pointed out that at the time of recombination of hydrogen, the lithium was still largely ionized so that the chemistry would be initiated alternatively by radiative association to form the molecular ion LiH$^+$.

Fig. 1 illustrates the set of chemical reactions that determine the abundances of LiH and LiH$^+$. Recent calculations of the LiH abundances (Puy *et al.* 1992) suggest that LiH first appeared in the expanding Universe at a red shift z of 400, but the chemistry is still very uncertain. In primordial collapsing clouds, there is less uncertainty because LiH will be formed by the three body reaction

$$Li + H + H \rightarrow LiH + H$$

and all the lithium should be incorporated into LiH. Attempts to detect LiH have not been successful. I turn now to H$_2$ of which there have been many observations.

4. Emission from H$_2$

Molecular hydrogen has rotation-vibration levels in the ground electronic state which radiate in the infrared through electric quadrupole transitions. The infrared lines were first observed towards the Becklin-Neugebauer object in Orion (Gautier *et al.* 1976). They have been observed subsequently in a wide range of objects in the Galaxy and in external galaxies. The observations of Orion have been interpreted as emission from shock-heated thermal gas. The derived excitation temperatures ranged between 1000K and 3000K, suggestive of bow shocks (Smith, Brand & Moorhouse 1991). A recent study by Parmar, Lacy & Achterman (1994) of emission from pure rotational transitions supports a two-shock model. Observations of the emission lines have been used to map the structure of shocked regions (Sugai *et al.* 1994).

A convenient diagnostic of shocked thermal gas is the intensity ratio of the 2-1S(1)/ 1-0S(1) lines. At 2000K, the ratio is 0.13 (Shull & Beckwith 1982). In other objects, significant enhancements of the ratio occur. For example in the planetary nebula Hubble 12, the ratio is 0.5 (Dinerstein *et al.* 1988; Ramsey *et al.* 1993). The value 0.5 is consistent with the excitation by ultraviolet pumping of a diffuse gas (Black & Dalgarno 1976; Black & van Dishoeck 1987; Sternberg 1988). Further evidence is provided by other line ratios and more generally by the detection of emission from high-lying vibrational levels. Extended emission of the $v = 6$-4 transition at $1.601\mu m$ has been observed from the Orion A molecular cloud which is attributable to ultraviolet fluorescent pumping (Luhman *et al.* 1994).

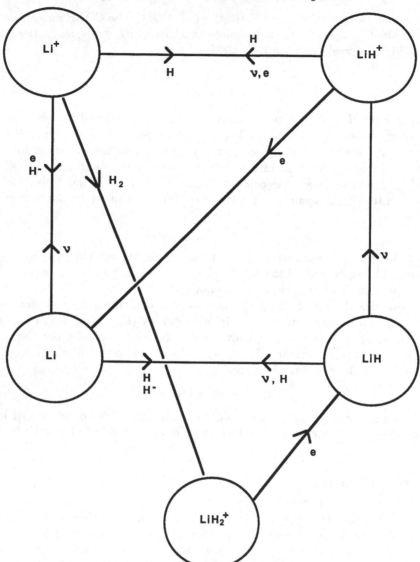

FIGURE 1. The chemistry of Li in the early Universe

Infrared H_2 emission has been detected in external galaxies. Table 1 contains the $2\text{-}1S/1\text{-}0S(1)$ intensity ratio for some starburst and Seyfert galaxies. The ratios tend to lie between the shock ratio of 0.13 and the ultraviolet fluorescence ratio of 0.54. A mix of fluorescence and shocks is a plausible interpretation but an alternative explanation is the fluorescence pumping of a dense gas (Sternberg & Dalgarno 1989; Burton, Hollenbach & Tielens 1990). As the density increases, the gas tends to thermal equilibrium, and the observed ratios can be reproduced by a gas density of about 10^5 cm^{-3}. Many observational data that have been interpreted to indicate shocked gas may in fact be dense gas subjected to intense ultraviolet radiation fields.

There is a third mechanism. A gas of molecular hydrogen subjected to X-rays will produce an infrared emission spectrum. The fast electrons released in photoionization excite electronic levels of H_2 which cascade into the rotation-vibration levels. The associ-

Galaxy	Type	Ratio
NCG 253	starburst	0.13 ± 0.06
IRAS 05189-2524	Sy 1.8	0.20 ± 0.26
NGC 3227	Sy 1.2	0.32 ± 0.12
NGC 3256	starburst	0.39 ± 0.06
NCG 6240	merger-starburst	0.22 ± 0.06

TABLE 1. The intensity ratio 2-1S/1-0S(1) in several galaxies[†]

[†] From Mouri (1994)

ated ultraviolet spectrum is essentially that observed (Feldman *et al.* 1993; Trafton *et al.* 1994) in Jovian auroras, which is also due to bombardment by fast electrons. Detailed calculations of the infrared emission spectrum have been carried out (Gredel & Dalgarno 1994). To distinguish the three mechanisms, additional diagnostic probes will be needed.

5. Shock chemistry

Slow shocks heat the gas but not so much that molecules are thermally dissociated. The heated gas has a quite different chemistry from that of cold molecular clouds. The chemistry is dominated by exothermic and endothermic reactions with H_2 such as

$$O + H_2 \rightleftarrows OH + H$$

$$OH + H_2 \rightleftarrows H_2O + H$$

The changes in composition are determined by the temperature of the shock-heated gas, and hence the shock velocity, and by the H/H_2 ratio, which together determine the fractions of oxygen that are taken up into OH and H_2O. The abundance of HCO^+ has been used as a diagnostic of a shocked gas, though some theoretical models predict it is enhanced and others that it is diminished. The production is enhanced by contributions from the sequence

$$C^+ + H_2 \rightarrow CH^+ + H$$
$$CH^+ + H_2 \rightarrow CH_2^+ + H$$
$$CH_2^+ + H_2 \rightarrow CH_3^+ + H$$

followed by

$$CH_2^+ + O \rightarrow HCO^+ + H$$
$$CH_3^+ + O \rightarrow HCO^+ + H_2$$

but the destruction depends on the H/H_2 ratio. If it is low so that the oxygen sequence ends up with much of the oxygen in H_2O, then HCO^+ is efficiently removed by

$$HCO^+ + H_2O \rightarrow H_3O^+ + CO$$

and the abundance of HCO^+ is decreased. If the H/H_2 is high, the oxygen ends up with OH and the abundance of HCO^+ is increased. However, the abundance of other species like SO should be increased by the sequence

$$S + H_2 \rightarrow SH + H$$
$$SH + O \rightarrow SO + H.$$

In fast shocks, the molecules are dissociated. The heated gas then cools and recombines and during the cooling phase the chemical composition differs from that of a gas heated by a non-dissociative shock, perhaps most markedly by the presence of a substantial population of molecular ions at a warm temperature (Hollenbach & McKee 1989; Neufeld & Dalgarno 1989). Indeed, Turner (1992, 1994), has argued that the presence of SO^+ in considerable abundance, which he has found through observation of its emission lines near 106 and 209 Ghz in the supernova remnant IC433, is a definitive indicator of a dissociative shock. It appears to be a means of distinguishing non-dissociative and dissociatrive shocks but a substantial abundance of SO^+ is also produced in photon-dominated or photodissociation regions (PDRs).

6. Photon-dominated regions

The intense infrared emission from H_2 arising from ultraviolet pumping has demonstrated the existence of regions of the interstellar medium in which gas is irradiated by intense photon fields. The thermal structure of such regions has been studied by Tielens & Hollenbach (1985) and Sternberg & Dalgarno (1989). There occurs a warm zone in which endothermic reactions can occur. The radiation maintains a high degree of ionization into a depth corresponding to a hydrogen column density of 10^{22} cm^{-2} in the case of S^+. The S^+ ions react with OH to form SO^+

$$S^+ + OH \rightarrow SO^+ + H$$

which is removed by dissociative recombination and photodissociation. A large equilibrium abundance of SO^+ results in the region with visual extinction less than two magnitudes (Sternberg & Dalgarno 1994). Hence the observation of abundant SO^+ is not in itself clear evidence for a shock. Other indicators such as velocity differentials and asymmetric line profiles must also be present. The simultaneous presence of an enhanced abundance of SO may indicate a shock and is not predicted for a PDR.

7. Supernova 1987a

Carbon monoxide was identified in the ejecta of supernova 1987a about 100 days after the explosion. It is formed mostly by the radiative association process

$$C + O \rightarrow CO + h\nu.$$

The removal of CO by the helium ion

$$He^+ + CO \rightarrow He + C^+ + O$$

is very efficient and calculations (Liu, Dalgarno & Lepp 1992) show that the observed amount of CO cannot be produced if the helium gas is mixed into the inner region where carbon and oxygen are both found. The early escape of X-rays and γ-rays from the ejecta had indicated that considerable mixing of the lighter gases into the inner region must have occurred. The discovery of a large abundance of CO shows conclusively that the mixing was macroscopic, not microscopic, with the parcels of helium gas retaining their integrity, as they penetrated into the interior.

8. Ages

Theoretical studies of the chemical composition of dense molecular clouds have shown that the measured abundances of complex molecules cannot be reproduced by steady-state models but that time-dependent models meet with some success (Herbst & Leung

1989). Different molecules behave differently with time. Because the time-scale for the formation of CO, where most of the carbon ultimately resides, is longer than the time-scale for formation of the hydrocarbons, their abundances pass through maxima at intermediate times. The nitrogen-containing molecules in contrast tend to increase in abundance monotonically. If this picture is correct, it suggests that chemical composition may provide a means of following the evolution of diffuse gas into a dense core in which star formation occurs. Suzuki *et al.* (1992) have compared measured abundances of C_3H_2, C_2S and NH_3. They suggest that the several sources in which C_2S is very abundant relative to NH_3 are cores in an early stage of evolution. The cores have no associated IRAS sources and no signs of star formation.

Acknowledgment

This work was supported by the National Science Foundation, Division of Astronomical Sciences, under grant AST 93-01099.

REFERENCES

BELLINI, M., DE NATALE, P., INGUSCIO, M., FINK, E., GALLI, D. & PALLA, F. 1994 *ApJ* **424**, 507

BLACK, J. H. & DALGARNO, A. 1976 *ApJ* **203**, 132

BLACK, J. H. & VAN DISHOECK, E. F. 1987 *ApJ* **322**, 412

BURTON, M., HOLLENBACH, D. & TIELENS, A. G. G. M. 1990 *ApJ* **365**, 620

DALGARNO, A. & LEPP, S. 1985 in *Astrochemistry* (ed. S. P. Tarafdar & M. P. Varshni) Dordrecht: Reidel, 109

DE BERNARDIS, P., DUBROVICH, V., ENCRENZAZ, P., MAOLI, R., MASTRANTONIO, G., MELCHIORRI, B., MELCHIORRI, F., SIGNORE, M. & TANZILLI, P. E. 1993 *A&A* **269**, 1

DINERSTEIN, H. L., LESTER, D. F., CARR, J. S. & HARVEY, P. M. 1988 *ApJ* **327**, L27

FELDMAN, P. D. MCGRATH, M. A., MOOS, W. H. DURRANCE, S. T., STROBEL, D. F. & DAVIDSEN, A. F. 1993 *ApJ* **406**, 279

GAUTIER, T. N., FINK, V., TREFFERS, R. R. & LARSON, H. P. 1976 *ApJ* **207**, L129

GREDEL, R. & DALGARNO, A. 1994 *ApJ* in press.

HERBST, E. & LEUNG, C. M. 1989 *ApJS* **69**, 271

HOLLENBACH, D. & MCKEE, C. F. 1989 *ApJ* **342**, 306

LEPP, S. & SHULL, M. 1984 *ApJ* **280**, 465

LIU, W., DALGARNO, A. & LEPP, S. 1992 *ApJ* **396**, 679

LUHMAN, M. L., JAFFE, D. T., KELLER, L. D. & PAK, S. 1994 *ApJL* in press

MAOLI, R., MELCHIORRI, F. & TOSTI, D. 1994 *ApJ* **425**, 372

MOURI, H. 1994 *ApJ* **427**, 777

PARMAR, P. S., LACY, J. H. & ACHTERMANN, J. M. 1994 *ApJ* **420**, 786

PUY, D., ALECIAN, G., LeBOURLOT, J., LE ORAT, J. & PINEAU DES FORÊTS, G. 1992 *A&A* **267**, 337

RAMSAY, S. K., CRYSOSTOMOU, A., GEBALLE, T. R., BRAND, P. W. J. & MOUNTAIN, M. 1993 *MNRAS* **263**, 695

ROTH, K. C., MEYER, D. M. & HAWKINS, I. 1993 *ApJ* **413**, L67

SHULL, J. M. & BECKWITH 1982 *ARA&A* **20**, 163

SIGNORE, M., VEDRENNE, G., DE BERNADIS, P., DUBROVICH, V., ENCRENAZ, P., MAOLI, R., MASI, S., MASTRANTONIO, G., MELCHIORRI, B., MELCHIORRI, F. & TANZILLI, P. E. 1994 *ApJS* **92**, 535

SMITH, M. D., BRAND, P. W. J. L. & MOORHOUSE, A. 1991 *MNRAS* **248**, 451

STERNBERG, A. 1988 *ApJ* **332**, 400

STERNBERG, A. & DALGARNO, A. 1994 *ApJS* in press

STERNBERG, A. & DALGARNO, A. 1989 *ApJ* **338**, 197

SUGAI, H., USUDA, T., KATAZA, H., TANAKA, M., INOUE, M. Y., KAWABATA, H., TAKAMI, H. AOKI, T. & HIROMOTO, N. 1994 *ApJ* **420**, 746

TAUBER, J. A., SNELL, R. L., DICKMAN, R. L. & ZIURYS, L. M. 1994 *ApJ* **421**, 570

TIELENS, A. A. G. M. & HOLLENBACH, D. 1985 *ApJ* **291**, 722 and 747

TRAFTON, L. M., GÉRARD, J. C., MUNHOVEN, G. & WAITE, J. H. 1994 *ApJ* **421**, 816

TURNER, B. E. 1992 *ApJ* **396**, L107

TURNER, B. E. 1994 *ApJ* **430**, 727

Abundance Determinations

By MANUEL PEIMBERT

Instituto de Astronomía, UNAM, Apdo. Postal 70-264, México 04510 D.F., México

A review of some aspects related to the abundance determinations of planetary nebulae, galactic H II regions and extragalactic H II regions is presented. The effect of the temperature structure of gaseous nebulae on the abundance determinations is explored. The relevance of abundance determinations to some aspects of the study of the evolution of stars, galaxies and the universe is discussed.

1. Introduction

From the study of the emission lines produced in galactic and extragalactic gaseous nebulae it has been possible to derive abundances of H, He, C, N, O, Ne, S and Ar. The chemical composition of these gaseous nebulae is needed to understand their physical conditions as well as their evolution. These abundances are also paramount to constrain evolutionary models of stars, galaxies and the universe.

Reviews and textbooks on the physical processes taking place in ionized nebulae have been presented by many astronomers, classic ones are those by Seaton (1960), Aller (1984) and Osterbrock (1989).

Some abundances have been determined based on detailed photoionization models while most abundances have been determined based on simple empirical methods. The input of a photoionization model consists of: a) a stellar radiation field, b) an electron density distribution, $N_e(\mathbf{r})$, (which defines the geometry of the nebula, c) a dust distribution, $N_d(\mathbf{r})$, and d) abundance distributions, which in most cases have been assumed homogeneous. The output consists of: a) a set of line intensities, b) the electron temperature distribution, $T_e(\mathbf{r})$, and c) the ionization structure.

The input of an empirical method consists of a set of observed line intensities (in some cases also some continuum intensities). The output consists of quantities averaged over the observed volumes like: a) $\langle T_e \rangle$, usually derived from the auroral to nebular intensity ratios of [O III] and [N II] lines, b) $\langle N_e \rangle$, usually derived from nebular intensity ratios of the [O II], [S II] and [Ar IV] lines, c) a set of average ionic abundances. Not all the ionization stages are observed, therefore to derive the total abundances, ionization correction factors, i_{cf}, have to be estimated based on photoionization models or on observations at different distances from the ionizing stars. Some of the i_{cf} are almost model independent.

There are discrepancies among different photoionization models of a given object. There are also discrepancies between results derived from photoionization models and empirical methods. Apparently, the main source of the discrepancy is due to the temperature structure of the gaseous nebulae.

In §2 I will discuss the temperature structure of gaseous nebulae and its relevance in abundance determinations. Some recent results on planetary nebulae, galactic H II regions and extragalactic H II regions are reviewed in §3, 4 and 5 respectively. The conclusions on the physical conditions of gaseous nebulae and on the constraints for the evolution of stars, galaxies and the universe are presented in §6.

2. Temperature structure

To a second order approximation we can characterize the temperature structure of a gaseous nebula by two parameters: the average temperature, T_0, and the root mean square temperature fluctuation, t, given by

$$T_0(N_e, N_i) = \frac{\int T_e(\mathbf{r}) N_e(\mathbf{r}) N_i(\mathbf{r}) dV}{\int N_i(\mathbf{r}) N_e(\mathbf{r}) dV}, \tag{2.1}$$

and

$$t^2 = \frac{\int (T_e - T_0)^2 N_e N_i dV}{T_0^2 \int N_e N_i dV}, \tag{2.2}$$

respectively, where N_e and N_i are the electron and the ion densities of the observed emission line and V is the observed volume (Peimbert 1967; Osterbrock 1970).

To determine T_0 and t^2 we need two different methods to derive T_e: one that weights preferentially the high temperature regions and one that weights preferentially the low temperature regions. For example the temperature derived from the ratio of the [O III] $\lambda\lambda$ 4363, 5007 lines, $T_e(4363/5007)$, and the temperature derived from the ratio of the Balmer continuum to $I(H\beta)$, $T_e(\mathrm{Bac}/H\beta)$, that are given by

$$T_e(4363/5007) = T_0 \left[1 + \frac{1}{2} \left(\frac{90800}{T_0} - 3 \right) t^2 \right], \tag{2.3}$$

and

$$T_e(\mathrm{Bac}/H\beta) = T_0 (1 - 1.70 t^2), \tag{2.4}$$

respectively. It is also possible to use the intensity ratio of a collisionally excited line of an element $p + 1$ times ionized to a recombination line of the same element p times ionized, this ratio is independent of the element abundance and depends only on the electron temperature. By combining this ratio with a temperature determined from the ratio of two collisionally excited lines like $T_e(4363/5007)$ it is also possible to derive T_0 and t^2.

To determine abundance ratios, $T_e(4363/5007)$ has been used very often under the assumption that $t^2 = 0.00$. In the presence of temperature variations however, the use of $T_e(4363/5007)$ yields O abundances that are smaller than the real ones (e.g. Peimbert 1967; Peimbert & Costero 1969). In general, abundance ratios derived from the ratio of a collisionally excited line to a recombination line are underestimated, while those derived from the ratio of two collisionally excited lines with similar excitation energies or from the ratio of two recombination lines, are almost independent of t^2. Nevertheless for some applications like the determination of the primordial helium abundance, which is based on the ratio of recombination lines, the errors introduced by adopting a $t^2 = 0.00$ value are very small but non-negligible.

From observations it has been found that $0.01 \le t^2 \le 0.09$, with typical values around 0.04; while from photoionization models of chemically and density homogeneous nebulae it has been found that $0.00 \le t^2 \le 0.02$, with typical values around 0.005. An explanation for the differences in the t^2 values predicted by photoionization models and those found by observations has to be sought.

3. Planetary nebulae

There have been many reviews on the abundances of PN (e.g. Pottasch 1984; Peimbert 1990; Clegg 1993; Aller 1994). Harrington et al. (1982) have found from photoionization models of PN t^2 values of 0.02 or less; in general, the lower the degree of ionization of the nebula the lower the t^2 value.

From the comparison of $T_e(\text{Bac}/H\beta)$ with $T_e(4363/5007)$, Peimbert (1971) obtained $\langle t^2 \rangle = 0.053$ for three PN. From a similar comparison Liu & Danziger (1993) obtained a large spread in t^2 values with a representative value of 0.03. Dinerstein, Lester & Werner (1985), determined T_0, N_e and t^2 for six PN from the [O III] lines at $\lambda\lambda$ 4363, 5007, 52 μm and 88 μm and found $\langle t^2 \rangle = 0.04$.

Under the assumption that $t^2 = 0.00$ the C^{++} abundances derived from the $3d^2D$ - $4f^2F^0$ $\lambda4267$ recombination line of C^+ are in general higher than those derived from the [C III] $3s^2\ ^1S_0$ - $3s3p^3\ P_2$ $\lambda1907$ and C III] $3s^2\ ^1S_0$ - $3s3p^3\ P_1$ $\lambda1909$ collisionally excited lines, the difference can be as high as an order of magnitude. General discussions of this problem have been given in the literature (e.g. Torres-Peimbert, Peimbert & Daltabuit 1980; Barker 1982; French 1983; Kaler 1986; Peimbert 1989; Rola & Stasinska 1994). Three different ideas have been advanced to explain these differences: a) the presence of an unknown mechanism, like stellar continuum resonance fluorescence, strengthening λ 4267, however no evidence for the presence of an unknown mechanism has been found, b) an overestimation of the weak lines, when this occurs, in addition to the overestimation of λ 4267, the overestimate of the weak auroral lines produces a higher electron temperature that combined with the strong $\lambda\lambda$ 1907, 1909 lines yields a low C^{++} abundance increasing the discrepancy between both determinations; this effect should not be present in observations of high quality, and c) the presence of spatial temperature variations.

Peimbert, Storey & Torres-Peimbert (1993) based on O^+ recombination lines of multiplets 1, 2 and 10 derived for NGC 6572 an O^{++}/H^+ abundance ratio a factor of 1.55 higher than that derived from the [O III] $\lambda\lambda$ 4363 and 5007 lines, this result corresponds to $t^2 = 0.040 \pm 0.025$; the recombination coefficients were computed by Storey (1994) in LS coupling. Liu et al. (1994) based on O^+ recombination lines of multiplets 1 and 2 in LS coupling and multiplets 10, 12, 19 and 20 in intermediate coupling derived for NGC 7009 an O^{++}/H^+ ratio a factor of 4.7 higher than that derived from forbidden lines under the assumption of $t^2 = 0.00$; to reconcile both O^{++}/H^+ values, a $t^2 = 0.098$ is needed. Similarly Liu et al. (1994) based on recombination lines derived abundances for C^{++}, C^{+++}, N^{++} and N^{+++} that give C/H and N/H ratios a factor of 6.1 and 4.1 times larger than those derived from collisionally excited lines implying t^2 values of 0.086 and 0.072, respectively. Liu et al. (1995) derived for a group of six PN an $\langle O/H \rangle$ ratio which is a factor of 2.2 higher than the one derived from forbidden lines, this result corresponds to $\langle t^2 \rangle \sim 0.05$; the $\langle O/H \rangle$ value is in excellent agreement with the solar photospheric value. Kingsburgh & López (1995) from O^+ recombination lines find for NGC 6543 an O^{++}/H^+ ratio which is a factor of three higher than the one derived from collisionally excited lines, which implies that $t^2 = 0.08$ and an O/H ratio two times higher than the solar one.

There are three possible explanations for the large t^2 values observed: a) erroneous $T_e(4363/5007)$ and t^2 determinations due to the presence of high density clumps (Viegas & Clegg 1994), b) the presence of chemical abundance inhomogeneities (Peimbert 1989) and c) the deposition of kinetic energy by shocks or by subsonic turbulence produced by mass loss from the central star.

Clumps with $N_e \geq 10^6 \text{cm}^{-3}$ embedded in a lower density medium with $N_e \sim 10^4 \text{cm}^{-3}$, are needed for the first explanation. This possibility might work for some objects but not for all. A two orders of magnitude increase in the density is likely to produce a drop in the ionization degree and a high [O II] $I(7320 + 7330)/I(3726 + 3729)$ ratio, which at least for NGC 7009 is not observed (Peimbert 1971). Moreover if this effect is present, it will also produce real temperature variations because the heating and cooling of these clumps will be different to that of the lower density medium.

There are some objects with C/H rich clumps like A30, A78 and NGC 4361 (e.g. Jacoby & Ford 1983; Torres-Peimbert, Peimbert & Peña 1990). For NGC 4361 Torres-Peimbert et al. have been able to compute a photoionization model with a large t^2 value that reconciles the C abundances derived from recombination and collisionally excited lines.

Probably the most important explanation is the third one. The central stars of PN are loosing mass at velocities in the 1000 to 4000 km s^{-1} range. This stellar wind interacts with material that was ejected previously at about 20 km s^{-1} producing shocks (e.g. Kwok et al. 1978; Kahn 1989; Balick 1989, 1994; Cuesta, Phillips & Mampaso 1994).

Most PN show expansion velocities of the shell in the 20 to 30 km s^{-1} range. Type I PN show components, including outer lobes, expanding with velocities in the 70 to 800 km s^{-1} range (e.g. Webster 1978; Meaburn & Walsh 1980; Sabbadin, Capellaro & Turatto 1987, Weller & Heathcote 1989; López et al. 1991). Middlemass, Clegg & Walsh (1989, 1991) and Manchado & Pottasch (1989) have reported higher T_e and lower O/H values for large faint halos relative to the brighter central part of several PN. Middlemass et al. interpret these results in terms of shocked filamentary regions, with the energy input coming from the winds produced by the central stars. The O/H differences could be due to different t^2 values and not to real O/H differences.

Rowlands, Houck & Herter (1994) find that the highly ionized regions of NGC 6302 and NGC 6537 show higher T_e values than those predicted by photoionization models, therefore they conclude that the difference is due to mechanical energy input produced by shocks. Balick et al. (1994) find fast low ionization emission regions, FLIERS, in several PN, among them NGC 7009; FLIERS show the ionization structure expected of bow shocks. Bohigas (1994) finds from line intensity ratios that there are regions of NGC 6302 dominated by shocks and regions dominated by photoionization.

Peimbert & Torres-Peimbert (1987a) find from λ 4267 a C^{++}/H$^+$ ratio which is a factor of seven higher than that derived by Dufour (1984) for NGC 2818, a type I PN. This difference, again, probably is due to the adoption of $t^2 = 0.00$, and implies that t^2 for this object is large. Similar differences are found for other type I PN.

If the N/O abundance ratio is derived from the [N II]/[O II] line ratio and the O abundance from collisionally excited lines under the assumption that $t^2 = 0.00$, the presence of large temperature variations would go in the direction of lowering O/H and increasing N/O. Therefore the low O/H values and the N/O versus O/H anticorrelation present in a group of type I PN, derived under the assumption that $t^2 = 0.00$ (Peimbert & Torres-Peimbert 1983; Peimbert 1990 and references therein), could be due to increasing t^2 values with decreasing O/H, without the need of invoking contamination of the nebulae due to ON cycling in the central stars.

3.1. *The helium abundance and temperature variations*

The $N(\mathrm{He}^+)/N(\mathrm{H}^+)$ ratios can be derived from equations of the type

$$\frac{N(\mathrm{He}^+)}{N(\mathrm{H}^+)} = \frac{\alpha(\mathrm{H}^0, H\beta)}{\alpha(\mathrm{He}^0, \lambda_{\mathrm{nm}})} \frac{\lambda_{\mathrm{nm}}}{4861} \frac{I(\lambda_{\mathrm{nm}})_R}{I(H\beta)}, \tag{3.5}$$

where the effective recombination coefficients, α, for hydrogen and helium have been computed by Hummer & Storey (1987) and Smits (1994), and $I(\lambda_{nm})_R$ is the pure recombination intensity that has to be obtained from the observed intensity, $I(\lambda_{nm})$. Radiative transfer effects and collisions from the 2^3S level affect $I(\lambda_{nm})$ and have to be estimated.

The collisions to recombinations ratio of a helium line is given by

$$\frac{I(\lambda_{nm})_C}{I(\lambda_{nm})_R} = \frac{N(2^3S)}{N(He^+)} \frac{\kappa(\lambda_{nm})}{\alpha(\lambda_{nm})}, \tag{3.6}$$

where κ is the effective collisional coefficient that depends strongly on T_e and

$$\frac{N(2^3S)}{N(He^+)} = \frac{5.62 \times 10^{-6} t_4^{-1.19}}{1 + 3130 t_4^{-0.5} N_e^{-1}}, \tag{3.7}$$

where ionizations from the 2^3S level have been neglected (Kingdon & Ferland 1995).

The latest estimates of the $I(\lambda_{nm})_C/I(\lambda_{nm})_R$ values for the different helium lines are those by Kingdon and Ferland (1995) based on the 29–state ab initio computation for collisions to He^0 states with $n \leq 5$ by Sawey & Berrington (1993) and the helium recombination coefficients by Smits (1994).

Robbins (1968) and Robbins & Bernat (1973) have computed the effect that atomic absorption has on the He I line intensity ratios. Robbins has used as a parameter the He I λ 3889 optical depth, $\tau(3889)$, for the triplet series. From the computations by Robbins and Cox & Daltabuit (1971) and the ratio of two He I lines it is possible to determine $\tau(3889)$ and consequently the effect of the radiation transfer on the triplet lines. A similar procedure can be followed for the singlet lines. It is found that the radiation transfer effect is almost negligible for $\lambda\lambda$ 4472, 5876 and 6678; alternatively it is large for $\lambda\lambda$ 3889, 7065 and 10830.

The He^+/H^+ values derived from different helium lines, based on equations (3.5) and (3.6) for $t^2 = 0.00$, do not agree for a given object, particularly for those PN with high N_e and T_e values (e.g. Peimbert & Torres-Peimbert 1987a, b; Peña et al. 1995). The differences imply that the collisional effects have been overestimated. This problem has at least four solutions: a) the He line intensities have not been properly measured, b) there is an unknown process depopulating the 2^3S level, c) the density has been overestimated (see equation 3.7), d) the temperature has been overestimated, i.e. $t^2 \neq 0.00$.

Even if λ 10830 is affected by telluric absorption (Kingdon & Ferland 1991), I consider that possibility a) above plays a minor role in well observed objects. Possibility b) suggested by Peimbert & Torres-Peimbert (1987a, b) has been studied by Clegg and Harrington (1989) who find that photoionization can reduce the $N(2^3S)$ population by as much as 25% in compact optically-thick PN; alternatively for the vast majority of the observed PN and for giant extragalactic H II regions the effect is very small and can be neglected. Possibility c) could be important for objects with $N_e \leq 3000$ cm^{-3}, but for PN with $N_e \gg 3000$ cm^{-3} is not important (see equation 3.7). Finally, possibility d) will be explored further.

In Figure 1 we present the He^+/H^+ abundances for the type I PN Hu 1-2 (Peimbert, Luridiana & Torres-Peimbert 1995b) based on three He I lines that are almost unaffected by radiative transfer effects. The observations correspond to the average of three different regions of the nebula. The temperature at which the three lines reach the same He^+/H^+ ratio is about 13 000 K, considerably smaller than that given by $\langle T_e \rangle$ (4363/5007) that amounts to 18 800 $\pm$ 600 K; this result implies a very large t^2 value. The density for the observed regions of Hu 1-2, $\langle N_e \rangle = 4\,900$ cm^{-3}, is higher than the critical density and errors in N_e possibly do not play a role in explaining the discrepancies in the He^+/H^+

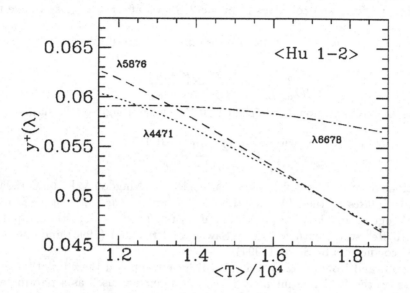

FIGURE 1. $N(\mathrm{He}^+)/N(\mathrm{H}^+) = y^+(\lambda)$ versus $\langle T_e \rangle$ diagram for the type I PN Hu 1-2, where $\langle T_e \rangle$
stands for the average of three different regions of the nebula (Peimbert et al. 1995b).

determinations. Peña et al. (1995) from a similar study of N66 also find that lower T_e
and N_e values than those given by [O III], [O II] and [Ar IV] lines are needed to derive
the same $\mathrm{He}^+/\mathrm{H}^+$ abundances from the $\lambda\lambda$ 4472, 5876 and 6678 lines.

The $I(3889)/I(4472)$, $I(7065)/(4472)$ and $I(10830)/(4472)$ ratios depend on $\tau(3889)$
and T_e. The T_e affects weakly the recombination coefficients but strongly the collisional
excitation effects from the $2^3\mathrm{S}$ level. The relationship between $\tau(3889)$ and T_e for any
line ratio is derived by comparing the observations with the computations by Robbins
(1968). The three line ratios depend on different functions of $\tau(3889)$ and T_e, therefore
the combination of two line ratios will provide us with a unique pair of $\tau(3889)$ and T_e
values.

In Figure 2 we present a $\tau(3889)$ versus T_e diagram for NGC 7009 (Peimbert et al.
1995b) where we have adopted $N_e = 6\,000$ cm^{-3}. From the $I(3889)/I(7065)$, $I(3889)$
$/I(10830)$ and $I(7065)/I(10\,830)$ crossings we obtain T_e values of 8 000 K, 6 700 K and
6 300 K respectively. For this object $T_e(4363)/(5007)$ is equal to 10 000 K, the differences
between $T_e(4363)/(5007)$ and the crossing temperatures are mainly due to the t^2 value
(which is similar to that derived by Liu et al. 1994); while the smaller T_e values derived
from the two $I(10830)$ crossings relative to that derived from the $I(3889)/I(7065)$ crossing
probably is due to telluric absorption and dust destruction inside NGC 7009 of λ 10830
photons (Clegg & Harrington 1989; Kingdon & Ferland 1991, 1993).

4. Galactic H II regions

Recent reviews on abundances of galactic H II regions are those by Peimbert (1993)
and Wilson & Rood (1994).

Photoionization models with uniform chemical composition predict t^2 values around

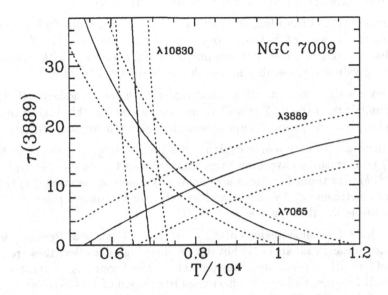

FIGURE 2. $\tau(3889)$ versus T_e diagram for NGC 7009. The solid lines stand for the $I(\lambda)/I(4472)$ ratio, the dotted lines to the right and to the left at a given $\tau(3889)$ correspond to ratios 10% higher and 10% lower than observed, respectively (Peimbert et al. 1995b).

Object	t^2(Bac/FL)	t^2(C^{++}R/C)	t^2(O^{++}R/C)
Orion	0.047 ± 0.015 (1,2)	0.055 ± 0.013 (6)	0.038 ± 0.011 (7)
Orion	0.024 ± 0.012 (2,3)	0.053 ± 0.013 (7)	0.050 ± 0.010 (9)
M8	0.034 ± 0.017 (4)	0.046 ± 0.010 (8)	$\ldots$
M17	0.041 ± 0.019 (5)	$\ldots$	0.037 ± 0.017 (7)

TABLE 1. Mean temperature variations in galactic H II regions. (1) Peimbert 1967; (2) Peimbert & Torres-Peimbert 1977; (3) Simpson 1973; (4) Sánchez & Peimbert 1991; (5) Peimbert et al. 1992; (6) Walter, Dufour & Hester 1992; (7) Peimbert et al. 1993a; (8) Peimbert et al. 1993b; (9) Peimbert et al. 1995a.

0.01 (e.g. Garnett 1992; Gruenwald & Viegas 1992). Baldwin et al. (1991) produced a photoionization model for the Orion nebula, in which grains and gas are well mixed, that included photoelectric heating and cooling of the gas by grain ionization and grain collisions respectively; for this model Peimbert et al. (1993a) find that $t^2 = 0.004$.

In Table 1 I present observational t^2 values for galactic H II regions of the solar vicinity based on three methods: *a*) the combination of $T_e(\text{Bac}/H\beta)$ and $T_e(4363/5007)$ (see equations 2.3 and 2.4), *b*) the condition that $N(\text{C}^{++}$ recombination$)/N(\text{C}^{++}$, collisions$)$ is equal to one; *c*) the condition that $N(\text{O}^{++}$ recombination$)/N(\text{O}^{++}$, collisions$)$ is equal to one. In addition to the results presented in Table 1, by combining the observations at high radiofrequencies (H56α, H66α) and the [O III] observations at 88.35 μm with

$T_e(4363/5007)$, the following values of t^2 are obtained for the Orion nebula: 0.037, 0.011 and 0.035 (Berulis, Smirnov & Sorochenko 1975; Wilson, Bieging & Wilson 1979; Moorwood et al. 1978; Torres-Peimbert, Peimbert & Daltabuit 1980).

Walter & Dufour (1994) from longslit spectra of the Orion nebula find a decrease in $T_e(\text{Bac})$ moving to the west of θ^1 Ori C, from 8400 K at a distance of $40''$ to a low of 2800 K at a distance of $220''$; this decrease implies an increase in t^2 with distance from θ^1 Ori C reaching values considerably higher than any reported to date.

From the previous discussion it follows that typical t^2 values in galactic H II regions of the solar vicinity are in the 0.03 to 0.05 range, values considerably higher than those predicted by photoionized chemical homogeneous models with constant density.

As for PN there are three possible explanations for the large t^2 values: a) that the $T_e(4363/5007)$ temperatures have been overestimated due to clumps and regions with $N_e \sim 10^6$ cm^{-3}, b) that the H II regions are chemically inhomogeneous and c) that there is deposition of kinetic energy by shocks and by subsonic turbulence produced by mass loss from stars inside the H II regions.

The Orion nebula does show regions of $N_e \sim 10^6$ cm^{-3} (Bautista, Pradhan & Osterbrock 1994; Bautista & Pradhan 1994) but due to their high densities, these regions are of low degree of ionization, probably in the partially ionized zone that separates the H II region from the H I region, and do not affect the distribution of $T_e(4363/5007)$. Also the determination of $N_e(3727/7325)$ indicates the presence of density fluctuations but the fraction of the emission due to regions with $N_e \geq 10^5$cm^{-3} is negligible (e.g. Peimbert & Torres-Peimbert 1977).

Cunha & Lambert (1992) and Cunha (1993) have divided the B stars of the Orion association into four subgroups according to age. They have found that the younger the subgroup the higher the O/H ratio and that the spread in the O/H ratio in one age group reaches 0.5 dex. Inhomogeneities of O/H in the 0.3 dex to 0.5 dex range in a given nebula can produce t^2 values in the 0.03 to 0.04 range.

There is a large body of evidence for the presence of high velocity components and shock waves inside shell H II regions (e.g. Dufour 1989 and references therein) and inside typical H II regions (e.g. Castañeda 1988; Clayton 1990; Massey & Meaburn 1993). These high velocity components are due to stellar winds from massive stars embedded in the nebulae.

Temperature fluctuations also play a paramount role in the determination of $\Delta Y/\Delta Z$. M17 is the H II region of the solar vicinity with the best He/H determination (Peimbert et al. 1992). This determination combined with a $t^2 = 0.04$ and the pregalactic helium abundance, Y_p, of 0.230 derived from O–poor extragalactic H II regions (e.g. Pagel et al. 1992, Pagel 1995) yields $\Delta Y/\Delta Z = 2.5 \pm 0.5$. From M17, $Y_p = 0.230$ and t^2 0.00 a $\Delta Y/\Delta Z = 4.59$ is derived.

The large $\Delta Y/\Delta Z$ and the small O/H values for $t^2 = 0.00$ found for O–poor H II regions led Maeder (1992, 1993) to suggest that there is a mass, $m(BH)$, above which massive stars would end their evolution without enriching the ISM at the time of the SN explosion. Carigi (1994) and Peimbert, Sarmiento & Colín (1994b), based on the stellar evolution models by Maeder (1992) find that it is possible to explain the $\Delta Y/\Delta Z$ value of 2.5 ± 0.5 ($t^2 = 0.04$) for the solar vicinity without the need to suppress the Z enrichment of massive stars at the end of their evolution. Alternatively for $t^2 = 0.00$, the mechanism suggested by Maeder becomes necessary to explain the $\Delta Y/\Delta Z$ ratio but would enter in contradiction with observational constraints such as $\Delta C/\Delta O$.

5. Extragalactic H II regions

Recent reviews on abundances derived from extragalactic H II regions are those by Shields (1990) and Dinerstein (1990). Sets of photoionization models that can be used for the study of extragalactic H II regions have been presented by Stasinska (1990) and Gruenwald & Viegas (1992).

As for PN and galactic H II regions there is plenty of evidence for the presence of mechanical energy input in addition to photoionization in giant extragalactic H II regions.

Many giant H II regions show line profiles that indicate supersonic velocities (e.g. Meaburn 1984; Iye, Ulrich & Peimbert 1987; Castañeda, Vilchez & Copetti 1990a, b). Moreover probably all giant extragalactic H II regions present supersonic velocities since Chu and Kennicutt (1994) find that all the LMC H II regions observed by them –30– present supersonic velocity dispersions; the most spectacular structures in 30 Doradus are several fast expandind shells which according to them appear to be produced, at least partially, by SNR. These supersonic velocities should be produced by stellar activity inside H II regions.

There are plenty of arguments in favor of exploding supernovae inside H II regions and of winds produced by O stars and WR stars. Five out of eleven known supernovae of Type Ib exploded in H II regions (Wheeler, Harkness & Capellaro 1987). Five out of eight oxygen-rich supernova remnants are found inside H II regions (van den Bergh 1988). Peimbert, Sarmiento & Fierro (1991) estimate that half of the O stars will explode as SN inside H II regions. Rosa & Mathis (1987) have found an O/H rich region inside 30 Doradus that they attribute to mass loss from a massive star inside the nebula.

The difficulty of measuring $I(\lambda 4363)$ (or any other direct temperature indicator) led Pagel et al. (1979) to propose an empirical method based on the ratio of the nebular oxygen lines to $H\beta$, $R_{23} \equiv ([\text{O II}]\lambda 3727 + [\text{O III}]\lambda\lambda 4959, 5007)/H\beta$, to determine T_e and the O/H ratio. There are significant differences between the calibration of Pagel's method to determine O/H ratios for extragalactic H II regions based on models (McCall, Rybski & Shields 1985; Dopita & Evans 1986; McGaugh 1991) and the calibrations based on observations (Edmunds & Pagel 1984; Torres-Peimbert et al. 1989). The differences are in the 0.2 dex to 0.4 dex range and could be mainly due to the presence of temperature inhomogeneities over the observed volume (Campbell 1988; Torres-Peimbert et al. 1989; McGaugh 1991). While the calibration based on models adjusts the observed R_{23} values with those derived from photoionization models (with $\langle t^2 \rangle \sim 0.005$), the calibration based on observations adjusts the observed R_{23} values with the abundances derived from $T_e(4363/5007)$ under the assumption that $t^2 = 0.00$. Peimbert et al. (1991) have found that by adding a supernova remnant to an H II region ionized by 100 O stars the $([\text{O III}]\lambda 4363)/H\beta$ ratio varies by a large factor while the R_{23} ratio is almost unaffected. From this result it follows that in the presence of shock waves Pagel's method is an excellent tool to derive the O/H ratio as long as it is calibrated with the R_{23} ratio predicted by photoionization models and not with the observed $T_e(4363/5007)$ value under the assumption that $t^2 = 0.00$.

Pagel et al. (1992), based on abundances derived from $T_e(4363/5007)$ under the assumption that $t^2 = 0.00$, found that H II regions with WR features show a larger scatter in the N versus Y diagram than those without WR features, they suggested that the scatter could be due to Y and N selfenrichment. Esteban and Peimbert (1995) from modelling the chemical evolution of H II regions due to selfenrichment, based on the stellar evolution modls by Maeder (1992), find that the Y and N enrichment produce displacements in the N versus Y diagram that are parallel to the fit, obtained by Pagel et al., for the H II regions without WR features. Esteban & Peimbert suggest that the

Y and N overabundances might be illusory and could be due to higher t^2 values than for H II regions without WR features.

From a selected group of well observed H II regions, where those with WR features were omitted, Pagel et al. (1992) found $\Delta Y/\Delta O = 10.2 \pm 3.5$ by assuming that: *a*) $t^2 = 0.00$, *b*) no O trapped in dust grains and *c*) no deviations from case B of the hydrogen lines. Pagel et al. recomended a $\Delta Y/\Delta O \sim 6.7$ by considering: *a*) an increase of 0.04 dex due to $t^2 \neq 0.00$, *b*) 0.08 dex of O trapped in dust grains and *c*) possible deviations of the H lines from case B for the metal richer H II regions.

Carigi et al. (1994) from a sample of 10 well observed extragalactic H II regions derived $\Delta Y/\Delta O = 7.1 \pm 1.6$ for $t^2 = 0.00$ without considering the amount of O in dust grains nor deviations from case B. The main difference in the $\Delta Y/\Delta O$ values between Carigi et al. and Pagel et al. (1992) is due to the use by Carigi et al. of recent results for the two metal poorest objects in their sample: IZw 18 and UGC 4483 (Skillman & Kennicutt 1993; Skillman et al. 1994). Carigi et al. derived $\Delta Y/\Delta O = 4.5 \pm 1.0$ for $t^2 = 0.035$ and adopting an increase of 0.04 dex in O due to the fraction trapped in dust grains.

To explain the large $\Delta Y/\Delta O$ ratios derived from samples of galaxies it has been proposed that a galactic outflow of O–rich material is present (Aparicio et al. 1988; Lequeux 1989; Pilyugin 1993; Tosi 1994; Peimbert et al. 1994a). This outflow is characterized by γ, the fraction of the mass ejected by supernovae to the intergalactic medium without mixing with the interstellar gas.

Carigi et al. (1994) from chemical evolution models of irregular galaxies are able to fit the $\Delta Y/\Delta O(t^2 = 0.035)$, C/O and $(Z$–C–O)/O observational restrictions for $\gamma = 0.23$; for larger values of γ they can not adjust the C/O and $(Z$-C-O)/O constraints while for smaller values of γ they can not adjust the $\Delta Y/\Delta O$ ratio. The C/O and $(Z$–C–O)/O ratios are almost independent of t^2 while $\Delta Y/\Delta O$ depends strongly on t^2. Larger values of t^2 will permit to adjust all the observational constraints for a smaller value of γ.

An additional argument in favor of large t^2 values of O–poor H II regions is provided by González-Delgado et al. (1995) who find a considerable smaller T_e from the Paschen continuum to Hα ratio than $T_e(4363/5007)$.

The most spectacular result of the study of abundances derived from gaseous nebulae is the determination of the primordial or pregalactic helium abundance, Y_p, that has cosmological implications (e.g. Peimbert & Torres-Peimbert 1974, 1976; Boegsgaard & Steigman 1985; Pagel et al. 1992; Pagel 1995 and references therein). Y_p is derived from the intensity ratios of He to H recombination lines in O–poor extragalactic H II regions assuming that

$$Y = Y_p + O(\Delta Y/\Delta O), \qquad (5.8)$$

where Y and O are the helium and oxygen mass fractions of each H II region and $\Delta Y/\Delta O$ –assumed to be constant– is the ratio of the Y and O enrichments of the interstellar medium after the galaxies were formed. The H and He recombination line intensities to a first approximation are inversely proportional to T_e, to a second approximation they show small differences in their T_e dependence (e.g. Hummer & Storey 1987; Smits 1994). The differences are such that for $t^2 = 0.04$ the Y_p value becomes about 0.003 smaller than for $t^2 = 0.00$. Alternatively for those objects for which collisional excitations from the 2^3S level are considerable a $t^2 \neq 0.00$ will increase the derived Y value (see Figure 1). The densities of most extragalactic H II regions are typically around 100 cm^{-3}, considerably smaller than the critical density (see equation 3.7); consequently the collisional effects are practically negligible for these objects, therefore $t^2 \neq 0.00$ will reduce the derived Y_p value.

6. Conclusions

The best determinations of abundance ratios are those that do not depend, or depend weakly, on T_e. Therefore some of the best determinations are those derived from the ratio of two recombination lines because, to a very good approximation, the temperature dependence of the recombination lines is very similar. Unfortunately, with the exception of the He lines, the brighter recombination lines of the most abundant elements – C, N and O – are typically about three orders of magnitude fainter than $H\beta$ in objects with solar abundances. Consequently only a small number of high quality determinations has been carried out based on recombination lines of C, N and O. Therefore, for less abundant elements, metal poor gaseous nebulae or faint gaseous nebulae, one has to rely on collisionally excited lines to derive abundances relative to H.

The intensity of a collisionally excited line depends on the Boltzmann factor, $-\Delta E/kT_e$, where ΔE is the energy difference between the excited level from which the line originates and the ground level. For UV lines and visual auroral lines, ΔE is in the $4 - 13$ eV range, for visual nebular lines in the $1.3 - 4$ eV range, and for far IR lines in the $0 - 1$ eV range. Since in gaseous nebulae $kT_e \sim 1$ eV, the UV and auroral lines depend strongly on T_e, the visual nebular lines depend moderately on T_e, and the IR lines are almost independent of T_e (but often their ratio relative to a H line depends strongly on N_e, see Simpson et al. 1995 and references therein). This dependence, in the presence of temperature variations over the observed volumes, produces significant errors in the abundance determinations if the nebulae are assumed to be isothermal.

More high quality abundance determinations derived from recombination lines are needed to advance in our knowledge of the chemical composition of gaseous nebulae and to estimate the effect of the temperature structure on the abundances derived from collisionally excited lines.

The photoionization models of gaseous nebulae have to be modified to include deposition of mechanical energy. Moreover further exploration of photoionization models with inhomogeneous density and chemical abundances should be carried out. Empirical abundance determination methods based on the intensity of nebular lines should be calibrated by considering objects with t^2 determinations; or if there are no suitable calibrators with t^2 determinations, the method should be calibrated with photoionization models by matching the nebular lines instead of the auroral to nebular line ratios.

From H II regions with good determinations of t^2 values it follows that to adopt a $t^2 \approx 0.04$ is a reasonable approximation to use for those objects without a high quality t^2 determination. PN show a wider range of t^2 values than H II regions; typical values for PN are in the $0.03 \leq t^2 \leq 0.06$ range.

Some results due to the adoption of abundances derived with $t^2 \neq 0.00$ follow:

a) O/H values for disk PN are similar to those of the sun removing the difference between the abundances of young stars and type II PN (those that have kinematic properties typical of intermediate population I stars). There is still the standing problem of NGC 6543 and NGC 7009 that show O/H abundances two times higher than solar; more abundance determinations, based on recombination lines, are needed to find out which is the fraction of O/H rich PN and to explain their presence in the context of stellar or galactic chemical evolution.

b) The O/H values of solar vicinity H II regions become similar to those of the sun and young stars of the solar vicinity, removing a long standing discrepancy.

c) The discrepancy among the He^+/H^+ abundance ratios derived from different He I lines of PN with high N_e and $T_e(4363/5007)$ values, after considering the effects of collisional excitation and absorption from the 2^3S He I level, dissapears.

d) By determining the abundances of type I PN considering their high t^2 values, the O/H ratios become similar to those of stars recently formed, their N/O ratios become smaller and there is no need to invoke the presence of ON cycling products in the nebular shells.

e) The $\Delta Y/\Delta Z$ ratio derived from M17 and Y_p can be explained by chemical evolution models of the solar vicinity without the need to invoke a mass cut-off above which massive stars do not enrich the interstellar medium with heavy elements during the SN event.

f) The scatter in the Y versus N diagram present in O–poor extragalactic H II regions with WR features could be due to different t^2 values for each object and not to He and N pollution by WR stars.

g) The smaller $\Delta Y/\Delta O$ ratios are in better agreement with chemical evolution models of irregular galaxies that fit other observational constraints and imply that O–rich galactic outflows are not as important as previously thought.

h) For those H II regions and PN where collisional excitations from the 2^3S level are thought to be important a $t^2 \neq 0.00$ will decrease the contribution to the He line intensities due to collisions and a higher Y value will be derived. Alternatively for O–poor extragalactic H II regions with negligible collisional effects from the 2^3S level of He I (objects with low N_e), a $t^2 = 0.04$ affects the Y_p determination in the sense of reducing Y_p by about 0.003.

It is a pleasure to aknowledge important and constructive suggestions on this subject with Pedro Colín, César Esteban, Valentina Luridiana, Miriam Peña, Silvia Torres-Peimbert & Antonio Sarmiento. I also like to thank the staff of the Space Telescope Science Institute, and in particular Bob Williams and Mario Livio, for their invitation to participate in this conference and for their warm hospitality. Over the years I have had the privilege of enjoying inspiring and fruitful academic relationships with Don Osterbrock and Mike Seaton.

REFERENCES

ALLER, L.H. 1984. *Physics of Thermal Gaseous Nebulae*. Reidel.

ALLER, L.H. 1994 *Ap. J.* **432**, 427.

APARICIO, A., GARCÍA-PELAYO, J.M. & MOLES, M. 1988 *A. & A. S. S.* **74**, 375.

BALDWIN, J.A., FERLAND, G.J., MARTIN, P.G., CORBIN, M.R., COTA, S.A., PETERSON, B.M. & SLETTEBAK, A. 1991 *Ap. J.* **374**, 580.

BALICK, B. 1989. In *Planetary Nebulae*, IAU Symposium 131 (ed. S. Torres-Peimbert) p. 83. Kluwer.

BALICK, B. 1994 *Ap. & S. S.* **216**, 13.

BALICK, B., PERINOTTO, M., MACCIONI, A., TERZIAN, Y., HAJIAN, A. 1994 *Ap. J.* **424**, 800.

BARKER, T. 1982 *Ap. J.* **253**, 167.

BAUTISTA, M.A., PRADHAN, A.K. & OSTERBROCK, D.E. 1994 *Ap. J.* **432**, L135.

BAUTISTA, M.A. & PRADHAN, A.K. 1994 *Ap. J.* Submitted.

BERULIS, J.J., SMIRNOV, G.T. & SOROCHENKO, R.L. 1975 *Soviet Ap. Letters* **1**, 9, 28.

BOESGAARD, A.M. & STEIGMAN, G. 1985 *Ann. Rev. Astr. Ap.* **23**, 319.

BOHIGAS, J. 1994 *A. & A.* **288**, 617.

CAMPBELL, A. 1988 *Ap. J.* **335**, 644.

CARIGI, L. 1994 *Ap. J.* **424**, 181.

CARIGI, L., COLÍN, P., PEIMBERT, M. & SARMIENTO, A. 1994 *Ap. J.* Submitted.

CASTAÑEDA, H.O. 1988 *Ap. J. S.* **67**, 93.

CASTAÑEDA, H.O., VÍLCHEZ, J.M. & COPETTI, M.V.F. 1990a *Rev. Mexicana Astron. Astrof.* **21**, 231.

CASTAÑEDA, H.O., VÍLCHEZ, J.M. & COPETTI, M.V.F. 1990b *Ap. J.* **365**, 164.

CHU, Y.H. & KENNICUTT, R.C. 1994 *Ap. & S.S.* **216**, 253.

CLAYTON, C.A. 1990 *M.N.R.A.S.* **246**, 712.

CLEGG, R.E.S. 1993. In *Planetary Nebulae*, IAU Symp. 155, (ed. R. Weinberger & A. Acker), p. 549. Kluwer.

CLEGG, R.E.S. & HARRINGTON, J.P. 1989 *M.N.R.A.S.* **239**, 869.

COX, D.P. & DALTABUIT, E. 1971 *Ap. J.* **167**, 257.

CUESTA, L., PHILLIPS, J.P. & MAMPASO, A. 1994 *Ap. & S.S.* **216**, 25.

CUNHA, K. & LAMBERT, D.L. 1992 *Ap. J.* **399**, 586.

CUNHA, K. 1993 *Rev. Mexicana Astron. Af.* **27**, 111.

DINERSTEIN, H.L. 1990. In *The Interstellar Medium in Galaxies* (ed. H.A. Thronson & J.M. Shull), p. 257, Kluwer.

DINERSTEIN, H.L., LESTER, D.F. & WERNER, M.W. 1985 *Ap. J.* **291**, 561.

DUFOUR, R.J. 1984 *Ap. J.* **287**, 341.

DUFOUR, R.J. 1989 *Rev. Mexicana Astron. Af.* **18**, 87.

DOPITA, M.A. & EVANS, I.N. 1986 *Ap. J.* **307**, 431.

EDMUNDS, M.G. & PAGEL, B.E.J. 1984 *M.N.R.A.S.* **211**, 507.

ESTEBAN, C. & PEIMBERT, M. 1995 *A. & A.* In press.

FRENCH, H.B. 1983 *Ap. J.* **273**, 214.

GARNETT, D.R. 1992 *A.J.* **103**, 1330.

GONZÁLEZ-DELGADO, R.M., PÉREZ, E., TENORIO-TAGLE, G., VÍLCHEZ, J.M., TERLEVICH, E., TERLEVICH, R.J., TELLES, E., RODRÍGUEZ-ESPINOSA, J.M., MAS-HESSE, M., GARCÍA-VARGAS, M.L., DÍAZ, A.I., CEPA, J. & CASTAÑEDA, H.O. 1995 *Ap. J.* In press.

GRUENWALD, R.B. & VIEGAS, S.M. 1992 *Ap. J. Suppl.* **78**, 153.

HARRINGTON, J.P., SEATON, M.J., ADAMS, S. & LUTZ, J.H. 1982 *M.N.R.A.S.* **199**, 517.

HUMMER, D.G. & STOREY, P.J. 1987 *M.N.R.A.S.* **224**, 801.

IYE, M., ULRICH, M.H. & PEIMBERT, M. 1987 *A. & A.* **186**, 84.

JACOBY, G.H. & FORD, H.C. 1983 *Ap. J.* **266**, 298.

KAHN, F.D. 1989. In *Planetary Nebulae*, IAU Symposium 131 (ed. S. Torres-Peimbert) p. 411. Kluwer.

KALER, J.B. 1986 *Ap. J.* **308**, 337.

KINGDON, J. & FERLAND, G.J. 1991 *P.A.S.P.* **103**, 752.

KINGDON, J. & FERLAND, G.J. 1993 *Ap. J.* **403**, 211.

KINGDON, J. & FERLAND, G.J. 1995 *Ap. J.* In press.

KINGSBURGH, R.L. & LÓPEZ, J.A. 1995. In preparation.

KWOK, S., PURTON, C.R. & FITZGERALD, M.P. 1978 *Ap. J.* **219**, L125.

LEQUEUX, J. 1989. In *Evolution of Galaxies-Astronomical Observations*, (ed. I. Appenzeller, H.J. Habing & P. Lena), p. 147. Springer.

LIU, X.W. & DANZIGER, J. 1993 *M.N.R.A.S.* **263**, 256.

LIU, X.W., STOREY, P.J., BARLOW, M.J. & CLEGG, R.E.S. 1994 *M.N.R.A.S.* In press.

LIU, X.W., STOREY, P.J., BARLOW, M.J. & CLEGG, R.E.S. 1995. Poster paper this conference.

LÓPEZ, J.A., FALCÓN, L.H., RUIZ, M.T. & ROTH, M. 1991 *A. & A.* **241**, 526.

MAEDER, A. 1992 *A. & A.* **264**, 105.

MAEDER, A. 1993 *A. & A.* **268**, 833.

MANCHADO, A. & POTTASCH, S.R. 1989 *A. & A.* **300**, 300.

MASSEY, R.M., MEABURN, J. 1993 *M.N.R.A.S.* **262**, L48.

McCALL, M.L., RYBSKI, P.M. & SHIELDS, G.A. 1985 *Ap. J. S.* **57**, 1.

McGAUGH, S.S. 1991 *Ap. J.* **380**, 140.

MEABURN, J. 1984 *M.N.R.A.S.* **211**, 521.

MEABURN, J. & WALSH, J.R. 1980 *M.N.R.A.S.* **191**, 5p.

MIDDLEMASS, D., CLEGG, R.E.S. & WALSH, J.R. 1989 *M.N.R.A.S.* **239**, 1.

MIDDLEMASS, D., CLEGG, R.E.S., WALSH, J.R. & HARRINGTON, J.P. 1991 *M.N.R.A.S.* **251**, 284.

MOORWOOD, A.F.M., BALUTEAU, J.P., ANDEREGG, M., CORON, N. & BIRAUD, Y. 1978 *Ap. J.* **224**, 101.

OSTERBROCK, D.E. 1970 *Q.J.R.A.S.* **11**, 199.

OSTERBROCK, D.E. 1989. *Astrophysics of Gaseous Nebulae and Active Galactic Nuclei.* University Science Books.

PAGEL, B.E.J. 1995. In *The Light Element Abundances* (ed. P. Crane), in press. Springer.

PAGEL, B.E.J., EDMUNDS, M.G., BLACKWELL, D.E., CHUN, M.S. & SMITH, G. 1979 *M.N.R.A.S.* **193**, 219.

PAGEL, B.E.J., SIMONSON, E.A., TERLEVICH, R.J. & EDMUNDS, M.G. 1992 *M.N.R.A.S.* **255**, 325.

PEIMBERT, M. 1967 *Ap. J.* **150**, 825.

PEIMBERT, M. 1971 *Bol. Obs. Tonantzintla y Tacubaya* **6**, 29.

PEIMBERT, M. 1989. In *Planetary Nebulae*, IAU Symp 131 (ed. S. Torres-Peimbert) p. 577. Kluwer.

PEIMBERT, M. 1990 *Rep. Prog. Phys.* **53**, 1559.

PEIMBERT, M. 1993 *Rev. Mexicana Astron. Af.* **27**, 9.

PEIMBERT, M., COLÍN, P. & SARMIENTO, A. 1994a. In *Violent Star Formation from 30 Doradus to QSOs*, (ed. G. Tenorio-Tagle) p. 79. Cambridge University Press.

PEIMBERT, M. & COSTERO, R. 1969 *Bol. Obs. Tonantzintla y Tacubaya* **5**, 3.

PEIMBERT, M., ESTEBAN, C., TORRES-PEIMBERT, S. & COSTERO, R. 1995a. In preparation.

PEIMBERT, M., LURIDIANA, V. & TORRES-PEIMBERT, S. 1995b. In preparation.

PEIMBERT, M., SARMIENTO, A. & COLÍN P. 1994b *Rev. Mexicana Astron. Af.* **28**, 181.

PEIMBERT, M., SARMIENTO, A. & FIERRO, J. 1991 *P.A.S.P.* **103**, 815.

PEIMBERT, M., STOREY, P.J. & TORRES-PEIMBERT, S. 1993a *Ap. J.* **414**, 626.

PEIMBERT, M. & TORRES-PEIMBERT, S. 1974 *Ap. J.* **193**, 327.

PEIMBERT, M. & TORRES-PEIMBERT, S. 1976 *Ap. J.* **203**, 581.

PEIMBERT, M. & TORRES-PEIMBERT, S. 1977 *M.N.R.A.S.* **179**, 217.

PEIMBERT, M. & TORRES-PEIMBERT, S. 1983. In *Planetary Nebulae, IAU Symposium 103* (ed. D.R. Flower), p. 233. Reidel.

PEIMBERT, M. & TORRES-PEIMBERT, S. 1987a *Rev. Mexicana Astron. Af.* **14**, 540.

PEIMBERT, M. & TORRES-PEIMBERT, S. 1987b *Rev. Mexicana Astron. Af.* **15**, 117.

PEIMBERT, M., TORRES-PEIMBERT, S. & RUIZ, M.T. 1992 *Rev. Mexicana Astron. Af.* **24**, 155.

PEIMBERT, M., TORRES-PEIMBERT, S. & DUFOUR, R.J. 1993b *Ap. J.* **418**, 760.

PEÑA, M., PEIMBERT, M., TORRES-PEIMBERT, S., RUIZ, M.T. & MAZA, J. 1995 *Ap. J.* In press.

PILYUGIN, L.S. 1993 *A. & A.* **277**, 42.

POTTASCH, S.R. 1984 *Planetary Nebulae.* Reidel.

ROBBINS, R.R. 1968 *Ap. J.* **151**, 511.

ROBBINS, R.R. & BERNAT, A.P. 1973 *Mémoires Societé Royale des Sciences de Liége, 6ᵉ série, tome V*, 263.

ROLA, C. & STASINSKA, G. 1994 *A. & A.* **282**, 199.

ROSA, M. & MATHIS, J.S. 1987 *Ap. J.* **317**, 163.

ROWLANDS, N., HOUCK, J.R. & HERTER, T. 1994 *Ap. J.* **427**, 867.

SABBADIN, F., CAPPELLARO, E. & TURATTO, M. 1987 *A. & A.* **182**, 305.

SÁNCHEZ, L.J. & PEIMBERT, M. 1991 *Rev. Mexicana Astron. Af.* **22**, 285.

SAWEY, P.M.J. & BERRINGTON, K.A. 1993 *Atomic Data and Nuclear Data Tables* **55**, 81.

SEATON, M.J. 1960, *Rep. Prog. Phys.* **23**, 313.

SHIELDS, G.A. 1990 *Ann. Rev. A.A.* **28**, 525.

SIMPSON, J.P. 1973 *P.A.S.P.* **85**, 479.

SIMPSON, J.P., COLGAN, S.W.J., RUBIN, R.H., ERICKSON, E.F. & HAAS, M.R. 1995 *Ap. J.* In press.

SKILLMAN, E.D. & KENNICUTT, R.C. 1993 *Ap. J.* **411**, 655.

SKILLMAN, E.D., TERLEVICH, R.J., KENNICUTT, R.C., GARNETT, D.R. & TERLEVICH, E. 1994 *Ap. J.* **431**, 172.

SMITS, D.P. 1994 *M.N.R.A.S.* In press.

STASINSKA, G. 1990 *A. & A.S.S.* **83**, 501.

STOREY, P.J. 1994 *A. & A.* **282**, 999.

TORRES-PEIMBERT, S., PEIMBERT, M. & DALTABUIT, E. 1980 *Ap. J.* **238**, 133.

TORRES-PEIMBERT, S., PEIMBERT, M. & FIERRO, J. 1989 *Ap. J.* **345**, 186.

TORRES-PEIMBERT, S., PEIMBERT, M. & PEÑA, M. 1990 *A. & A.* **233**, 540.

TOSI, M. 1994. Preprint.

VAN DEN BERGH, S. 1988. In *Galactic and Extragalactic Star Formation* (ed. R.E. Pudritz & M. Fich), p. 381. Kluwer.

VIEGAS, S.M. & CLEGG, R.E.S. 1994 *M.N.R.A.S.* In press.

WALTER, D.K. & DUFOUR, R.J. 1994 *Ap. J.* **434**, L29.

WALTER, D.K., DUFOUR, R.J. & HESTER, J.J. 1992 *Ap. J.* **397**, 196.

WEBSTER, B.L. 1978 *M.N.R.A.S.* **185**, 45p.

WELLER, W.G. & HEATHCOTE, S.R. 1989. In *Planetary Nebulae*, IAU Symposium 131 (ed. S. Torres-Peimbert) p. 180. Kluwer.

WHEELER, J.C., HARKNESS, R.P. & CAPELLARO, E. 1987. *13 Texas Symposium on Relativistic Astrophysics* (ed. M.P. Ulmer) p. 402. World Scientific.

WILSON, T.L., BIEGING, J. & WILSON, W.E. 1979 *A. & A.* **71**, 205.

WILSON, T.L. & ROOD, R.T. 1994 *Ann. Rev. A.A.* **32**, 191.

Astrophysical Gamma Ray Emission Lines

By REUVEN RAMATY[1] AND
RICHARD E. LINGENFELTER[2]

[1]Laboratory for High Energy Astrophysics, Goddard Space Flight Center, Greenbelt, MD
20771, USA

[2]Center for Astrophysics and Space Sciences, University of California San Diego, La Jolla, CA
92093, USA

We review the wide range of astrophysical observations of gamma ray emission lines and we
discuss their implications. We consider line emission from solar flares, the Orion molecular cloud
complex, supernovae 1987A and 1991T, the supernova remnants Cas A and Vela, the interstellar
medium, the Galactic center region and several Galactic black hole candidates. The observations
have important, and often unique, implications on particle acceleration, star formation, processes
of nucleosynthesis, Galactic evolution and compact object physics.

1. Introduction

Gamma ray lines are the signatures of nuclear and other high energy processes occurring in a wide variety of astrophysical sites, ranging from solar flares and the interstellar medium to accreting black holes and supernova explosions. Their measurement
and study provide direct, and often unique, information on many important problems
in astrophysics, including particle acceleration, star formation, nucleosynthesis and the
physics of compact objects.

The physical processes that produce astrophysical gamma ray emission lines are nuclear
deexcitation, positron annihilation and neutron capture. Excited nuclear levels can be
populated by the decay of long-lived radioactive nuclei as well as directly in interactions
of accelerated particles with ambient gas. Nuclear deexcitation lines following radioactive
decay have been seen from supernova 1987A (Matz et al. 1988; Tueller et al. 1990; Kurfess et al. 1992), from the supernova remnants Cas A (Iyudin et al. 1994) and Vela (Diehl
et al. 1995), and the interstellar medium (Mahoney et al. 1984; Share et al. 1985; Diehl
et al. 1994; 1995). The observation of such line emission provides unique information
on processes of nucleosynthesis. Nuclear deexcitation lines following accelerated particle
interactions have been observed from solar flares (Chupp et al. 1973; Rieger 1989; Chupp
1990) and recently from the Orion molecular cloud complex with the COMPTEL imaging
spectrometer on the Compton Gamma Ray Observatory (CGRO, Bloemen et al. 1994).
The observed Orion lines, at 4.44 MeV from ^{12}C and 6.13 MeV from ^{16}O, cannot result
from processes of nucleosynthesis since there are no significant long lived radioactive
isotopes that decay into the excited states of these nuclei. The lines must therefore be
produced contemporaneously by accelerated particles.

The accelerated ions which produce the deexcitation lines also produce neutrons and
positrons. The neutrons can be captured by various nuclei. Capture on hydrogen produces deuterium and 2.223 MeV line photons. This line has been extensively observed
from solar flares (e.g., Chupp 1990). Capture on ^{56}Fe produces ^{57}Fe and a variety of
gamma ray lines, the most important of which are at 7.645 and 7.631 MeV corresponding to captures into the ground state and first excited state of ^{57}Fe. The observability of
the neutron capture lines requires a region in which the ambient density is high enough
to allow the capture of the neutrons before they decay and the opacity low enough to

allow the escape of the line emission. Such a site is provided by the solar photosphere. Neutron capture lines have not yet been seen from other astrophysical sites, except for unconfirmed lines from a transient source (Ling *et al.* 1982) which were interpreted as redshifted lines from neutron capture on both hydrogen and iron (Lingenfelter, Higdon, & Ramaty 1978).

Positron-electron annihilation leads to the 0.511 MeV line provided that the temperature of the annihilation site is sufficiently low; otherwise the annihilation radiation is broadened and blueshifted, and at temperatures approaching $m_e c^2/k$ it is eventually smeared into a continuum. The annihilation radiation can also be redshifted, leading to line emission below 0.511 MeV, if the annihilation site is sufficiently close to a neutron star or black hole. A narrow line at precisely 0.511 MeV (with line centroid error of only a few tenths of keV and width less than 3 keV) has been observed on many occasions with Ge detectors from the direction of the Galactic center (Leventhal, MacCallum, & Stang 1978; Gehrels *et al.* 1991; Smith *et al.* 1993). This line has also been observed with the OSSE instrument on CGRO (Purcell *et al.* 1993). The line centroid and width require that the positrons annihilate in the interstellar medium, at considerable distances from compact objects.

Line features at energies just below 0.511 MeV have been seen from Galactic black hole candidates (*e.g.*, Gilfanov *et al.* 1994; Smith *et al.* 1993; Briggs *et al.* 1995). Their origin, however, is not clear. They have generally been attributed to redshifted annihilation radiation, although they could also be due to the Compton down scattering of collimated high energy continuum (Skibo, Dermer, & Ramaty 1994). Another line produced by Compton scattering is that at $\sim$0.2 MeV. This line, seen from the black hole candidate Nova Muscae (Goldwurm *et al.* 1992; Sunyaev *et al.* 1992) and unidentified objects in the direction of the Galactic center (Leventhal & MacCallum 1980; Smith *et al.* 1993), has been identified with Compton backscattered annihilation radiation (Lingenfelter & Hua 1991). However, the feature could also result from Compton down scattering of higher energy continuum in a jet (Skibo *et al.* 1994).

Cyclotron absorption features in intense (teragauss) magnetic fields have been observed from X-ray binaries and gamma ray bursts. As here we deal only with emission lines, we shall not discuss these features. Instead, we refer the reader to our previous review (Ramaty & Lingenfelter 1994).

The plan of the present article is as follows: in section §2 we deal with the line emission from accelerated particle interactions and we discuss solar flares and Orion; in §3 we treat the lines from processes of nucleosynthesis and we discuss the supernova and ^{26}Al observations; in §4 we deal with Galactic positron annihilation; in §5 we discuss gamma ray line emissions from black hole candidates; and we present our conclusions in §6.

2. Accelerated particle interactions

The interactions of accelerated particles with ambient matter produce a variety of gamma ray lines following the deexcitation of excited nuclei in both the ambient matter and the accelerated particles. Deexcitation gamma ray lines can be broad, narrow or very narrow, depending on their widths. Broad lines are produced by accelerated C and heavier nuclei interacting with ambient H and He. The broadening of these lines (widths ranging from a few hundreds of keV to an MeV) is due to the motion of the accelerated heavy particles themselves. Narrow lines are produced by accelerated protons and α particles interacting with ambient He and heavier nuclei. The broadening in this case (widths ranging from a few tens of keV to around 100 keV), is due to the motion of the heavy targets which recoil with velocities much lower than those of the projectiles.

The ^{7}Li and ^{7}Be deexcitation lines at 0.478 and 0.429 MeV, produced by α particle interactions with ambient He, are also considered as narrow. Very narrow lines result from excited nuclei which have slowed down and stopped due to energy losses before emitting gamma rays. The broadening of these lines is due only to the bulk motion of the ambient medium (widths around a few keV or less for the interstellar medium).

There are two distinct processes which can lead to very narrow line emission: deexcitation of heavy nuclei embedded in dust grains (Lingenfelter & Ramaty 1976), and deexcitation of excited nuclei populated by long lived radionuclei. In the case of the dust, the excitations are due to protons and α particles. The best example of a very narrow grain line is that at 6.129 MeV from ^{16}O. The mean life of the corresponding nuclear level, 1.2×10^{-11} s, is long enough to allow the excited nucleus to stop in grain material before the gamma ray is emitted. In contrast, the 4.438 MeV line of ^{12}C cannot be very narrow because the lifetime of the corresponding level, 2.9×10^{-14} s, is too short to allow the excited nucleus to stop. In addition to a relatively long lifetime, the production of very narrow grain lines also requires grains that are large enough. Their size distribution and the amount of O locked up in dust will then determine the ratio of the very narrow to narrow 6.129 MeV line fluxes. A ratio of about 1/3 is a reasonable average.

Long lived radionuclei produced by accelerated particle bombardment can stop in ambient gas before they decay thereby producing excited nuclei essentially at rest. The most important such radionuclei are ^{55}Co($\tau_{1/2} = 17.5$h), ^{52}Mn($\tau_{1/2} = 5.7$d), ^{7}Be($\tau_{1/2} = 53.3$d), ^{56}Co($\tau_{1/2} = 78.8$d), ^{54}Mn($\tau_{1/2} = 312$d), ^{22}Na($\tau_{1/2} = 2.6$y), and ^{26}Al($\tau_{1/2} = 0.72$my), all of which can be produced in accelerated particle interactions, for example ^{56}Fe(p,n)^{56}Co. Unlike the very narrow grain lines which are produced almost exclusively by accelerated protons and α particles, very narrow lines from long lived radioactivity can result from both these interactions and interactions due to accelerated heavy nuclei. When the interactions are predominantly due to heavy ion interactions (as might be the case for Orion, see below), the only narrow features in the spectrum are those from the long lived radioisotopes. To produce a very narrow 0.847 MeV line from ^{56}Co it is necessary to stop a ~ 10 MeV/nucleon ^{56}Co in less than about 100 days, and this requires that the density of the ambient medium exceed about 2×10^4 cm^{-3}. Such densities may be present in dense molecular clouds. Clearly the discovery of very narrow lines would provide unique information on the density of the medium in which the lines are formed.

A theoretical spectrum, (Fig. 1 from Ramaty 1995) illustrates some of the above line features. The calculated spectra were obtained by using a nuclear deexcitation line code which employs a large number of nuclear reaction cross sections and allows calculations to be performed for a variety of compositions, accelerated particle spectra and interaction models (Ramaty, Kozlovsky, & Lingenfelter 1979; Murphy *et al.* 1991). The top panel of Fig. 1 shows a deexcitation line spectrum obtained by assuming a solar photospheric composition for the ambient medium and cosmic ray source composition for the accelerated particles. Both broad and narrow lines, as well as very narrow lines from long lived radionuclei, are present. (Very narrow lines from dust, however, are not included). The ^{12}C complex around 4.44 MeV clearly shows the narrow line superimposed on its broad counterpart. The bottom panel of the figure shows the spectrum obtained by suppressing the accelerated protons and α particles. The narrow lines are obviously absent in this case. However, we now can clearly see the very narrow lines from the long lived radionuclei. To allow the shortest lived radionucleus to stop before it decays, we assumed that the ambient density exceeds 2×10^6 cm^{-3}.

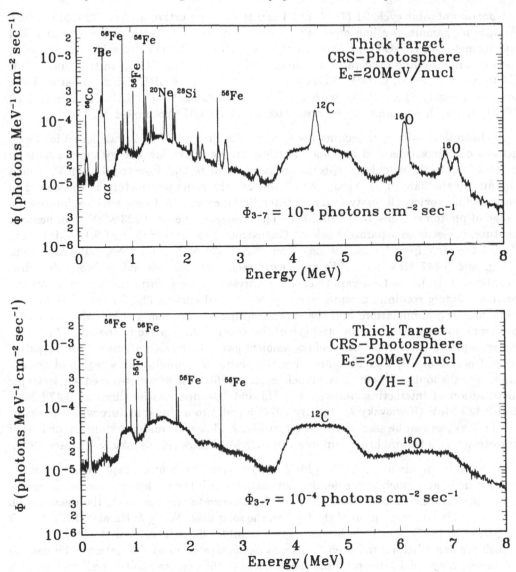

FIGURE 1. Calculated gamma ray deexcitation spectra; upper panel—a combined narrow and broad line spectrum; lower panel—a broad line spectrum showing the very narrow lines from long lived radionuclei (from Ramaty 1995).

As mentioned in the Introduction, gamma ray lines from accelerated particle interactions were seen from the flaring Sun and the Orion Complex. We first consider the flares.

2.1. *Solar flares*

Gamma ray lines from solar flares were first observed in 1972 with the NaI scintillator on OSO-7 (Chupp *et al.* 1973). But it was not until 1980 that routine observations of gamma ray lines and continuum became possible with the much more sensitive NaI spectrometer on the Solar Maximum Mission (SMM, Rieger 1989; Chupp 1990). The SMM detector operated successfully until 1989, making important observations during both the declining portion of solar cycle 21 (1980–1984) and the rising portion of cycle 22 (1988–1989). Additional gamma ray line observations during cycle 21 were carried out with a CsI spectrometer on HINOTORI (Yoshimori 1990). During the peak of solar cycle 22 solar flare gamma ray line observations were carried out with the CGRO instruments OSSE (Murphy *et al.* 1993), COMPTEL (Ryan *et al.* 1993), and EGRET (Schneid *et al.* 1994, see also Ramaty *et al.* 1994), with the Phebus instrument on GRANAT (Barat *et al.* 1994), and with a gamma ray spectrometer on YOHKOH (Yoshimori *et al.* 1994).

A theoretical solar flare gamma ray spectrum is shown in Fig. 2, calculated for interactions of flare accelerated ions and electrons having power law energy spectra with an index of -3.5; the calculated spectra are normalized to the observed 4–7 MeV gamma ray flux in the flare of 27 April 1981. These are the same parameters as those used in our previous review (Ramaty & Lingenfelter 1994), except that here we have limited the range of photon energies to 0.1–10 MeV. The strongest line, at 2.223 MeV from neutron capture, is discussed separately below. The narrow deexcitation lines at 6.129 MeV from ^{16}O, 4.438 MeV from ^{12}C, 1.779 MeV from ^{28}Si, 1.634 MeV from ^{20}Ne, 1.369 MeV from ^{24}Mg, and 0.847 MeV from ^{56}Fe are clearly seen. The corresponding broad deexcitation lines, together with a variety of other unresolved lines, form the excess above the bremsstrahlung continuum represented by the dashed curve. The line at 0.511 MeV is from positron annihilation, and the excess continuum just below 0.511 MeV is due to positronium annihilation. The strength of this continuum relative to the 0.511 MeV line depends primarily on the density of the ambient gas. These calculations are for a positronium fraction of 0.9, which requires that the positrons annihilate in a region of density lower than about 10^{15} cm^{-3} (Crannell *et al.* 1976). Since we assumed an isotropic distribution of interacting particles the ^{7}Li and ^{7}Be deexcitation lines at 0.478 MeV and 0.429 MeV (Kozlovsky & Ramaty 1974) blend into a single feature which peaks at $\sim$0.45 MeV, as can be seen in the insert in Fig. 2. However, under certain conditions of anisotropy, this feature breaks up into two individual lines (Kozlovsky & Ramaty 1977).

As already mentioned, the 2.223 MeV line is formed by neutron capture on ^{1}H in the photosphere, at a much larger depth than that at which the nuclear reactions take place. Consequently, the ratio of the 2.223 MeV line fluence to the fluence in the deexcitation lines depends on the position of the flare on the solar disk (Wang & Ramaty 1974; Hua & Lingenfelter 1987). The ratio becomes quite small for flares at or near the limb, and can vanish for flares behind the limb. This was demonstrated most dramatically by gamma ray observations of a flare on 1 June 1991 located at 10° degrees behind the limb for which the 2.223 MeV line was absent while the deexcitation lines were still seen (Barat *et al.* 1994). Evidently, a considerable fraction of the nuclear reaction occurred in the corona, at a site which was visible even though the location of the optical flare was occulted. While this 'classical' behavior is reassuring, (that the 2.223 MeV should be formed in the photosphere and hence strongly attenuated from flares located at or behind the limb was predicted theoretically, Wang & Ramaty 1974), there is another observation of a flare about 10 ± 5 degrees behind the limb for which only the 2.223 MeV line was seen (Vestrand & Forrest 1993). This flare, on 29 September 1989, was one of the largest on record, having produced a multitude of emissions including protons up to 25 GeV (Swinson & Shea 1990). Because of the very strong expected attenuation, the observed 2.223 MeV line must have been produced by charged particles interacting on the visible hemisphere of the Sun. It was suggested (Cliver, Kahler & Vestrand 1993) that these particles were accelerated by a coronal shock over a large volume thereby producing an extended gamma ray emitting region visible from the Earth even if the optical flare was behind the limb.

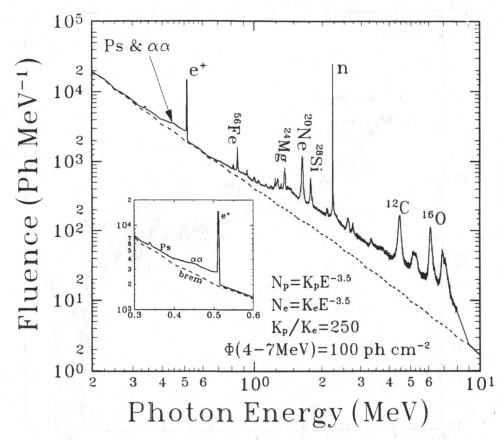

FIGURE 2. A theoretical solar flare gamma ray spectrum showing the strongest expected nuclear deexcitation lines.

The ratio of the fluence of the bremsstrahlung continuum to that in the lines was used to determine the electron-to-proton ratio (e/p) for the accelerated particles which interact at the Sun (Ramaty *et al.* 1993). The derived values of e/p were found to be generally larger than the corresponding e/p's obtained from observations of solar flare particle events in interplanetary space. Such interplanetary e/p observations have been used to distinguish two classes of solar flare particle events (Cane, McGuire & von Rosenvinge 1986; Reames 1990). Impulsive events, for which the associated soft X-ray emission is of relatively short duration, have large e/p ratios; gradual events, for which the soft X-rays last longer, exhibit smaller values of e/p. The gamma ray results reveal comparable or even higher values of e/p than the impulsive events in space, regardless of whether the flare is impulsive or gradual. This result suggests that the particles which are trapped at the Sun and produce the gamma rays, and the particles observed in interplanetary space from impulsive flares are accelerated by the same mechanism, probably stochastic acceleration due to gyroresonant interactions with plasma turbulence. Relativistic electron acceleration by such turbulence, particularly whistler waves, can be quite efficient (Miller & Steinacker 1992). On the other hand, for gradual events the particles are thought to be accelerated from cooler coronal gas probably by a shock which is not expected to accelerate electrons efficiently.

In Fig. 3 we show a solar flare gamma ray spectrum observed with the NaI spectrometer on SMM (Murphy *et al.* 1990). The ^{16}O, ^{12}C and ^{20}Ne lines can be clearly seen, but the

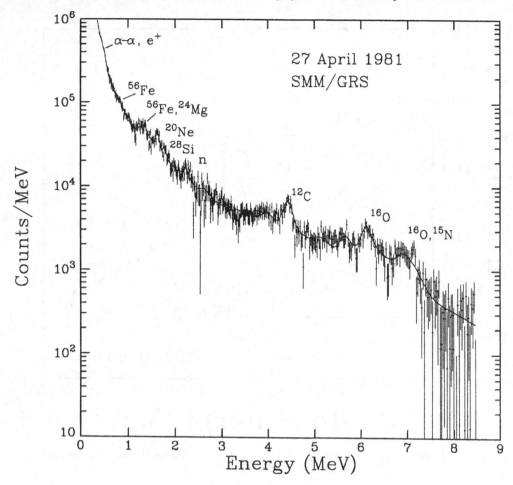

FIGURE 3. Observed solar flare gamma ray line spectrum fitted with theoretical curve (from Murphy *et al.* 1991).

^{24}Mg line is not resolved from several neighboring features (compare with Fig. 1). The annihilation line, possibly accompanied by a positronium continuum, and the α–α lines produce a broad feature above the strong bremsstrahlung continuum. The 2.223 MeV line is weak because of the location of this flare at the limb of the Sun.

The spectrum shown in Fig. 3 was used to determine abundances in both the ambient solar atmosphere and the accelerated particle population (Murphy *et al.* 1991). For the accelerated particles the results indicate that the abundances of the heavy elements, in particular Mg and Fe, are significantly enhanced (relative to C and O) in comparison to their abundances in either the photosphere or the corona. Similar enhancements are observed in the abundances of the accelerated particles observed from impulsive flares (Reames 1990). This supports the conclusion mentioned above that the particles responsible for gamma ray production and the particles observed in interplanetary space from impulsive flares are accelerated by the same mechanism. It has been shown (Miller and Viñas 1993) that stochastic acceleration by plasma turbulence produce these enhancements (Reames, Meyer & von Rosenvinge 1994).

For the ambient gas, the Mg, Si and Fe abundances relative to C and O are consistent with coronal abundances but enhanced in comparison with photospheric abundances. The enhanced Mg, Si and Fe abundances (elements with low first ionization potential, FIP) could be understood in terms of a charge dependent ambient gas transport process from the photosphere to the chromosphere and corona which favors the collisionally ionized, low FIP elements in the photosphere (Meyer 1985). Indeed, the enhancement of the low FIP elements in the corona is rather well established from various observations, in particular solar flare accelerated particle observations. The abundance of Ne determined from the gamma ray data is problematic. Based primarily on the strong 1.634 MeV ^{20}Ne line (Fig. 2), the gamma ray data yielded (Murphy *et al.* 1991) a Ne to O ratio that is more than a factor of 2 larger than the coronal Ne/O which is thought to be consistent with the local galactic Ne abundance (Meyer 1989). The photospheric Ne abundance is not known, so in principle it is possible that the gamma ray derived abundance would be representative of the photosphere. This point of view, however, is challenged by other astronomical data which would favor a photospheric Ne abundance about equal to its coronal value (Meyer 1989). It has been proposed that the Ne enhancement could be due to photoionization by soft X-rays (Shemi 1991), an interpretation which predicts that S should also be enhanced. Both the Ne and S enhancements have been confirmed by soft X-ray observations (Schmelz 1993), but only in one flare. Thus the issue of the gamma ray derived Ne abundance remains unresolved, awaiting new observations and their analysis.

Another isotope whose photospheric abundance is not well known is ^{3}He. It was first pointed out by Wang & Ramaty (1974) that ^{3}He in the photosphere could capture as much as one half of the neutrons, so that the flux and time profile of the 2.223 MeV line, resulting from the capture of the other half on ^{1}H, would strongly depend on the photospheric ^{3}He abundance. Observations (Prince *et al.* 1983) of the time dependent flux of the 2.223 MeV line were used (Hua & Lingenfelter 1987) to derive a photospheric ^{3}He/H $\simeq 2 \times 10^{-5}$, which is sufficiently low to be consistent with the ^{3}He abundance expected solely from primordial nucleosynthesis without requiring much mixing in the solar atmosphere.

Thus, we see that gamma ray emission lines from the Sun are providing a wealth of new information on both abundances and the acceleration of energetic particles in the solar flares.

2.2. *The Orion complex*

The recent discovery of gamma ray emission lines from the Orion giant molecular cloud complex has now revealed exciting new particle acceleration processes in this nearest region of recent star formation that have very important implications for light element nucleosynthesis. Gamma ray line emission in the 3 to 7 MeV range was observed from the Orion complex (Fig. 4a) with COMPTEL (Bloemen *et al.* 1994). The radiation shows (Fig. 4b) emission peaks near 4.44 and 6.13 MeV, consistent with the deexcitation of excited states in ^{12}C and ^{16}O produced by accelerated particle interactions. Moreover, the intensity of these lines is roughly two orders of magnitude greater than that expected from irradiation by low energy cosmic rays with energy density equal to that of the local Galactic cosmic rays (Ramaty *et al.* 1979). Thus this emission requires that the ambient matter in Orion, both gas and dust, is undergoing bombardment by an unexpectedly intense, locally accelerated, population of energetic particles.

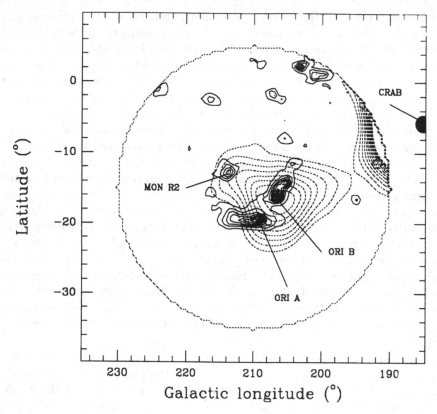

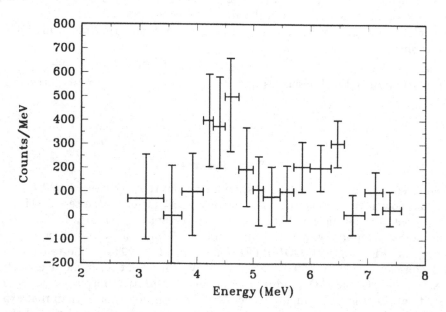

FIGURE 4. The observed location of the nuclear line emission in the Orion star formation region and the spectrum of the line emission (from Bloemen *et al.* 1994).

2.2.1. *Implication of gamma ray emission lines*

We have explored (Ramaty, Kozlovsky & Lingenfelter 1995a,b) the implications of these observations on the composition, energy spectra and total power of the accelerated particles in the Orion region. As we will discuss in detail, we have found that the ratio of the measured flux in the 3 to 7 MeV band compared to limits on the 1 to 3 MeV band place very strong constraints on the composition of the accelerated particles, requiring significant enrichment in C and O relative to the heavier elements in order not to produce too much line emission in the 1 to 3 MeV band. At the same time, the overall energetics and efficiency of the gamma ray line production require significant enrichment of C and O relative to H and He. The lack of enhanced flux at higher (>100 MeV) energies also requires that the accelerated particles have a much softer energy spectrum than that of the Galactic cosmic rays. But the spectrum of the accelerated particles should not be too soft in order to maintain an acceptable overall energetic efficiency.

We have calculated the gamma ray line emission rate together with the power deposited by the accelerated particles during gamma ray production. In the thick target model which we employ, the ratio of the photon production rate to the power deposited by the accelerated particles depends only on the composition and spectrum of the accelerated particles and on the composition and state of ionization of the ambient medium. In Fig. 5 (from Ramaty *et al.* 1995b) we show the deposited power at Orion for a neutral ambient medium with solar system abundances and for various accelerated particle compositions as a function of the spectral parameter E_0 of the accelerated particles. The spectrum

$$N_i(E) = K_i E^{-1.5} e^{-E/E_0}, \tag{2.1}$$

where E is energy per nucleon and the K_i's are proportional to the assumed abundances, is predicted by acceleration to nonrelativistic energies near a strong shock (compression ratio $r = 4$) with the effects of a finite acceleration time or a finite shock size taken into account by the exponential turnover (*e.g.*, Ellison & Ramaty 1985). We have carried out calculations for E_0 in the range 2–100 MeV/nucl; E_0, however, should not exceed about 30 MeV/nucl, because otherwise the accelerated particles would contribute to pion production and thus be in conflict with the high energy gamma rays observed from Orion with EGRET; these data only require a cosmic ray flux equal to that observed locally near Earth (Digel, Hunter & Mukherjee 1995).

The curves in Fig. 5 correspond to various assumed accelerated particle compositions (Ramaty *et al.* 1995b): SS—solar system (Anders & Grevesse 1989), CRS—cosmic ray source (Mewaldt 1983), SN35—the ejecta of a 35 M⊙ supernova (Weaver & Woosley 1993), WC—the late phase wind of a Wolf-Rayet star of spectral type WC (Maeder & Meynet 1987); GR—pick up ions resulting from the breakup of interstellar dust. Concerning the dust, in analogy with the anomalous component of the cosmic rays observed in interplanetary space (*e.g.*, Adams *et al.* 1991), we considered the effects of the pick up of ions onto a magnetized high speed wind (*e.g.*, the ejecta of supernovae or the winds of massive stars). As the ions acquire considerable energy during the pick up process, they form a seed population that is much more easily accelerated than the rest of the ambient plasma. In the solar system the pick up ions are interstellar neutrals that penetrate into the solar cavity where they are ionized. For Orion we proposed that the incoming matter is essentially neutral dust that is broken up by evaporation, sputtering or other processes. The GR composition, therefore, has no He and Ne and is relatively poor in H. The SS and CRS compositions will thus produce combined broad and narrow line spectra, while the WC and GR compositions will lead to essentially pure broad line spectra. The SN35 will still have a weak narrow line component. We see from In Fig. 5 that the gamma ray line production is energetically most efficient for large values of E_0 and accelerated

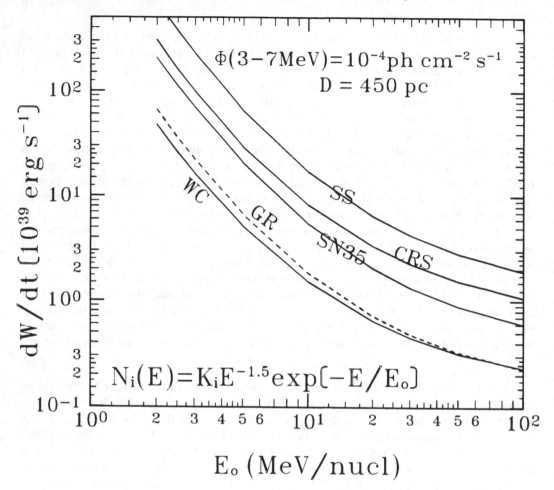

$$\Phi(3-7\,\text{MeV})=10^{-4}\text{ph cm}^{-2}\,\text{s}^{-1}$$
$$D = 450 \text{ pc}$$

$$N_i(E)=K_iE^{-1.5}\exp[-E/E_o]$$

$E_o\,(\text{MeV/nucl})$

FIGURE 5. Calculated energy deposition rates in Orion by accelerated particles of various compositions in a neutral ambient medium with solar photospheric composition, as functions of the accelerated particle spectral parameter E_0 (from Ramaty *et al.* 1995b).

particle compositions that are poor in protons and α particles (*i.e.*, the GR and WC compositions). Thus, just from energetic arguments, the broad line spectra are preferred, even though the COMPTEL gamma ray data can so far not distinguish between pure broad line spectra and combined broad and narrow line spectra (*i.e.*, spectra produced by heavy projectiles as well as protons and α particles, Ramaty *et al.* 1995a).

For the GR and WC compositions and $E_0 = 30$ MeV/nucl, the deposited power is 4×10^{38} erg s^{-1} and the ionization rate is 0.3 $M_\odot$ yr^{-1} or $\zeta = 10^{-13} M_5^{-1}$ (H atom)$^{-1}$ s^{-1}, where M_5 is the total irradiated neutral H mass in units of 10^5 solar masses. Setting the ionization rate equal to the recombination rate,

$$\zeta n_{\text{H}} = \alpha_r n_e^2\,, \qquad (2.2)$$

where $\alpha_r \simeq 10^{-11}(T/100\text{K})^{-0.7}$ (Bates & Dalgarno 1962), and n_{H} and n_e are the neutral H and electron densities, respectively, we obtain an equilibrium ionization fraction of about $0.1M_5^{-1/2}n_{\text{H}}^{-1/2}(T/100\text{K})^{0.35}$. We see that for a large enough irradiated mass, and temperatures and H densities typical of dense molecular clouds, the irradiated cloud can remain essentially neutral. The total deposited power depends of course on the duration of the irradiation. For example, if the irradiation lasts $\sim 10^5$ years, the total energy requirement would be 1.2×10^{51} ergs, equal to the total mechanical output of a supernova.

Just such a supernova, occurring $\sim 80,000$ years ago in the OB association at $l = 208°$ and $b = -18°$, the same location as the center of the gamma ray line source, was suggested by Burrows *et al.* (1991) from analyses of the X-ray emission from the Orion-Eridanus bubble.

In addition to being energetically efficient, the WC and GR compositions have the advantage of predicting gamma ray spectra which are consistent with the upper limit set by the COMPTEL observations on the emission in the 1–3 MeV range. We have shown (Ramaty *et al.* 1995b) that the discrepancy between this limit and the CRS prediction is greater than 3σ, and that the predictions of the SS and SN35 compositions are also inconsistent at greater than 2σ. On the other hand, both the GR and WC compositions yielded 1–3 MeV fluxes which are lower than the COMPTEL upper limit. However, while the WC composition predicts practically no emission in the 1–3 MeV region, the GR composition predicts significant broad line emission in this region, due to Mg, Si and Fe. The reduction in the overall 1–3 MeV emission for the GR case is caused by the absence of 1.634 MeV line due to the lack of Ne in grains.

The nature of the acceleration mechanism in Orion is still very poorly understood. Acceleration by the shocks associated with the winds of young O and B stars was proposed by Nath & Biermann (1994), while Bykov & Bloemen (1994) proposed that the acceleration is due to the shocks produced by colliding stellar winds and supernova explosions. We have proposed (Ramaty *et al.* 1995b) that the pick up ions resulting from the breakup of interstellar grains could be the dominant injection process to any of these acceleration mechanisms. We have also emphasized the comparison with solar flares (Ramaty 1995; Ramaty *et al.* 1995a). The solar flare gamma ray spectra show much higher ratios of 1–3 MeV to 3–7 MeV fluxes than does Orion. It was shown (Murphy *et al.* 1991) that this enhanced emission below 3 MeV is, in part, due to the enrichment of the flare accelerated particle population in heavy nuclei. Such enrichments are routinely seen in direct observations of solar energetic particles from impulsive flares (*e.g.*, Reames, Meyer & von Rosenvinge 1994). These impulsive flare events are also rich in relativistic electrons. On the other hand, in gradual events the composition is coronal and the electron-to-proton ratio is low. As we have pointed out above, the acceleration in impulsive events is thought to be due to gyroresonant interactions with plasma turbulence while in gradual events it is the result of shock acceleration. The fact that the ratio of bremsstrahlung-to-nuclear line emission in Orion is very low lends support to the shock acceleration scenario. Moreover, as we noted above, these shocks may be powered by an $\sim 80,000$ year old supernova of a massive star in the OB association which is also responsible for the Orion-Eridanus bubble.

2.3. *The relationship to light isotope production*

That cosmic ray spallation is important to the origin of the light isotopes ^{6}Li, ^{7}Li, ^{9}Be, ^{10}B, and ^{11}B has been known for over two decades (Reeves, Fowler & Hoyle 1970). It was shown that cosmic rays with flux equal to the observed flux near Earth, interacting with interstellar matter prior to the formation of the solar system, can produce the observed solar system abundances of ^{6}Li, ^{9}Be and ^{10}B (Meneguzzi, Audouze & Reeves 1971; Mitler 1972). Since low energy cosmic rays with spectra similar to that of the accelerated particles in Orion also produce these isotopes, the Galactic inventories of ^{6}Li, ^{9}Be and ^{10}B can set limits on the total Galactic irradiation by such low energy cosmic rays. In Fig. 6 (from Ramaty *et al.* 1995b) we show the ratio of ^{9}Be production

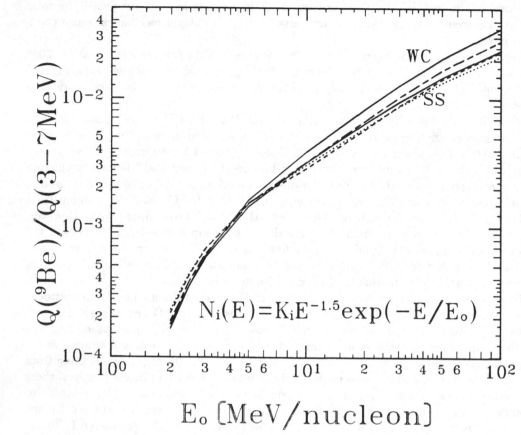

FIGURE 6. Production ratios of ^{9}Be to 3–7 MeV nuclear deexcitation photons for various accelerated particle compositions as functions of the accelerated particle spectral parameter E_0 (from Ramaty *et al.* 1995b).

to 3–7 MeV deexcitation photon production. As both the 3–7 MeV photons and the ^{9}Be are produced predominantly in interactions involving C and O, this ratio is practically independent of composition. Adopting a total Galactic ^{9}Be inventory of 10^{57} atoms and assuming that currently there are N_{orr} regions in the Galaxy with the same level of low energy cosmic ray activity as Orion (*i.e.*, producing 3–7 MeV nuclear gamma rays at a rate $Q_{orr}(3-7\text{MeV}) = 2.3 \times 10^{39}$ s^{-1}), we have that

$$N_{orr}Q_{orr}(3-7\text{MeV})[Q(^9Be)/Q(3-7)\text{MeV}]T_{irr} < 10^{57}\text{atoms,} \qquad (2.3)$$

where the quantity in square brackets is from Fig. 6 and T_{irr} is the total duration of the irradiation. Taking $Q(^9Be)/Q(3-7\text{MeV}) = 8 \times 10^{-3}$ (from Fig. 6) and $T_{irr} = 3 \times 10^{17}$ s, the Galactic age, we get that $N_{orr} < 200$. Eq. (2.3) assumes the irradiation is constant in time and ignores the destruction of ^{9}Be in stars. Using a similar argument, but employing the Galactic inventory of B, Reeves & Prantzos (1995) obtained $N_{orr} < 100$. They obtained a lower value because the B inventory of 10^{58} atoms that they used corresponds (for the solar system B/Be) to a ^{9}Be inventory which is lower by a factor of 3 than the value we used.

The upper limit on the number of currently active 'Orion-like' regions can be used to set limits on the 3–7 MeV nuclear line emission from the central regions of the Galaxy. For a given type of emission (*e.g.*, the 3–7 MeV nuclear line emission), the relationship between the flux from the solid angle defined by the central radian of Galactic longitudes and all latitudes, Φ_{crad}, and the total Galactic photon luminosity, Q_G, can be written as

$$\Phi_{crad}(3-7\text{MeV}) = \xi 10^{-46} Q_G(3-7\text{MeV}) = \xi 10^{-46} N_{orr} Q_{orr}(3-7\text{MeV}), \qquad (2.4)$$

where ξ is obtained by integrating the photon source distribution along all the lines of sight within the above solid angle. For a variety of assumed Galactic source distributions ξ was found to range from about 0.6 to 1.7 (Skibo 1993). Thus, for $N_{orr} < 200$, $\Phi_{crad}(3-7\text{ MeV}) < 5 \times 10^{-5}\xi$ photons cm^{-2} s^{-1}, which is lower by at least a factor of 3 than the upper limit on $\Phi_{crad}(3-7\text{ MeV})$ obtained from SMM data (Harris, Share & Messina 1995). We note, however, that the limit on N_{orr} obtained from ^{9}Be would become higher if a significant amount of ^{9}Be is destroyed by incorporation into stars. On the other hand, the limit would be lower if the rate of irradiation in the early Galaxy was much higher than the average. We clearly need more observations of nuclear gamma ray lines to map out the distribution of low energy cosmic rays in the Galaxy.

While Galactic cosmic ray spallation can account for the ^{6}Li, ^{9}Be and ^{10}B, such cosmic rays cannot account for the abundances of ^{7}Li and ^{11}B. The ^{11}B excess is probably produced by spallation, either by low energy cosmic rays (Casse, Lehoucq & Vangioni-Flam 1995) or by neutrinos in supernovae (Woosley *et al.* 1990). Here we discuss some of the issues related to Li.

^{7}Li production in standard big bang nucleosynthesis models leads to ^{7}Li/^{1}H $\sim 10^{-10}$, in agreement with the ^{7}Li abundance in extremely metal deficient Pop II stars (see Reeves 1994 and references therein). This cosmological ^{7}Li, however, is insufficient to account for the ^{7}Li abundance in Pop I stars and meteorites, where ^{7}Li/^{1}H $\sim 10^{-9}$. The excess ^{7}Li is thought to be produced in stars, most likely AGB stars, although neutrino induced spallation in supernovae was also considered (Woosley *et al.* 1990). Thus, while cosmic ray spallation leads to ^{7}Li/^{6}Li $\simeq 1.4$ (Reeves 1994), the enhanced meteoritic ^{7}Li/^{6}Li of 12.3 is generally understood as being due to these additional ^{7}Li sources. That the enhancement cannot be due to low energy cosmic rays with spectrum given by eq. (2.1) can be seen from Fig. 7, which shows our calculations of ^{7}Li/^{6}Li as a function of E_0 for the 5 assumed compositions. For E_0 around 30 MeV/nucl, ^{7}Li/^{6}Li is 1.5 independent of composition and essentially equal to the cosmic rays ratio.

Rather than producing the high meteoritic ^{7}Li/^{6}Li, strong localized irradiation by low energy cosmic rays could significantly lower this ratio in selected molecular clouds that have undergone 'Orion-type' irradiation episodes (Lemoine, Ferlet & Vidal–Madjar 1995; Reeves & Prantzos 1995). This argument is suggested by the recent observations of Lemoine *et al.* (1995) which indicate that in the direction of ζ Oph there are two absorbing clouds (A and B) with different Li isotopic ratios, $(^7\text{Li}/^6\text{Li})_A \simeq 8.6$ and $(^7\text{Li}/^6\text{Li})_B \simeq 1.4$. However, the energetic and ionization implications of the required massive low energy cosmic ray irradiation have not yet been examined.

2.3.1. *The relationship to ^{26}Al production*

Evidence for the existence of freshly nucleosynthesized ^{26}Al in the interstellar medium is provided by the observations of the 1.809 MeV line from various directions in the Galaxy, as we discuss in detail below (see §3.3). Evidence for the presence of live ^{26}Al at the time of the formation of the solar system was obtained from the analysis of meteoritic material (Lee, Papanastassiou & Wasserburg 1977). Motivated by the Orion gamma ray line observations, Clayton (1994) suggested that both the general Galactic ^{26}Al and the protosolar ^{26}Al could result from low energy cosmic ray spallation. However, the production ratio of ^{26}Al to 3–7 MeV line emission from such spallation reactions, $< 10^{-2}$ (Ramaty *et al.* 1995a), combined with the upper limit on the Galactic 3–7 MeV line flux

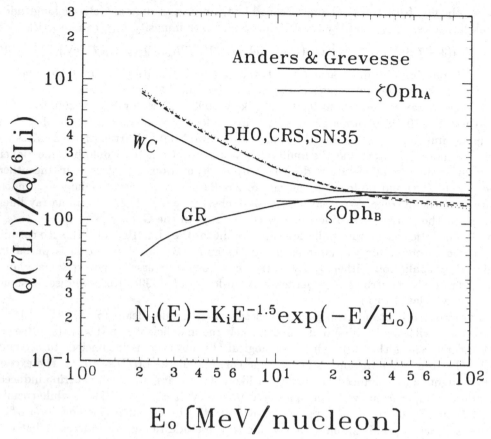

FIGURE 7. Production ratios of ^{7}Li to ^{6}Li for various accelerated particle compositions as functions of the accelerated particle spectral parameter E_0. Shown are the meteoritic ratio (Anders & Grevesse 1989) and the two values for clouds in the direction of ζ oph (Lemoine *et al.* 1995).

set by the SMM data (Harris *et al.* 1995), shows that low energy cosmic rays produce at best only 1% of the total Galactic ^{26}Al.

Although Clayton & Jin (1995) now recognize that it is not possible to produce the Galactic ^{26}Al by spallation, they still argue that the protosolar ^{26}Al could have resulted from an episode of low energy cosmic ray irradiation similar to that currently taking place in Orion. We have shown (Ramaty *et al.* 1995a), however, that the COMPTEL Orion observations, and in particular the upper limit on the 1–3 MeV emission, imply a significantly lower ^{26}Al production rate than that estimated by Clayton (1994) to suggest that the protosolar ^{26}Al/^{27}Al of 5×10^{-5} could have resulted from low energy cosmic ray bombardment. That argument, however, depended on the total irradiated mass and on the level of protosolar low energy cosmic ray activity which could have been different from that in Orion.

More recently by comparing the ^{9}Be and ^{26}Al yields, we have shown (Ramaty *et al.* 1995b) that an independent limit can be set on possible protosolar spallation production of ^{26}Al that does not depend of these parameters. In Fig. 8 we show the ratio of ^{26}Al to ^{9}Be production. The solid bar is the ratio implied by the protosolar value of ^{26}Al/^{27}Al and the SS ^{27}Al/^{9}Be. We see that for $E_0 > 10$ MeV/nucl (lower values are energetically very inefficient) the SS ^{9}Be abundance limits the contribution of particle bombardment

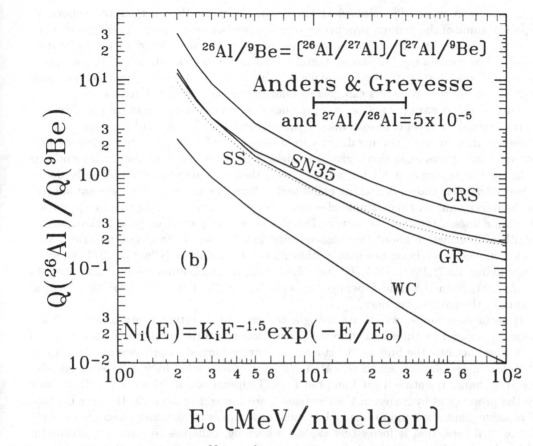

FIGURE 8. Production ratios of ^{26}Al to ^{9}Be for various accelerated particle compositions as functions of the accelerated particle spectral parameter E_0 (from Ramaty *et al.* 1995b). The horizontal bar represents the protosolar ^{26}Al abundance and the probable E_0 range.

to ^{26}Al/^{27}Al to less than 3 to 10% for compositions consistent with the 1–3 MeV flux limits. Moreover, if we take into account the facts that most of the ^{9}Be had already been produced prior to the formation of the protosolar nebula, and that some of the ^{26}Al must have decayed during irradiation, these limits become much lower. This shows that it is practically impossible to produce the protosolar ^{26}Al by accelerated particle bombardment, quite independent of the amount of material irradiated.

3. Gamma ray lines from nucleosynthesis

Because the sites of explosive nucleosynthesis are optically thick to gamma-rays, only the delayed gamma ray line emission from the decay of synthesized radionuclei can be observed from such sites that become at least partially transparent on time scales less than the radioactive decay mean lives. Such gamma ray lines from the decay of ^{56}Co and other freshly synthesized radionuclei in supernovae were predicted by Clayton, Colgate & Fishman (1969). The most intense lines are from ^{56}Ni$\rightarrow$^{56}Co$\rightarrow$^{56}Fe decay, followed by those from ^{57}Co$\rightarrow$^{57}Fe and ^{44}Ti$\rightarrow$^{44}Sc $\rightarrow$^{44}Ca. Such radionuclei, carried outward in the expanding supernova envelope, decay and then deexcite by gamma ray line emission. At early times, the gamma rays interact with the material in the supernova and are Compton scattered down to X-ray energies which are photoelectrically absorbed and their energy is eventually released at longer wavelengths. However, as the supernova expands, some of the gamma rays begin to escape without scattering. These gamma ray lines are Doppler-broadened by the velocity spread of the radionuclei in the expanding nebula. The gamma ray line shapes therefore reflect the velocity distribution within the supernova, modified by the opacity along the line of sight, and their measurement with high resolution spectrometers can give us information on this distribution.

Thus, observations of these gamma ray lines provide the most direct means of testing current models of both explosive nucleosynthesis and the dynamics of supernova ejecta. These gamma ray lines, coming directly from the decay of radioactive nuclei freshly produced in supernova explosions, give a straight forward measure of the nucleosynthetic yields of the supernovae. Being nuclear lines, their emission rates are directly determined from their known branching ratios and radioactive half lives; they are not subject to the uncertainties in the estimated excitation rates that complicate the interpretation of atomic lines. Moreover, because of Doppler broadening and Compton attenuation, a wealth of information about the mass-velocity distribution of the ejecta, and the distribution of nucleosynthetic products within it, can be obtained (Clayton 1974; Chan & Lingenfelter 1987; 1988; 1991; Gehrels, Leventhal & MacCallum 1987; Ruiz-Lapuente *et al.* 1993) from the time dependent observations of the intensity ratios and spectral shapes of the gamma ray lines.

Type Ia supernovae, which are thought to occur in accreting white dwarfs and are optically defined by their lack of hydrogen envelopes, are the most luminous sources of such lines, because the bulk of the star is explosively burned to produce nearly $1M_\odot$ of ^{56}Ni, leaving a relatively small ($< 1M_\odot$) overlying envelope to obscure the emission. On the other hand, the other Type I and the Type II supernovae, which are optically defined by the presence of hydrogen in their envelope, are thought to occur in the core collapse of massive giant stars and are much less luminous sources of such lines, because the bulk of the ^{56}Ni core, that is formed by explosive burning, collapses to form a neutron star and only a small amount (typically $< 0.1M_\odot$) is ejected; furthermore the gamma rays from its decay are obscured for a much longer time by the massive (typically $> 10M_\odot$) overlying envelope.

The most intense gamma ray lines (with branching ratios) are those from ^{56}Co$\rightarrow$^{56}Fe decay (Huo *et al.* 1987) at 0.8468 MeV (99.9%), 1.2383 MeV (68.4%), 2.5985 MeV (17.4%), 1.7714 MeV (15.5%), and 1.0378 MeV (14.1%). The ^{56}Co results from the decay of ^{56}Ni which is the parent nucleus produced in supernovae. ^{56}Co decays (Huo *et al.* 1987) with a mean life of 111.3 days, on a time scale comparable to that required for the supernova ejecta to become transparent to the gamma rays. The decay of ^{56}Ni also gives important gamma ray lines, 0.1584 MeV (98.8%), 0.8119 MeV (86.0%), 0.7500 MeV (49.5%), 0.2695 MeV (36.5%) and 0.4804 MeV (36.5%), but its much shorter mean life of 8.8 days allows the lines to be detectable for a shorter period of time only in the most rapidly expanding and least massive supernovae. Another important radioisotope is ^{57}Co, resulting from the decay of ^{57}Ni produced by neutron capture onto ^{56}Ni. ^{57}Co decays with a longer mean life of 392 days, emitting two major lines (Burrows & Bhat 1986) at 0.1221 MeV (85.9%) and 0.1365 MeV (10.3%), which can be used to study the neutron flux during nucleosynthesis (Clayton 1974).

The principal lines from ^{56}Co and ^{57}Co have been observed from the Type II supernova 1987A in the nearby Large Magellanic Cloud, and these gamma ray line observations have already provided important information on the nucleosynthetic yield and dynamics of that Type II supernova, as we discuss below. Similar observations of gamma ray lines from extragalactic Type Ia supernovae by spectrometers on the CGRO and the planned INTEGRAL (Winkler 1994) should provide critical tests of current models of the nucleosynthetic yield and dynamics of these supernovae. The detection with COMPTEL of the ^{56}Co lines from the Type Ia supernova SN1991T in NGC4527 at a distance of about 13 Mpc has recently been reported (Dan Morris, oral presentation, 17 Texas Symposium, 1994). The expected fluxes of the principal lines from ^{56}Co decay are such that they should be detectable (Chan & Lingenfelter 1991) with INTEGRAL from roughly half of the Type Ia supernovae in the Virgo Cluster. The calculations of the gamma ray line profiles have also indicated how sensitive the line shapes are to the mass-velocity distribution in the supernova ejecta. Thus, future gamma ray line measurements will be used as tests of supernova models and as diagnostics of supernova structure.

3.1. *The Type II supernova 1987A*

The occurrence of the Type II supernova 1987A in the Large Magellanic Cloud, at a distance of about 50 kpc from the Earth, the brightest supernova seen in nearly 400 years, has given us an unprecedented opportunity to directly study supernovae through their gamma ray emission. This supernova explosion occurred (Arnett *et al.* 1989) in the star Sk −69 202, a B3 I supergiant with a mass of about $16M_\odot$ at the time of the explosion. This supernova has been studied over the full spectrum from radio to gamma rays, and extensive observations of the time dependent gamma ray line intensities and profiles have been made with a variety of spectrometers on balloons and spacecraft (*e.g.*, Matz *et al.* 1988; Tueller *et al.* 1990). Comparisons of these observations with predicted line intensities as a function of time (Chan & Lingenfelter 1987; Gehrels, MacCallum & Leventhal 1987; Bussard, Burrows & The 1989) have confirmed the production $0.075M_\odot$ of nucleosynthetic ^{56}Co which has been independently derived from the bolometric luminosity (Arnett *et al.* 1989).

The gamma ray observations have also shown that the ^{56}Co is extensively mixed in the ejecta. Early calculations (Chan & Lingenfelter 1987) of the gamma ray line emission expected from existing models (Weaver & Woosley 1980a,b) of supernovae in a $15M_\odot$ star suggested that the line emission from ^{56}Co decay would not be visible until nearly 600 days after the explosion if the ^{56}Co was confined to the innermost layers of the supernova ejecta. The subsequent detection (Matz *et al.* 1988) of the ^{56}Co lines at 0.847 and 1.238 MeV within 200 days after the explosion showed that substantial mixing had occurred in the ejecta (Chan & Lingenfelter 1988; Gehrels *et al.* 1988), raising the ^{56}Co to high levels in the ejecta where the obscuration by overlying matter was much lower. This has led to extensive modifications of the supernova models to explore the causes and effects of mixing on both the explosive nucleosynthesis and the ejecta dynamics (*e.g.*, Nomoto *et al.* 1988; Pinto & Woosley 1988; Bussard *et al.* 1989).

The measured 0.847 and 1.238 MeV line fluxes (Tueller *et al.* 1990) after the supernova explosion have been compared (see Fig. 9) with calculations (Pinto & Woosley 1988) for the mixed model 10HMM, yielding a reasonably good fit. This model is for a 16 $M_\odot$ star, having a $6M_\odot$ helium core and 10 $M_\odot$ blue supergiant envelope, which explodes ejecting the matter above the silicon shell and explosively synthesizes the 0.075 $M_\odot$ of ^{56}Ni that is required to account for the bolometric light curve. In this model the ^{56}Ni has been mixed out through the helium core into the envelope in order to account for the early gamma ray line observations. However, as can be seen in Fig. 9, there is still a significant difference between expected model and the observed profile of the 0.847 MeV line measured 613 days after the explosion, suggesting that even more extensive mixing and asymmetric ejection are required to account for the observed line shape, because the line predicted for the 10HMM model is much too narrow, and too blueshifted to be an acceptable fit. Clearly further study is needed of the dynamics of the mixed and asymmetric ejecta in Type II supernovae.

SN1987A 847 LINE

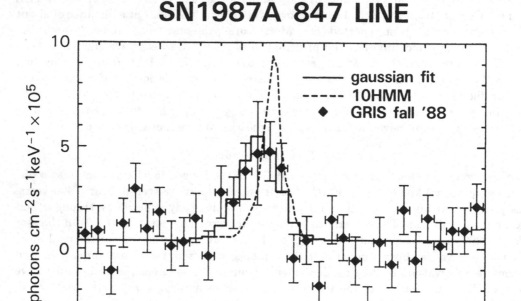

FIGURE 9. The 0.847 MeV line observed (Tueler *et al.* 1988) with a high resolution detector from Supernova 1987A compared with calculations (Pinto & Woosley 1988) for the mixed model 10HMM, suggesting significant asymmetry in the ejecta.

The 0.122 MeV line from the decay of ^{57}Co, resulting from neutron capture on freshly synthesized ^{56}Ni, has been detected with OSSE (Kurfess *et al.* 1992). The observed flux of about 10^{-4} photons cm^{-2} s^{-1} suggests that the production ratio of ^{57}Ni/^{56}Ni is about 1.5 times the solar abundance ratio of ^{57}Fe/^{56}Fe. This is quite consistent with current model calculations (*e.g.*, Thielemann, Hashimoto & Nomoto 1990).

3.2. ^{44}Ti decay lines from Cas A and other young supernovae

On a longer time scale, ^{44}Ti decays with a mean life of anywhere between 78 years (Frekers *et al.* 1983) and 96 years (Alburger & Harbottle 1990) to ^{44}Sc, producing gamma ray lines (Lederer & Shirley 1978) at 67.9 (100%) and 78.4 keV (98%); ^{44}Sc subsequently decays with a 5.7 hr meanlife to ^{44}Ca, producing a line at 1.157 MeV (100%). The relatively long lifetime of ^{44}Ti, together with its lower nucleosynthetic yield, make these lines too weak to observe from extragalactic supernovae using current detectors. However, this longer life should allow us to observe these lines from Galactic supernovae for several hundred years after the explosion and thus use them to discover the most recent supernovae in our Galaxy. Historical records have allowed us to identify only 2 or 3 nearby Galactic supernovae within the past 300 years. However, the estimated (van den Bergh & Tammann 1991) Galactic supernova rate of $8.4h^2$ per 100 years, gives an expected number of between 6 and 24 Galactic supernovae within the last 300 years as-

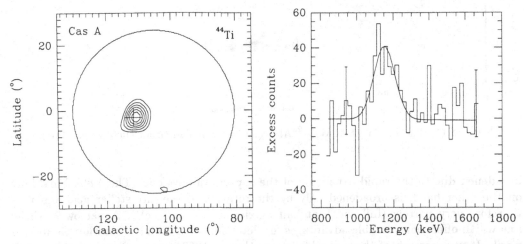

FIGURE 10. The ^{44}Ti line emission from the young supernova remnant Cas A; left panel—the location of the line emission; right panel—the spectrum of the line emission (from Iyudin *et al.* 1994).

suming $0.5 < h < 1$, or $50 < H_o < 100$ km s^{-1} Mpc^{-1}. The detection of the ^{44}Ti decay lines thus can enable us to discover the locations of all of these other recent supernovae, which were optically obscured, and from their gamma ray locations we can then look in radio and other bands to identify their remnants.

Line emission at 1.157 MeV has recently been measured (Iyudin *et al.* 1994) with COMPTEL from the youngest known Galactic supernova remnant, Cas A (see Fig. 10). At an estimated (Ilovaisky & Lequeux 1972) distance of 2.8 kpc and estimated (Fesen & Becker 1991) age of ~ 310 yr, the observed ^{44}Sc line flux of $(7.0 \pm 1.7) \times 10^{-5}$ photons cm^{-2} s^{-1} corresponds to an initial ^{44}Ti mass of about 1.4 to $3.2 \times 10^{-4} M_\odot$, given the present uncertainty in the ^{44}Ti meanlife. Such a yield is quite consistent with that expected from current models of Cas A as a Type Ib supernova from the core-collapse of a $\sim 20 M_\odot$ Wolf Rayet star (Ensman & Woosley 1988; Shigeyama *et al.* 1990).

Assuming a comparable ^{44}Ti yield in the much more frequent Type II supernovae, COMPTEL may be able to locate several other younger but more distant Type II supernovae, and the planned INTEGRAL should discover the most recent dozen or two in our Galaxy.

3.3. *Gamma ray line emission from ^{26}Al decay*

On a much longer time scale, gamma ray line radiation at 1.809 MeV results from the decay of ^{26}Al (mean life 1.07×10^6 years) into the first excited states of ^{26}Mg. Because of this long lifetime and an encouraging theoretical ^{26}Al yield in supernovae (Schramm 1971), we proposed (Ramaty & Lingenfelter 1977) that this line should be the nucleosynthetic line with the best prospects for detection. The same idea was also independently proposed (Arnett 1977). The long lifetime is important for at least three reasons: it allows the accumulation of ^{26}Al in the interstellar medium, thereby ensuring steady line emission; it allows the escape of the nucleosynthetic ^{26}Al from its production site before it decays, independent of the complications introduced by the dynamics of the region in which the explosive nucleosynthesis takes place; and it guarantees that the width of the line is going to be very narrow. Gamma ray lines from shorter lived isotopes, such as the 0.847 MeV line of ^{56}Co (mean life 111 days), can be observed from a relatively close source (*e.g.*, supernova 1987A) only for a short period of time and they are significantly

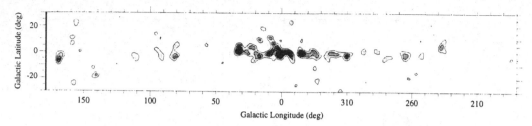

FIGURE 11. The sky in 1.809 MeV ^{26}Al decay line emission (from Diehl *et al.* 1995).

broadened due to the rapid expansion of the supernova envelope. The 1.809 MeV line, on the other hand, is broadened only by the rotation of the interstellar gas, and it is therefore very narrow (full width at half maximum about 3 keV). A narrow intrinsic line width offers considerable advantages for detection with a high resolution Ge instrument. It was because of these considerations that in 1977, two years before the launch of HEAO-3, we proposed that the 1.809 MeV line should be detectable.

The 1.809 MeV line was indeed the first nucleosynthetic gamma ray line to be observed, showing that nucleosynthesis is an ongoing process in the Galaxy at the present epoch. The 1.809 MeV line was first detected (Mahoney *et al.* 1984) from the direction of the Galactic center with the HEAO-3 high resolution Ge spectrometer and was subsequently confirmed by observations (Share *et al.* 1985) with the NaI detector on SMM, and balloon borne Ge instruments (*e.g.*, MacCallum *et al.* 1987). Most recently, imaging observations (Diehl *et al.* 1994, 1995) of the 1.809 MeV line have been carried out with COMPTEL. As can be seen in Fig. 11, the COMPTEL data reveal a broad, patchy longitude distribution, that is very different from that of the 0.511 MeV line which is strongly peaked at the Galactic center (§4). This result clearly demonstrates that the two line emissions have quite different origins.

Estimates (Mahoney *et al.* 1984; Skibo, Ramaty & Leventhal 1992) of the total amount of ^{26}Al that resides in the Galaxy range from about 1.7 to $3M_\odot$, depending on assumed model for the Galactic distribution of ^{26}Al, the distance to the Galactic center and the exact values of the 1.809 MeV line fluxes implied by the observations. It should be emphasized that the derivation of a photon flux from data obtained by wide field of view detectors such as SMM and HEAO-3 does depend on the assumed longitude and latitude distribution of the 1.809 MeV line emission.

Although the requirement of 1.7 to 3 $M_\odot$ of ^{26}Al exceeds the originally predicted (Ramaty & Lingenfelter 1977) yield from supernovae, recent increases in the estimates of supernova yields of ^{26}Al and of the present rate of Type II supernova occurrence now suggest that such supernovae could in fact produce ^{26}Al at close to the observed rate. In particular, revisions (Woosley & Weaver 1986) of the reaction rates in core collapse models have substantially increased the calculated yield of ^{26}Al to $\sim 6.9 \times 10^{-5} M_\odot$ in the ejecta of a Type II supernova. More recent studies (Woosley *et al.* 1990) of hitherto neglected neutrino induced nucleosynthesis suggest a further doubling of the yield of ^{26}Al in such supernovae. Such a yield combined with the higher recent estimate (van den Bergh & Tammann 1991) of the Galactic Type II supernova rate of $6.1h^2$ per 100 years, gives an expected Galactic nucleosynthesis rate of $(4 - 8)h^2 M_\odot$ of ^{26}Al per 10^6 years. Thus Type II supernovae alone could account for all of the observed Galactic ^{26}Al production for any $h > 0.5$, or $H_o > 50$ km s^{-1} Mpc^{-1}. The calculated ^{26}Al yields for Type I and other supernovae are negligible by comparison (Nomoto, Thielemann & Yokoi 1984). Other possible sources are novae and Wolf-Rayet stars. However, recent calculations attribute no more than $0.5 M_\odot$ of ^{26}Al per 10^6 years to Galactic novae (Prantzos 1991), and $0.2 - 0.3 M_\odot$ per 10^6 years to Wolf-Rayet stars (Paulus & Forestini 1991). These objects, therefore, appear to produce less than $1/2$ of the ^{26}Al required to account for the observations. Moreover, the COMPTEL maps of the 1.809 MeV line intensity show an enhancement at the Vela supernova remnant (Fig. 11), supporting a supernova origin of the ^{26}Al, and not at γ Vel which would have been expected if Wolf Rayet stars were the source.

In addition to the Galactic longitude and latitude distribution of the 1.809 MeV line emission, information on the origin of the ^{26}Al could also be obtained from studies of the shape of the line. Doppler shifts of the line centroid energy of as much as 0.5 keV are expected due to Galactic rotation, depending on the distribution of the ^{26}Al and the direction of observation (Skibo & Ramaty 1991). All of these studies will allow a much better understanding of the origin of the radioactive Al, and hence of the chemical evolution of the Galaxy.

4. Galactic positron annihilation radiation

The 0.511 MeV line is perhaps the most important line in gamma ray astronomy. Its study began in 1970 when a line-like feature around 0.47 MeV, assumed to be due to positron annihilation, was observed from the direction of the Galactic center with a balloon borne NaI detector (Johnson, Harnden, & Haymes 1972). Various schemes were proposed to account for the redshift. Ramaty, Borner & Cohen (1973) suggested that the observed line was due to gravitationally redshifted annihilation radiation produced on the surfaces of neutron stars, while Leventhal (1973) pointed out that the convolution of a spectrum consisting of the 0.511 MeV line and the accompanying positronium continuum with the response function of a detector with poor energy resolution would lead to the apparent redshift. It was not until 1977, that the line energy was accurately determined with a Ge spectrometer (Leventhal, MacCallum, & Stang 1978). The observed line center energy, 510.7 ± 0.5 keV, clearly established that the radiation was due to the annihilation of positrons. In this and all subsequent detections with Ge spectrometers (Riegler *et al.* 1981; Gehrels *et al.* 1991; Leventhal *et al.* 1993; Smith *et al.* 1993), the line width was found to be very narrow (full width at half maximum < 3 keV) and the line center energy to be at 0.511 MeV within errors less than a keV (Fig. 12). Reviewing the possible origins for the line emission (Ramaty & Lingenfelter 1979), we pointed out that the most likely sources for the positrons responsible for the observed annihilation radiation were the radionuclei ^{56}Co, ^{44}Ti and ^{26}Al resulting from various processes of Galactic nucleosynthesis (see also Clayton 1973).

This point of view, however, was challenged by the subsequent HEAO-3 result, that the 0.511 MeV line flux from the direction of the Galactic center had varied on a time scale shorter than $1/2$ year (Riegler *et al.* 1981). Even though the significance of this result was weakened by a different analysis of the HEAO-3 data (Mahoney, Ling & Wheaton 1994), confirmation for the 0.511 MeV time variability was provided by a series of observations carried out with balloon borne Ge detectors from 1977 through 1984. Whereas strong 0.511 MeV line emission was seen in the 1977 and 1979 flights, only upper limits were obtained in 1981 and 1984. The statistical significance of the implied variability was about 3σ. Since variability on time scales of a few years or less can not be expected if the positrons result from multiple nucleosynthetic events, we proposed that the bulk of the positrons should be produced at or near a single compact object of less than a light year in diameter, most likely a black hole in the Galactic center region (Ramaty, Leiter & Lingenfelter 1981). We subsequently suggested (Lingenfelter & Ramaty 1982)

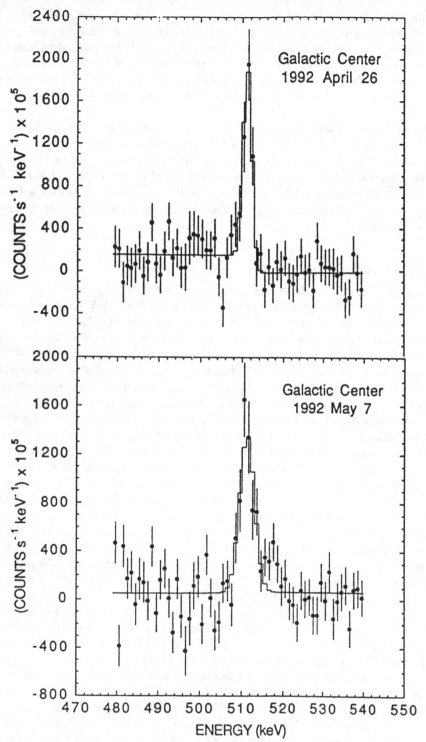

FIGURE 12. The 0.511 MeV positron annihilation line observations from the region of the Galactic center (from Leventhal *et al.* 1993).

that the annihilating positrons could be produced by $\gamma\gamma$ pair production of the higher energy ($> m_e c^2$) continuum photons which showed an even greater intensity variation simultaneous with the annihilation radiation. From the observed continuum flux and the estimated (Galactic center) distance to the source, the required rate of pair production limited the size of the source region, less than about 10^9 cm. On the basis of this result, we pointed out that the source probably was a stellar mass black hole that did not necessarily reside at the dynamical center of the Galaxy.

The possibility of positron production near black holes was strengthened by the discovery of a broad line at ~ 0.4 MeV from the X-ray source 1E1740.7-2942, located at an angular distance of $0.9°$ from the Galactic center (Bouchet *et al.* 1991; Sunyaev *et al.* 1991; see also Gilfanov *et al.* 1994). This transient line emission was observed on 13–14 October 1990 with the imaging gamma ray spectrometer SIGMA. Similar line emission from an unidentified source or sources in the direction of the Galactic center was also observed with non imaging Ge detectors flown on balloons in 1977 (Leventhal *et al.* 1978; Leventhal & MacCallum 1980) and 1989 (Smith *et al.* 1993), as well as from another unidentified source 12° away from the Galactic center with HEAO-1 (Briggs *et al.* 1995). We discuss these further in §5. If these 0.4 MeV line features are due to gravitationally redshifted positron annihilation produced near black holes, it is reasonable to assume that a fraction of the positrons could escape and annihilate at large distances from the hole to produce a line at precisely 0.511 MeV (Ramaty *et al.* 1992). If the pair production near the hole is time variable, the 0.511 MeV line flux will also vary in time, although the time scale of the latter variability would depend on the density of the medium in which the positrons annihilate, $\sim 10^5/n(\text{cm}^{-3})$ yrs (Guessoum, Ramaty, & Lingenfelter 1991).

Numerous observations of 0.511 MeV line emission from the Galactic center and the Galactic plane were carried out with OSSE on CGRO. This observatory was launched in 1991, and there are published 0.511 MeV line observations from July 1991 to October 1992 (Purcell *et al.* 1993). OSSE is a non-imaging NaI instrument with a relatively small field of view ($11°4 \times 3°8$). Most of the OSSE observations were carried out with the detector pointing at or close to the Galactic center. These observations suggest that the 0.511 MeV line emission is strongly concentrated toward the Galactic center. None of the OSSE observations show significant time variability, but they allow 3σ limits on weekly variations of $\pm60\%$ and daily variations of $\pm120\%$ relative to the observed flux of $(2.5 \pm 0.3) \times 10^{-4}$ photons cm^{-2} s^{-1}. Thus the question of the variability of the 0.511 MeV line flux must await future observations.

Models for the 3 dimensional Galactic distribution of annihilation radiation based on the OSSE and other 0.511 MeV line observations have been developed (Skibo 1993; Ramaty, Skibo & Lingenfelter 1994). The calculated longitude distributions (integrated over all Galactic latitudes) for two of the models are shown in Fig. 13. The dashed curve is based on the nova distribution of Higdon & Fowler (1989) which consists of two morphological parts, a disk and a spheroid centered at the Galactic center. Because novae occur on accreting white dwarfs that eventually become Type Ia supernovae, and because such supernovae could be important positron sources, this distribution should provide a reasonable starting point for the analysis. But as demonstrated by Skibo (1993) and Ramaty, Skibo & Lingenfelter (1994), this distribution cannot account for the very strong concentration of the 0.511 MeV line emission at the Galactic center ($P \sim 10^{-6}$, Skibo 1993). The solid curve was obtained by adding to the nova distribution an additional spheroid, which, for simplicity was assumed to have the same shape as the spheroid associated with the nova distribution itself. We refer to the nova part and the additional spheroidal part of the total distribution as the Galactic plane and the central Galactic

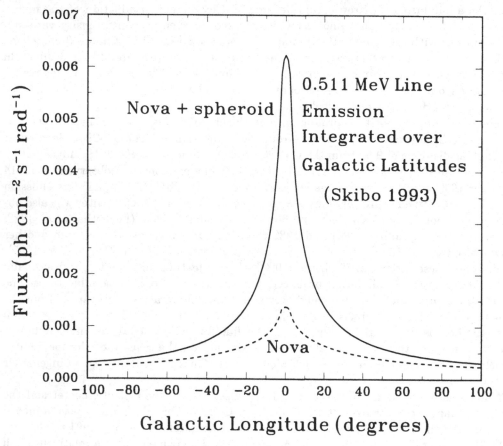

FIGURE 13. Galactic longitude profiles of the 0.511 MeV line emission (from Skibo *et al.* 1993).

components, respectively. By allowing the ratio of positron production in these two components to vary, a good fit to the data was obtained (P $\sim$ 0.6, Skibo 1993) for a central Galactic–to–Galactic plane positron production ratio of 2.6. Normalization to the 0.511 MeV line observations also determines the absolute productions, 2.6×10^{43} e$^+$ s^{-1} and 1×10^{43} e$^+$ s^{-1}, for the two components respectively.

The bulk of the positrons responsible for the Galactic plane component could result from the decay of ^{56}Co, ^{44}Sc and ^{26}Al produced in various Galactic processes of nucleosynthesis (*e.g.*, Lingenfelter, Chan & Ramaty 1993). The contribution of other processes (*e.g.*, cosmic ray interactions, pair production in pulsars) is quite small (Ramaty & Lingenfelter 1991). The total Galactic positron production rate from the decay of ^{26}Al was estimated (Skibo *et al.* 1992) to be about 0.2×10^{43} positrons s^{-1} (see also §3), *i.e.*, about 20% the total Galactic plane component. The rest of the positrons probably result from ^{56}Co and ^{44}Sc, where the total production rate of these isotopes scales with the present rate of Galactic iron nucleosynthesis and the relative contributions of these two isotopes depends on the positron escape fraction from the envelopes of Type Ia supernovae (Chan & Lingenfelter 1993). These two radionuclei are thought to be produced primarily in Type Ia supernovae and the distribution of their production rate is expected to follow that of Galactic novae.

The origin of the positrons in the central component is essentially unknown. It is possible that black hole candidates are a major contributor to positron production in the

Galactic center region at the present epoch. Because both the positron annihilation time and the distance that the positrons can travel from their sources to their annihilation site is strongly dependent on the properties of the medium in which they are trapped, both the spatial extent of annihilation radiation produced by a point source of positrons and the time dependence of the emission are highly uncertain. It is possible that the entire central Galactic component is fed by just 1E1740.7-2942. But additional sources may also contribute significantly (see Ramaty & Lingenfelter 1994 for more details). We might expect comparable enhancements of the 0.511 MeV emission in the direction of Nova Muscae and other candidate black hole sources, although the production of a narrow 0.511 MeV line depends on the existence of not only a positron source but also a sufficiently dense annihilation site. Clearly much better mapping of the Galactic annihilation radiation by the planned INTEGRAL and perhaps other missions is needed to resolve the question of its origin.

5. Line features from black hole candidates

As mentioned above, line-like emission features at $\sim$ 0.4 MeV have been observed from a number of sources assumed to be accreting black holes, and in several of these observations this line was accompanied by another line at $\sim$ 0.2 MeV. Line emission at both $\sim$ 0.48 MeV and $\sim$ 0.19 MeV were simultaneously observed with SIGMA from Nova Muscae during its outburst on 20 January 1991 (Goldwurm *et al.* 1992). Similar features at 0.496 MeV and $\sim$ 0.17 MeV and at $\sim$ 0.40 MeV and $\sim$ 0.16 MeV were observed with balloon-borne Ge spectrometers from an unidentified source, or sources, in the Galactic center region in 1977 (Leventhal & MacCallum 1980) and 1989 (Smith *et al.* 1993), respectively. Only the higher energy line at $\sim$ 0.40 MeV was observed with SIGMA from 1E1740.7-2942 during an outburst on 13–14 October 1990 (Bouchet *et al.* 1991; Gilfanov *et al.* 1994), and at $\sim$ 0.46 MeV with HEAO-1 from another source, possibly the low mass X-ray binary 1H1822–371, about 12° away from the Galactic center (Briggs *et al.* 1995). Two other transients exhibiting emission features were observed with SIGMA from 1E1740.7-2942 (Cordier *et al.* 1993; Gilfanov *et al.* 1994). However, for one of these, that on 19–20 January 1992, the flux reported by SIGMA is in conflict at more than 3σ with co-temporal CGRO OSSE and BATSE observations (Jung *et al.* 1995; Smith *et al.* 1995).

The $\sim$ 0.4 MeV line has frequently been assumed to be redshifted positron annihilation radiation. If the redshift is gravitational, the line must be formed around a compact object, presumably an accreting black hole, at distances varying from a few Schwarzschild radii for the 1E1740.7-2942 source, to more than 10 Schwarzschild radii for the 1977 source and Nova Muscae. The $\sim$ 0.2 MeV line, observed at the same time from the latter two sources, can been interpreted as Compton backscattered reflection of the annihilation feature from the inner edge of an optically thick accretion disk (Lingenfelter & Hua 1991; Hua & Lingenfelter 1993).

This second line results from the fact that Compton scattering of photons of energy E_0 into an angle θ will produce photons of energy

$$E_s = \frac{E_0}{1 + \frac{E_0}{m_e c^2}(1 - \cos\theta)}, \qquad (5.5)$$

so that, if the initial photons form a line at $E_0 \simeq m_e c^2$, then backscattered photons ($\cos\theta = -1$) will form another line at $E_s \simeq m_e c^2/3 = 0.17$ MeV. The intensity of the backscattered line is typically only $\sim$ 5% of that of the initial line, while the observed ratio of these lines is about 30% for Nova Muscae. However, the initial line can be

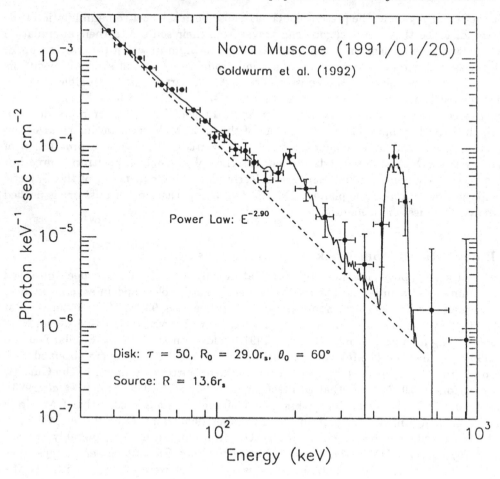

FIGURE 14. The observed (Goldwurm *et al.* 1992) spectrum from Nova Muscae during the outburst on 20 January 1991 compared with the Monte Carlo simulated (Hua & Lingenfelter 1993) spectrum of Compton scattered annihilation radiation for an accretion disk around a black hole, superimposed on a background power law with index $s = -2.90$ and viewed at an observing angle of $68°$ with respect to the axis of the accretion disk.

strongly attenuated by the accretion disk itself. Hua & Lingenfelter (1993) have shown that such scattering gives excellent agreement with the observed pairs of line-like features which were seen from both Nova Muscae (Fig. 14) and the 1977 source (Lingenfelter & Hua 1991). Moreover, assuming that the observing angle of $68 \pm 14°$ relative to the axis of the accretion disk, determined for Compton scattering in Nova Muscae, is the same as the inclination angle of the binary system, a central black hole mass of $5.6 \pm 1.3 \, M_\odot$ can be determined using the optically determined mass function (Remillard, McClintock, & Bailyn 1992). However, the temperature kT of the gas which backscatters the photons should not exceed about 10 keV since otherwise the backscattered feature will be broader than observed.

On the other hand, for the 1E1740.7-2942, 1989 and 1H1822–371 sources, the higher energy lines were observed at energies implying much larger redshifts. This would imply that pair production and annihilation occur at essentially the same physical site, which leads to a problem because the temperature required to produce the pairs greatly exceeds the upper limit on the temperature set by the width of the ~ 0.4 MeV line. This argument has been quantified by Maciolek-Niedzwiecky & Zdziarski (1994) who showed that the line centroid requires that the positrons annihilate in a region around 3 Schwarzschild radii from the hole and that the line width requires that the temperature in this region be ($kT \lesssim 50$ keV).

A different explanation of the redshift, assuming positron annihilation, is that the observed ~ 0.4 MeV line energy is due to the motion of the annihilating region (Misra & Melia 1993) where the pairs annihilate at the base of a jet emanating from the black hole. This explanation would allow the separation of the annihilation region from the pair production region. While this scenario could provide an explanation for the 1E1740.7-2942 observation, which revealed only the higher energy line, it might not explain the 1989 data, for which the lower energy line was more intense than that at higher energies; is not clear whether the annihilation line could be greatly attenuated relative to the backscattered line in such a jet geometry.

An alternative interpretation (Skibo *et al.* 1994) for both the ~ 0.2 and ~ 0.4 MeV line features is that they result form Compton scattering of high energy continuum photons in a jet. This can again be seen from Eq. (5.5). If $E_0 >> m_e c^2$, then $E_s \simeq m_e c^2/(1 - cos\theta)$, *i.e.*, energy of the scattered photon is independent of its initial energy and depends only on the scattering angle. Consequently, if the original spectrum were very flat, or consisted entirely of photons well above $m_e c^2$, the scattered photons would accumulate at E_s, producing a line-like feature. Skibo *et al.* (1994) showed that such Compton scattering in a double sided jet can produce the two lines around 0.4 and 0.2 MeV for jet bulk flow velocities β around 0.55 and observing angle cosines in the range 0.2 to 0.6. The required observing angle for Nova Muscae for this model is 68° (J. G. Skibo, private communication 1994), essentially the same as that for the backscatter model, so that the implied black hole mass is nearly identical. Also radio observations show that 1E1740.7-2942 is in fact associated with a double sided jet (Mirabel *et al.* 1992).

Lastly, a broad line in the 0.4–0.5 MeV can also be produced by interactions amongst α particles, $^4He(\alpha, p)^7Li^{*478}$ and $^4He(\alpha, n)^7Be^{*429}$ (Kozlovsky & Ramaty 1974). This feature has been seen from solar flares (§2). However, the expected line centroid, without any redshift, is at 0.45 MeV and this is in marginal conflict with the Nova Muscae observation, for which the line centroid was at 0.48 ± 0.02 MeV (Goldwurm *et al.* 1992), and definitely with the 1977 observation for which the line centroid was at 0.496 MeV. On the other hand, the observations (Martin *et al.* 1992; 1994) of high lithium abundances in the binary companions of the black hole candidates V404 Cygni, Cen X-4 and A0620, possibly due to α–α reactions in the accretion disks of the holes, do provide support for this model. Again Compton backscattering of these α–α photons from the inner region of an optically thick accretion disk could account for the $\lesssim 0.2$ MeV line emission.

Fortunately, there are tests to distinguish between the various models. The principal prediction of the annihilation/backscattering model is that the two lines should always be just below 0.511 MeV and around 0.2 MeV for all sources. This seems to be borne out by the current, albeit very limited, source population. On the contrary, the jet model in which both line are due to Compton scattering, does predict lines at various energies, including energies above 0.511 MeV. The observation of a broad line around 1 MeV from Cygnus X-1 (Ling & Wheaton 1989) may be due Compton scattering of high energy photons into viewing angles close to the forward direction. Polarization observations could also distinguish the models. For Compton scattering in jets, both lines should be polarized (see Skibo *et al.* 1994); however, no polarization is expected in either of these lines for the annihilation/backscattering model. (Unpolarized radiation undergoing Compton scattering becomes polarized, except in the backward direction.) The required polarization observations have not yet been carried out.

The α–α interaction model makes two important predictions. The first one concerns the isotopic ratio ^{7}Li/^{6}Li in the binary companions of the black hole candidates. A ratio around 1.5 would favor the α–α reactions; a significantly higher ratio would probably require another mechanism for producing the Li. The second one involves the very narrow delayed 0.478 MeV line from ^{7}Be decay. For an outburst similar to that of Nova Muscae on 20 January 1991, we expect a delayed flux of $\sim 4 \times 10^{-6}$ photons cm^{-2} s^{-1}, equal to the ratio of ^{7}Be to ^{7}Li production of ~ 1 (Ramaty *et al.* 1979) times the 10.4% branching ratio times the observed broad line flux of 6×10^{-3} photons/cm^{2}sec times its 0.5 day duration divided by the 77 day ^{7}Be meanlife. Such a narrow line flux could be detected by the planned INTEGRAL spectrometer.

6. Conclusions

As we have seen, the study of gamma ray line radiation indeed spans a broad range of astrophysical problems. In solar flares the line radiation addresses problems of particle acceleration and transport, particle trapping, magnetic field structure, plasma turbulence and solar atmospheric composition. Of all astrophysical sites, the richest gamma ray line spectra observed so far are those from the Sun.

Deexcitation gamma ray line emission has been observed from the Orion star formation region showing that the gas and dust in this molecular cloud complex are currently undergoing strong irradiation by low energy cosmic rays. Given that the duration of the irradiation is about 10^5 years, the time span since a possible recent supernova in Orion, the total energy in accelerated particles is about 10^{51} erg. The relationship of such irradiation to the question of the origin of the light elements is currently under investigation. However, neither the overall Galactic ^{26}Al nor the ^{26}Al that was present at the formation of the solar system could have originated in accelerated particle bombardment.

Gamma ray lines from processes of nucleosynthesis have been observed from a variety of sites. Observations of the ^{56}Co lines from supernova 1987A have shown that the supernova explosion is much more complex than previously thought. Specifically, the early appearance of the lines requires mixing, while the shapes of the lines, determined with a high resolution spectrometer, require asymmetric supernova ejecta. Recently, gamma ray line emission from ^{44}Ti decay has been observed from Cas A, opening the possibility of observing similar line emission from other young, and hitherto unknown, supernova remnants in our Galaxy. The 1.809 MeV line from ^{26}Al decay has been observed and imaged. The bulk of the 2 M$_\odot$ of ^{26}Al in the interstellar medium are thought to be due to Type II supernovae which exploded in the Galaxy in the last million years. A supernova origin is supported by the observation of enhanced 1.809 MeV line emission from the Vela supernova remnant.

Galactic annihilation radiation has been observed. The high resolution Ge detectors flown on balloons have shown that there is strong, very narrow line emission produced at precisely 0.511 MeV at or near the Galactic center. The recent OSSE mapping of the flux in this line has shown that there is much emission strongly concentrated within less than a few degrees from the Galactic center. We believe that this concentration points to a black hole origin, specifically pair production in photon-photon collisions near the hole and escape and annihilation of the positrons in the surrounding medium. In addition, observations of the diffuse Galactic annihilation radiation set important constraints on the current rate of iron nucleosynthesis in the Galaxy.

Transient emission lines from black hole candidates have been discovered. Such lines were seen from 1E1740.7-2942 during an outburst in 1990, from Nova Muscae during an outburst in 1991, and from several other unidentified sources. The interpretations involve redshifted pair annihilation, Compton backscattering, Compton downscattering of collimated high energy gamma ray continuum and possibly α–α interactions producing ^{7}Li and ^{7}Be. The various models make predictions that will be tested by future observations. Because of their low duty cycle, these transients can be best studied by continuously monitoring the Galactic plane with wide field of view gamma ray spectrometers. These lines are providing unique probes of the accretion disks and jets of black holes.

We wish to acknowledge Hans Bloemen and Roland Diehl for supplying us with postscript figures of the COMPTEL data. We also acknowledge financial support from NASA under Grant NAG5-2811.

REFERENCES

ADAMS, J. H., *et al.* 1991 *ApJ* **375**, L45.

ALBURGER, D. E., & HARBOTTLE, G. 1990 *Phys. Rev. C* **41**, 2320.

ANDERS, E., & GREVESSE, N. 1989 *Geochim. et Cosmochim. Acta* **53**, 197.

ARNETT, W. D. 1977 *Ann. NY Acad. Sci.* **302**, 90.

ARNETT, W. D., BAHCALL, J. N., KIRSHNER, R. P., & WOOSLEY, S. E. 1989, *ARAA* **27**, 629.

BARAT, C. *et al.* 1994 *ApJ* **425**, L109.

BATES, D. R., & DALGARNO, A. 1962 in *Atomic and Molecular Processes*, (ed. D. R. Bates) New York: Academic Press.

BLOEMEN, H., *et al.* 1994 *A&A* **281**, L5.

BOUCHET, L. *et al.* 1991 *ApJ* **383**, L45.

BRIGGS, M. S., GRUBER, D. E., MATTESON, J. L., & PETERSON, L. E. 1995 *ApJ* in press.

BURROWS, D. N., SINGH, K. P., NOUSEK, J. A., GARMIRE, G. P., & GOOD, J. 1993 *ApJ* **406**, 97.

BURROWS, T. W., & BHAT, M. R. 1986, *Nuclear Data Sheets*, **47**, 1.

BUSSARD, R. W., BURROWS, A., & THE, L. S. 1989 *ApJ* **341**, 401.

BYKOV, A., & BLOEMEN, H. 1994 *A&A* **283**, L1.

CANE, H. V., McGUIRE, R. E., & VON ROSENVINGE, T. T. 1986 *ApJ* **310**, 448

CASSE, M., LEHOUCQ, R., & VANGIONI-FLAM, E. 1995 *Nature* **373**, 318.

CHAN, K-W., & LINGENFELTER, R. E. 1987 *ApJ* **318**, L51.

CHAN, K-W., & LINGENFELTER, R. E. 1988 in *Nucl. Spectr. of Astrophysical Sources* (eds. N. Gehrels and G. H. Share) 110.

CHAN, K-W., & LINGENFELTER, R. E. 1991 *ApJ* **368**, 515.

CHAN, K-W., & LINGENFELTER, R. E. 1993 *ApJ* **405**, 614.

CHUPP, E. L. 1990 *Physica Scripta* **T18**, 15.

CHUPP, E. L., FORREST, D. J., HIGBIE, P. R., SURI, A. N., TSAI, C., & DUNPHY, P. P. 1973 *Nature* **241**, 333.

CLAYTON, D. D. 1973 *Nature Phys. Sci.* **244**, 1973.

CLAYTON, D. D. 1974 *ApJ* **188**, 155.

CLAYTON, D. D. 1994 *Nature* **368**, 222.

CLAYTON, D. D., COLGATE, S. A. & FISHMAN, G. J. 1969 *ApJ* **155**, 75.

CLAYTON, D. D., & JIN, L. 1995 preprint.

CLIVER, E. W, KAHLER, S. W., & VESTRAND, W. T. 1993 *23rd Internat. Cosmic Ray Conf. Papers* **3**, 91.

CORDIER, B. *et al.* 1993 *A&A* **275**, L1.

CRANNELL, C. J., JOYCE, G., RAMATY, R., & WERNTZ, C. 1976 *ApJ* **210**, 582.

DIEHL, R. *et al.* 1994 *ApJS* **92**, 429.

DIEHL, R. *et al.* 1995 *A&A* in press.

DIGEL, S. W., HUNTER, S. D. & MUKHERJEE, R. 1995 *ApJ* **441**, 270.

ELLISON, D. C., & RAMATY, R. 1985 *ApJ* **298**, 400.

ENSMAN, L. M., & WOOSLEY, S. E. 1988 *ApJ* **333**, 754.

FESEN, R. A., & BECKER, R. H. 1990 *ApJ* **371**, 621.

FREKERS, D. *et al.* 1983 *Phys. Rev.* **C28**, 756.

GEHRELS, N., BARTHELMY, S. D., TEEGARDEN, B. J., TUELLER, J., LEVENTHAL, M., & MACCALLUM, C. J. 1991 *ApJ* **375**, L13.

GEHRELS, N., LEVENTHAL, M., & MACCALLUM, C. J. 1987 *ApJ* **322**, 215.

GEHRELS, N., MACCALLUM, C. J., & LEVENTHAL, M. 1987 *ApJ* **320**, L19.

GILFANOV, M. *et al.* 1994 *ApJS* **92**, 411.

GOLDWURM, A. *et al.* 1992 *ApJ* **389**, L79.

GUESSOUM, N., RAMATY, R., & LINGENFELTER, R. E. 1991 *ApJ* **378**, 170.

HARRIS, M., SHARE, G. H. & MESSINA, D. C. 1995 *ApJ* in press.

HIGDON, J. C., & FOWLER, W. A. 1989 *ApJ* **339**, 956.

HUA, X-M., & LINGENFELTER, R. E. 1987 *ApJ* **319**, 555.

HUA, X-M., & LINGENFELTER, R. E. 1993 *ApJ* **416**, L17.

HUO, J. D., HU, D. L., ZHOU, C. M., HAN, X. L., HU, B. H., & WU, Y. D. 1987 *Nuclear Data Sheets* **51**, 1.

ILOVAISKY, S. A., & LEQUEUX, J. 1972 *A&A* **18**, 169.

IYUDIN, A. F. *et al.* 1994 *A&A* **284**, L1.

JOHNSON, W. N., HARNDEN, & R. C. HAYMES 1972 *ApJ* **172**, L1.

JUNG, G. V. *et al.* 1995 *A&A* in press.

KOZLOVSKY, B., LINGENFELTER, R. E., & RAMATY, R. 1987 *ApJ* **316**, 801.

KOZLOVSKY, B., & RAMATY, R. 1974 *ApJ* **191**, L43.

KOZLOVSKY, B., & RAMATY, R. 1977 *Astrophys. Letters* **19**, 19.

KURFESS, J. D. *et al.* 1992 *ApJ* **399**, L137.

LEDERER, C. M., & SHIRLEY, V. 1978 *Tables of Isotopes* (New York: Wiley).

LEE, T., PAPANASTASSIOU, D. A., & WASSERBURG, G. J. 1977 *ApJ* **211**, L107.

LEMOINE, M., FERLET, R., & VIDAL-MADJAR, A. 1995 *A&A* in press.

LEVENTHAL, M. 1973 *ApJ* **183**, L147.

LEVENTHAL, M., BARTHELMY, S. D., GEHRELS, N., TEEGARDEN, B. J., TUELLER, J., & BARTLETT, L. M. 1993 *ApJ* **405**, L25.

LEVENTHAL, M., & MACCALLUM, C. J. 1980 *Ann. NY Acad. Sci.* **336**, 248.

LEVENTHAL, M., MACCALLUM, C. J., & STANG, P. D. 1978 *ApJ* **225**, L11.

LING, J. C., MAHONEY, W. A., WILLET, J. B., & JACOBSON, A. S. 1982 in *Gamma Ray Transients and Related Astrophysical Phenomena* (eds. R. E. Lingenfelter, H. S. Hudson & D. M. Worrall) New York: AIP, 143.

LING, J. C., & WHEATON, W. A. 1989 *ApJ* **343**, L57.

LINGENFELTER, R. E., CHAN, K-W., & RAMATY, R. 1993 *Physics Reports* **227**, 133.

LINGENFELTER, R. E., HIGDON, J. C., & RAMATY, R. 1978 in *Gamma Ray Spectroscopy in Astrophysics* (eds. T. Cline & R. Ramaty) NASA, 252.

LINGENFELTER, R. E., & HUA, X-M. 1991 *ApJ* **381**, 426.

LINGENFELTER, R. E. & RAMATY, R. 1976 *ApJ* **211**, L19.

LINGENFELTER, R. E., & RAMATY, R. 1982 in *The Galactic Center* (eds. G. R. Riegler & R. D. Blandford) New York: AIP, 148.

MACCALLUM, C. J., HUTERS, A. F., STANG, P. D., LEVENTHAL, M. 1987 *ApJ* **317**, 877.

MACIOLEK-NIEDZWIECKY, A., & ZDZIARSKI, A. 1994 *ApJ* **436**, 762.

MAEDER, A. & MEYNET, G. 1987 *A&A* **182**, 243.

MAHONEY, W. A., LING, J. C., & WHEATON, W. A. 1994 *ApJS* **92**, 387.

MAHONEY, W. A., LING, J. C., WHEATON, W. A., & JACOBSON, A. S. 1984 *ApJ* **286**, 578.

MARTIN, E. L., REBOLO, R., CASARES, J., & CHARLES, P. A. 1992 *Nature* **358**, 129.

MARTIN, E. L., REBOLO, R., CASARES, J., & CHARLES, P. A. 1994 *ApJ* **435**, 791.

MATZ, S. M. *et al.* 1988 *Nature* **331**, 416.

MENEGUZZI, M., AUDOUZE, J., & REEVES, H. 1971 *A&A* **15**, 337.

MEWALDT, R. A. 1983 *Revs. Geophys. Space Phys.* **21**, 295.

MEYER, J-P. 1985 *ApJS* **57**, 151.

MEYER, J-P. 1989 *Cosmic Abundances of Matter*, (ed. C. J. Waddington) New York: AIP, 245.

MILLER, J. A., & STEINACKER, J. 1992 *ApJ* **399**, 284.

MILLER, J. A., & VIÑAS, A. 1993 *ApJ* **412**, 386.

MIRABEL, I. F., RODRIGUEZ, L. F., CORDIER, B., PAUL, J., & LEBRUN, F. 1992 *Nature* **358**, 215.

MISRA, R., & MELIA, F. 1993 *ApJ* **419**, L25.

MITLER, H. E. 1972 *Ap. and Sp. Sci.* **17**, 186.

MURPHY, R. J. *et al.* 1993 *23rd Internat. Cosmic Ray Conf. Papers*, 3, 99.

MURPHY, R. J., KOZLOVSKY, B., & RAMATY, R. 1990, *ApJ* **331**, 1029.

MURPHY, R. J., RAMATY, R., KOZLOVSKY, B., & REAMES, D. V. 1991 *ApJ* **371**, 793.

MURPHY, R. J., SHARE, G. H., LETAW, J. R., & FORREST, D. J. 1990 *ApJ* **358**, 298.

NATH, B. B. & BIERMANN, P. 1994 *MNRAS* **270**, L33.

NOMOTO, K., THIELEMANN, F-K. & YOKOI, K. 1984 *ApJ* **286**, 644.

NOMOTO, K. *et al.* 1988 in *Physics of Neutron Stars and Black Holes* (ed. Y. Tanaka) Tokyo: Universal Academy Press, 441.

PAULUS, G., & FORESTINI, M. 1991 in *Gamma Ray Line Astrophysics*, (eds. P. Durouchoux & N. Prantzos) New York: AIP, 183.

PINTO, P. A., & WOOSLEY, S. E. 1988 *Nature* **333**, 534.

PRANTZOS, N. 1991 in *Gamma Ray Line Astrophysics*, (eds. P. Durouchoux & N. Prantzos) New York: AIP, 129.

PRINCE, T. A. *et al.* 1983 *18th Internat. Cosmic Ray Conf. Papers* 4, 79.

PURCELL, W. R. *et al.* 1993 *ApJ* **413**, L85.

RAMATY, R. 1995 in *The Gamma Ray Sky with COMPTON GRO and SIGMA* (eds. M. Signore, P. Salati, & G. Vedrenne) Dordrecht: Kluwer, p. 279.

RAMATY, R., BORNER, G., & COHEN, J. M. 1973 *ApJ* **181**, 891.

RAMATY, R., KOZLOVSKY, B., & LINGENFELTER, R. E. 1979 *ApJS* **40**, 487.

RAMATY, R., KOZLOVSKY, B., & LINGENFELTER, R. E. 1995a *ApJ* **438**, L21.

RAMATY, R., KOZLOVSKY, B., & LINGENFELTER, R. E. 1995b *Ann. NY Acad. Sci.*, in press.

RAMATY, R., LEITER, D., & LINGENFELTER, R. E. 1981 *Ann. NY Acad. Sci.* **375**, 338.

RAMATY, R., LEVENTHAL, M., CHAN, K-W., & LINGENFELTER, R. E. 1992 *ApJ* 392, L63.

RAMATY, R., & LINGENFELTER, R. E. 1977 *ApJ* **213**, L5.

RAMATY, R., & LINGENFELTER, R. E. 1979 *Nature* **278**, 127.

RAMATY, R., & LINGENFELTER, R. E. 1991 in *Gamma Ray Line Astrophysics*, (eds. P. Durouchoux & N. Prantzos) New York: AIP, 67.

RAMATY, R., & LINGENFELTER, R. E. 1994 in *High Energy Astrophys.* (ed. J. M. Matthews) Singapore: World Scientific, 32.

RAMATY, R., MANDZHAVIDZE, N., KOZLOVSKY, B., & SKIBO, J. 1993 *Adv. Space Res.* **13**, No. 9(9), 275.

RAMATY, R., SKIBO, J. G., & LINGENFELTER, R. E. 1994 *ApJS* **92**, 393.

RAMATY, R., SCHWARTZ, R. A., ENOME, S., & NAKAJIMA, H. 1994 *ApJ* **436**, 941.

REAMES, D. V. 1990 *ApJS* **73**, 235.

REAMES, D. V., MEYER, J-P., & VON ROSENVINGE, T. T. 1994 *ApJS* **90**, 649.

REEVES, H. 1994 *Revs. Modern Physics* **66**, 193.

REEVES, H., FOWLER, W. A., & HOYLE, F. 1970 *Nature, Phys. Sci.* **226**, 727.

REEVES, H., & PRANTZOS, N. 1995 in *Light Element Abundances* (ed. Ph. Crane) in press.

REMILLARD, R. A., MCCLINTOCK, J. E., & BAILYN, C. D. 1992 *ApJ* **399**, L145.

RIEGER, E. 1989 *Solar Phys.* **121**, 323.

RIEGLER, G. R. *et al.* 1981 *ApJ* **248**, L13.

RUIZ-LAPUENTE, P., LICHTI, G. G., LEHOUCQ, R., CANAL, R., AND CASSE, M. 1993 *ApJ* **417**, 547.

RYAN, J. *et al.* 1993 *23rd Internat. Cosmic Ray Conf. Papers* **3**, 103.

SCHMELZ, J. T. 1993 *ApJ* **408**, 381.

SCHNEID, E. J. 1994 Paper Presented at the 1994 AAS Winter Meeting, Crystal City, Virginia.

SCHRAMM, D. N. 1971 *Ap. and Sp. Sci.* **13**, 249.

SHARE, G. H., KINZER, R. L., KURFESS, J. D., FORREST, D. J., & CHUPP, E. L. 1985 *ApJ* **292**, L61.

SHEMI, A. 1991 *MNRAS* **251**, 221.

SHIBAZAKI, N., & EBISUZAKI, T. 1988 *ApJ* **327**, L9.

SHIGEYAMA, T., NOMOTO, K., TSUJIMOTO, T., & HASHIMOTO, M. 1990 *ApJ* **361**, L23.

SKIBO, J. G. 1993 *Diffuse Galactic Positron Annihilation Radiation and the Underlying Continuum*, PhD Dissertation, University of Maryland.

SKIBO, J. G., DERMER, C., & RAMATY, R. 1994 *ApJ* **431**, L39.

SKIBO, J. G., & RAMATY, R. 1991 in *Gamma Ray Line Astrophysics*, (eds. P. Durouchoux & N. Prantzos) New York: AIP, 168.

SKIBO, J. G., RAMATY, R., & LEVENTHAL, M. 1992 *ApJ* **397**, 135.

SMITH, D. M. *et al.* 1993 *ApJ* **414**, 165.

SMITH, D. M., LEVENTHAL, M., CAVALLO, R., GEHRELS, N., TUELLER, J., & FISHMAN, G. 1995 *ApJ* submitted.

SWINSON, D. B., & SHEA, M. A. 1990 *Geophys. Rev. Lett.* **17**, 1073.

SUNYAEV, R. *et al.* 1987 *Nature* **330**, 227.

SUNYAEV, R. *et al.* 1991 *ApJ* **383**, L49.

SUNYAEV, R. *et al.* 1992 *ApJ* **389**, L75.

TEEGARDEN, B. J., BARTHELMY, S. D., GEHRELS, N., TUELLER, J., LEVENTHAL, M., & MACCALLUM, C. J. 1989 *Nature* **339**, 122.

TEEGARDEN, B. J., BARTHELMY, S. D., GEHRELS, N., TUELLER, J., LEVENTHAL, M., & MACCALLUM, C. J. 1991 in *Gamma Ray Line Astrophysics*, (eds. P. Durouchoux & N. Prantzos) New York: AIP, 116.

THIELEMANN, F.-K., HASHIMOTO, M., & NOMOTO, K. 1990 *ApJ* **349**, 222.

TUELLER, J., BARTHELMY, S., GEHRELS, N., TEEGARDEN, B. J., LEVENTHAL, M., & MACCALLUM, C. J. 1990 *ApJ* **351**, L41.

VAN DEN BERGH, S. & TAMMANN, G. A. 1991 *ARAA* **29**, 363.

VESTRAND, W. T., & FORREST, D. J. 1993 *ApJ* **409**, L69.

VON BALLMOOS, P., DIEHL, R., & SCHONFELDER, V. 1987 *ApJ* **318**, 654.

WANG, H. T., & RAMATY, R. 1974 *Solar Phys.* **36**, 129.

WEAVER, T. A., & WOOSLEY, S. E. 1980a *Ann. NY Acad. Sci.* **336**, 335.

WEAVER, T. A., & WOOSLEY, S. E. 1980b in *Supernova Spectra* (eds. R. Meyerott & G. H. Gillespie) New York: AIP, 15.

WEAVER, T. A. & WOOSLEY, S. E. 1993 *Physics Reports* **227**, 65.

WERNTZ, C., LANG, F. L., & KIM, Y. E. 1990 *ApJS* **73**, 349.

WINKLER, C. 1994 *ApJS* **92**, 327.

WOOSLEY, S. E., HARTMANN, D., HOFFMAN, R., D., & HAXTON, W. C. 1990 *ApJ* **356**, 272.

WOOSLEY, S. E., & PINTO, P. A. 1988 in *Nuclear Spectroscopy of Astrophysical Sources*, (eds. N. Gehrels & G. H. Share) New York: AIP, 98.

WOOSLEY, S. E. 1991 in *Gamma Ray Line Astrophysics*, (eds. P. Durouchoux & N. Prantzos) New York: AIP, 270.

WOOSLEY, S. E., & WEAVER, T. A. 1986 in *Nucleosynthesis and Its Implications on Nuclear and Particle Physics* (eds. J. Audouze & N. Mathieu) Dordrecht: Reidel, 145.

YOSHIMORI, M. 1990 *ApJS* **73**, 227.

YOSHIMORI, M. *et al.* 1994 *ApJS* **90**, 639.

Summary Remarks

By VIRGINIA TRIMBLE

Physics Department, University of California, Irvine, CA 92717 and Astronomy Department,
University of Maryland, College Park, MD 20742

"Reading maketh a full man, conference a ready man,
and writing an exact man."   F. Bacon

1. Highlights

For the reader who has only a couple of moments to spare, the strongest overall impressions from "analysis of emission lines" were (1) infrared and ultraviolet astronomy have merged with optical astronomy in their techniques and power, and no longer need to be considered separately (except that Dufour and Dinerstein do these things so well); (2) limited wavelength resolution keeps this from being the case yet in X-ray astronomy, though planned missions promise improvements (Mushotzky), while gamma ray emission, coming largely from nuclear rather than atomic processes, will continue to require very different approaches (Ramaty); (3) the enormous growth of detailed atomic data (Pradhan) and sophisticated techniques for handling the partial redistribution of photons across line profiles and other non-linear processes in radiative transfer (Hummer) means that current computing power is not yet able to implement the best calculations that we, in principle, know how to do, especially for intrinsically complex systems like supernovae (Pinto) and lumpy stellar winds (Drew); and (4) there is something reassuring about encountering a large body of astronomical endeavor to which it matters hardly at all whether or not the early universe was dominated by a Gaussian, Harrison-Zeldovich spectrum of adiabatic fluctuations in biased Cold Dark Matter.

2. Historical introduction and how to conclude a conference

The final remarks at the Osterbrock-Seaton symposium were assigned to Bob Kirshner, who was very much missed throughout the three days of the meeting. The task was reassigned to me midway through Monday, and I was instantly made aware of the truth of the remark attributed to Louis Pasteur (in translation) that "chance favors the prepared mind." If you look back at the original French, it says something more like "chance favors only the prepared mind," which was even more sobering, since I had, of course, collected none of the preliminary material with which I would normally have approached the task.

Good heavens, you ask, do concluding remarkers prepare their viewgraphs before the meeting? Absolutely—perhaps a third or half of them, including lists of questions with answers to be filled in during the talks. Do they even sometimes recycle their viewgraphs from meeting to meeting? Well no (though accusations have been made), except perhaps with humorous intent, to show that particular issues have been with us for a long time.

How does one prepare? First, look up something of the history of the subject. For instance, C. A. Young first spotted solar coronal emission lines in 1869. They were not fully identified until the work of Edlén (1942–43) and the cause, in the sense of why the corona is so hot, remains poorly understood down to the present (a harsh warning to gamma ray burst theorists of what may yet lie ahead 20 years after their phenomenon was

1909	4686	6563
1488	4151	6300
1068	4486	6552
1952	4862	6890
1640	4267	6667
1785	4101	6948
1560	4339	6716
2403	4363	6548
2992	5477	6584
3227	5822	6355
3727	5007	7009
3726	5949	7027
3729	5548	7378
3782	5577	7411
3516	5128	7097
3869	6741	

TABLE 1. Wavelengths of emission lines mentioned by workshop participants (Unfortunately contaminated by an assortment of NGC numbers)

discovered). But even earlier, in 1867, W. Huggins had recognized that galactic nebulae (meaning those more or less along the equator) separated cleanly into two groups, those with continuous spectra and those with bright line, gaseous spectra. Emission lines in comets were known at least from the advent of the great comet of 1882 onward.

The story for spiral nebulae turns out to be quite different from what I had supposed. Of course they have strong continua, known from the mid 19th century. Their redshifts and rotation curves today are measured from emission lines (optical or radio), and I had supposed this always to have been so. But when V. M. Slipher took the first spiral spectrogram revealing lines (Andromeda on 3–4 December 1912, from Lowell Observatory), it was the absorption features he spotted, recognized as sun-like, and used to find a large heliocentric velocity of 300 km/sec.

In the stellar realm, a surviving spectrogram of SN 1895B in NGC 5253 preserves its characteristic emission/absorption feature (but these were not fully recognized as such for another 70-some years). I failed to find, in a very cursory search, the first observer of (non-solar) stellar emission lines. But they were worth two pages (878–79) to Russell, Dugan, and Steward in 1926, when the name was just changing from bright lines to emission lines. Neither the "intrinsic" (gas hotter than photosphere) nor "extrinsic" (diffuse gas volume bigger than photosphere) causes mentioned by Drew was properly understood then. The authors bet on something like fluorescence or population inversions, even for hydrogen.

Which have been the most important and informative emission lines over the history of astronomy? The 1420 MHz line of neutral hydrogen, not much discussed here, surely comes first. Hβ (which we saw most often as part of the inexplicable Lyα/Hβ ratio) probably comes next. And the lowest lying transition of CO is perhaps third. A couple dozen others were mentioned at least once. The table lists them. It has, unfortunately, become contaminated by some NGC numbers. Most four digit numbers can, of course, be both, but are more famous as one than the other. A small prize will be given to the first three readers who correctly separate the table into emission lines and NGC objects likely to be of interest to workshop participants.

Additional pre-conference research involves looking up proceedings of previous meetings on similar topics, to see which important questions were left unanswered. The preliminary program also repays study. What have the invited speakers been working on lately? What are the potentials for interactions and disagreements among them, along the lines we saw here in the case of shock ionization vs. photo-ionization models. Which speakers sound like they might answer which of the old questions (for instance is Lyα/Hβ a radiative transfer problem? Hummer never quite told us.) The contributed or poster papers are a guide to the future because they reflect work in progress. What is the balance of topics, and how has it changed? An astronomer who thinks that the main use of spectral lines is abundance measurements would have been surprised by the present inventory. Finally, who is coming (and who is not, and why), and how are their names pronounced? I am a little disappointed that Ewa Szuszkiewicz did not present a poster.

If the conference honors particular colleagues, one has to explain one's connections with them, the host institution, and organizers. The mathematical concept of the Erdös number can be useful in this regard. Only Erdös himself has an Erdös number = 0. Those who have written papers with him have E = 1; those who have written a paper with someone who has written a paper with him have E = 2, and so forth. It is said that every living mathematician has E $\leq$ 3. It was clear from other speakers' remarks that nearly all workshop members have Seaton and/or Osterbrock numbers of at most two. My claims include having studied undergraduate stellar structure at UCLA in the same class as Joe Miller (now nominally Osterbrock's boss at Lick), having attempted to discriminate between Seaton's "long" and "short" distance scales for planetary nebulae (it doesn't matter what the answer was, the method was wrong); having had my Space Telescope proposals rejected; and having written the third PhD dissertation ever on the Crab Nebula (Bob Williams in 1967 beat me by a year, and Woltjer by a decade), with Bengt Strömgren, then AAS president, in the chair when I first talked about the work.

And then you have to decide how to organize all the neat, new items presented at the meeting, first for the remarks and then for the manuscript. Most symposia focus either on a class of objects (Be stars, earth-crossing asteroids, barred galaxies), or on methodologies (automated telescopes, very long baseline interferometry, Fabry-Perot spectroscopy). The natural organization of a summary is then in terms of whichever was not the focus. If it was a VLBI conference, you organize by objects, from the solar system outward. If it was an AGN conference, you organize by wavelengths and angular scales. This emission line workshop included both method-oriented and object oriented talks and posters. As a result, no very tidy structure immediately suggested itself, and I have not succeeded in finding one. An attempt has been made to mention every poster at least briefly, since they are not otherwise represented in this volume. Poster authors are given with initials; invited speakers without).

3. Logistics and mechanics of astrophysical living

3.1. *Atomic and molecular data*

Many astronomers have relatively little idea what is involved here ("Oh, you mean f-values and that sort of thing."), and rank the topic somewhere just above or below positions, proper motions, and parallaxes—necessary, of course, but preferably to be investigated by somebody else, who then is supposed to provide an error-free table of all the numbers we might need. Pradhan's talk made clear just how much more is involved. Besides f-values (or transition probabilities) we need data on energy levels, electron impact excitation and ionization, photo-ionization, electron-ion recombination, charge exchanges, and line broadening mechanisms, not to mention isotope shifts and hyperfine structure (T. Brage), and classification of lines in some scheme that allows us some chance to find the data we need (R. Williams).

Large groups in half a dozen places (London—the opacity project, Belfast, Vienna, Japan, Oak Ridge, NIST, JILA, Ohio State, ...) are at work assembling giant data bases for many elements and ions, supplemented by special-purpose programs (e.g. f-values for highly ionized iron, reported by J. You.) Three surprises were the shear numbers involved (23,871 transitions of Fe III!), that one has to start reading a whole new set of journals to keep up (the Atomic and Nuclear Data Tables, J. Phys. Chem., J. de Physique, Physica Scripta, Physics Reports, Advances in Atomic & Molecular Physics, J. Phys. B, and others, not just ApJ Supplements), and that nearly all the new results come from calculations, not laboratory measurements. It seems that most of the levels you can store on earth long enough to look at have already been done, though there are exceptions, like the intersystem A values for CII] etc. measured by W. Parkinson.

New relationships continue to show up. For instance, the $^6P_{3/2}$ and $^2P_{3/2}$ levels of a particular ultraviolet Fe II multiplet (191) mix, and the spectrum of CH Cyg may actually show the effects (S. Johansson). More serious, enhanced excitation of N II by a transition at 748 Å may mimic an enhanced nitrogen abundance in novae, star burst galaxies, and such (T. Tripp).

Closely related is the continuing effort to identifying correctly every emission feature seen in real astronomical objects. RR Tel (H. Duerbeck) seems to be the richest of the cataclysmic variables and NGC 7027 (D. Pequignot) of the planetary nebulae. The element inventory in PNe now extends well past the iron peak to include Kr, Xe, and Pb.

Interpretation of molecular spectra requires a different but equally extensive armamentarium of constants for level properties, transitions, dissociation and recombination coefficients, and so forth. Without these, the ratio of particular transitions in H_2 is not a diagnostic for temperature and heating mechanisms in interstellar clouds (including the probable need for a third excitation mechanism, penetration by X-rays); the amount of lithium expected from the big bang cannot be translated into a cooling rate due to LiH, LiH^+, and LiH_2^+; and many other strange and wondrous processes discussed by Dalgarno would be nearly meaningless. In a couple of cases, the very existence of important reactions is a recent discovery, for instance H (in n = 2) + $H_2 \rightarrow H_3^+$ in SN1987A and $CO + He^+ \rightarrow C^+ + O + He$ as a major threat to the CO seen there.

3.2. *Codes, algorithms, and calibrations*

The ideal numerical code (recipe, algorithm, ... I do not mean these as technical terms) is a transparent window between fully-understood physical processes and minimally-massaged data. Real life is less simple, even for mature programs like CLOUDY (Ferland) for photo-ionization studies and the corresponding treatment of shock ionization (Dopita). Disagreement here persists even about which process matters in active galaxies, supernova remnants, cooling flows, and so forth. Historical precedent suggests that the correct answer is going to be "both, please" (as Pooh said when asked whether he would like honey or condensed milk on his bread).

In the absence of the ideal, specific diagnostics (like EUV iron lines in the solar corona, B. Monsignori), special-purpose codes (like SUMA for AGNs, M. Contini), and simplifying assumptions (like the 5-level atom analysis package, Dufour, R. Shaw) remain of great importance. Other examples include an improvement on curve of growth (which nevertheless falls short of being a complete model atmosphere calculation) suitable for optically thick lines in luminous blue variables (R. Viotti and G. Muratorio), the expanded inventory of X-ray lines (318!) that must now be summed to match ASCA spectra of thermal sources (U. Hwang), and (my private, favorite example of the improbable) a Monte Carlo approach to photo-ionization (S. Och).

Not surprisingly, it is nearly always easier to start with a model and "predict" what you should see than to start with the data and work backwards (R. Sutherland on HII regions around OB associations; P. van Hoof on distance indicators for planetary nebulae); and model fitting is indeed now generally regarded as the proper way to approach such problems, and related ones like deconvolution of time histories of continuum and line emission to get sizes and structures of the emitting regions (Horne, J. Krolik).

The advent of HST data en masse has necessitated calibrations to ensure that its results can be tied and compared to those from other sources. The relationship between GHRS and IUE spectra seems to be under control (D. Walter), but that between FOS and ground based data still presents problems (E. Vassiliadis).

3.3. *Physical processes: New, newly recognized, newly constrained*

Two that I, at least, had never even heard of before are stimulated dielectronic recombination, potentially relevant to N III λ4640 in PNe (E. Oliva) and the scattering of radiation by Langmuir plasma turbulence, potentially relevant to line profile changes in luminous blue variables (G. Israelian).

Stimulated emission in general is, of course, well known, but as it moves into the far infrared (as, perhaps, in boron-like ions of C to S; J. Peng and M. Bautista), should we call it masing, lasing, irasing, or what? The first page of a poster announcing beaming of H_2O megamasers (J. Braatz) seemed briefly redundant, until it became clear that the advertised beaming was in the plane of the obscuring torus of active galactic nuclei, not what one would expect *a priori* at all!

Three additional processes now better defined are the partial redistribution of photons within optically thick lines, which affects level populations and ionization (I. Hubeny) as well as emergent line profiles (Hummer), the break-out of HII regions from galactic disks to produce low-density froth (G. Tenorio-Tagle) and entrainment (meaning the pick-up of additional gas by supersonic jets as they move through an ambient medium) in Herbig-Haro objects (J. Morse).

And, finally, three previously advertised processes, now seen or limited. First, the dependence of acceleration of particles into stellar coronae upon first ionization potential. This happens in the sun and is generally regarded as relevant to the difference between cosmic elemental abundances and those in the cosmic ray sources. But it does not happen in Procyon (M. Laming). Second, flashes or thermal pulses in the thin hydrogen and helium burning shells in ABG and post-AGN stars have been blamed for several kinds of rapid evolution (including that of FG Sge, the galloping giant). A pulse has apparently been caught in progress in the nucleus of the LMC PN N66 (S. Torres-Peimbert). The signature changes are in lines from the central star, not the PN (whose spectrum was constant through the observing period). A composition indicator of such pulses in the past in the C^{13}/C^{12} ratio in carbon stars (W. Rose). Third, one way to fuel an AGN sporadically is occasional tidal disruption of a passing star by the gravitational field of the central black hole. T. Storchi-Bergmann showed a double-peaked, broad Hα component in the nuclear spectrum of NGC 1097 that appeared between November 1991 and February 1994 and may be the result of such a disruption and the strewing of stellar gas into a disk or ring. If so, then the central continuum source should brighten, though not necessarily within the author's lifetime.

3.4. *Sources of ionization and excitation energy*

To first order, the competitors are photons beyond the Lyman limit and shocks. The diffuse ionized gas of the Milky Way, however, appears to need yet a third source, for which R. Dettmar proposes cosmic rays and/or magnetic reconnection. In the skirmishes between the two favorites, there exist some clean-cut cases, where we see independent evidence for the energy source. On the photo-ionization side, examples include photons from post-AGB stars in early type elliptical galaxies (L. Binette) and photons from the central star in Nova Cyg 1992, where the high ionization potential lines turned on at the same time as the EUV and X-ray continuum radiation (D. DePoy).

Shocks, on the other hand, can be tied to the ionization in IRAS galaxies that have nuclear winds (S. Veilleux). Slow, dense, non-dissociative (C-type) hydrodynamic shocks excite some H_2O maser emission (M. Kaufman). This latter poster generously took the trouble to explain the properties of C and J-type shocks sufficiently that one might guess the letters to stand for continuous and jump (without being quite sure just which variable is being described). But I think I have been told something of the sort, and forgotten it, several times before, and so am reduced to thinking of the names as analogous to S-band, X-band, etc. within radar and radio astronomy (where, it is said, they are deliberately meaningless and were chosen to confuse the enemy).

Among the various brands of active galaxies, the evidence for the source of ionization energy is less direct. A fair statement, however, seems to be that it is easier to fit most of the line data with photo- than with shock ionization (A. Marconi and E. Oliva). This is true also for the notorious $Ly\alpha/H\beta$ ratio, provided you remember to include effects of dust (M. Patoriza). Supporting evidence includes the correlation of line fluxes with X-ray luminosity in low redshift quasars (B. Wilkes) and the interpretation of NLR clouds as ionization bounded, rather than matter bounded (S. Komossa and H. Schulz).

4. Composition gradients, kinematics, and other items with both micro- and macro-structure

This section occupies the no-person's land between processes and object. For instance, Miller focused on results of polarimetry, but an amusing implication is that NGC 1068 will be a Type 1 Seyfert in only 10^6 yr (apply for your telescope time now!) and that the type classifications can, in general, be transitory, suggesting the classification Seyfert 3 ($= 2 + 1$).

Improved angular resolution from WFPC-2 reveals that 30 Doradus is a lot like large galactic HII regions in its fine structure (P. Scowen) and that, although its ionization is structured on scales of $1-4''$, the derived composition is satisfactorily stable as you move from large to small apertures (M. Rosa). Upon looking closely at these and other giant extragalactic HII regions, it becomes clear that supersonic turbulence is ubiquitous (M. Muqoz-Tuqon). High angular resolution is similarly important for answering the question "what does a face-on shock look like" in IC 443 (M. Richter).

Much better resolution could conceivably provide definitive evidence for (or against) the large temperature fluctuations (Peimbert's notorious t^2) needed to reconcile elemental abundances found from recombination lines with those found from collisionally excited lines in planetary nebulae (X. Liu). Also to be sought in the sub-arcsecond regime are the dense (10^{5-7} cm^{-3}) clumps in Orion described by M. Bautista and the even denser ones (10^{14-15} cm^{-3}) in nova ejecta mentioned by R. Williams.

You can sometimes cheat and use time and wavelength resolution to probe otherwise unavailable spatial scales. Horne's tracing out of the gas stream, hot spot, and so forth in

cataclysmic variables and the high (wavelength!) resolution spectroscopy of CVs reported
by U. Munari are sterling examples, though I continue to wish for a definitive answer on
whether the outbursts of dwarf novae commence with something happening to the donor
star or something happening to the disk!

At the other end of the length-scale continuum, dynamic range is often the limiting
factor for studying whole galaxies. For instance, the composition gradient in M101
persists clear out to r = 32 kpc (where the O/H and S/O ratios have dropped to a
tenth of their central values–D. Garnett)—but you have to be able to record very faint
optical emission to see that far out. The very extended, warped HI disk of dwarf irregular
NGC 3782 (P. Bo) is similarly a triumph in signal to noise ratio.

Dust normally reveals itself in broad band features (not the province of this workshop),
but it can also affect the strengths and ratios of emission lines if you know what is likely
to have started out on the far side of the dust. The effects don't always go the way you
expect (J. Shields on high metallicity HII regions) and can even change the sign of a
line's equivalent width (W. Chen on Lyman alpha).

An issue of considerable importance is the amount of absorption of visible light in the
disks of spiral galaxies. We look both up and down out of our own disk with not more
than 0.1^m or so of obscuration in V. But E. A. Valentijn (not at the workshop) has claimed
that typical disks are so opaque that their real luminosities have been underestimated
by factors of more than two, permitting interpretation of spiral rotation curves in terms
of a constant M/L ratio (translation, no dark matter needed). L. Peterson reported data
on HII regions in NGC 2403 that imply A_v is indeed 0.86 to 1.3^m (of which only 0.12^m
comes from the Milky Way along our particular line of sight). These HII regions and
their stars emit much of the blue light of a typical spiral, and apparently also contain
a disproportionate share of dust. Thus, while much of the disk may be transparent,
nevertheless much of the light may be absorbed. But it still has to come out somewhere
in the infrared, and ordinary spirals do not have $L_{IR}/L_{opt} \gg 1$. There still seems to be
something going on here that is not understood, and "more work is needed!"

5. Active galaxies

5.1. *Unified models?*

Active galaxies include, at least, the radio loud and quiet quasars (with subtypes like
steep-spectrum compact), radio galaxies (FR I and II and perhaps some other numbers),
BL Lacs and other OVVS, Seyferts 1 and 2 (and numbers in between), and LINERS. Just
how faint active galaxies can be is uncertain, but a search for dwarf Seyferts is underway
(L. Ho). Some objects combine traits of two or more classes, for instance Mkn 1066
(studied by R. Gelderman, though we wish very much we could blame D. Hastings of
MIT and C. Norman of ST ScI) which is a Seyfert at its core and a LINER (star burst
galaxy) further out.

Within the standard AGN model, many parameters can be adjusted to account for
the range of observed phenomena. These include the mass of the central black hole,
the rate of accretion on to it, the size and shape (thin/thick/toroidal/warped) of the
accretion disk, the velocity and opening angle of the jets collimated by it, the type of the
parent galaxy, and the orientation of all this structure relative to the observer. A truly
unified model needs to examine the effects of varying each of these independently. But
the phrase most often implies some focus on the effects of the orientation of the observer
relative to the jet and disk. I hasten to say that everyone believes orientation determines
some of what we see, and no one believes that it is the entire story.

A number of presentations bore on where in between these two extremes the truth lies. The importance of polarization data in revealing active nuclei hidden by misalignment was stressed by Miller. The technique can be extended from nearby Seyfert 2's to radio galaxies at redshift ≈ 1 and has revealed hidden QSOs in some of them (S. de Serego Alighieri). Another manifestation of orientation is that the emission line regions of some QSOs "see" more ionizing flux (big blue bump) than we do (S. Mathur), though the corresponding region of Seyfert NGC 4151 does not (Z. Tsvetanov). The emission line profiles you see will depend both on the opening angle of the cone of ionizing radiation and on your direction relative to that cone (S. Viegas). But the equivalent width of OVI in QSO's is anti-correlated with the ultraviolet and soft X-ray flux in a way that is probably intrinsic to the sources (W. Zhang).

Among radio galaxies, the optical line equivalent widths are correlated with radio power for both Fararoff-Riley I and II sources, but the correlations are different (in the sense of FR II's having 10 times as much line emission at a given radio power), and this again suggests intrinsic differences (E. Zirabel and S. Baum). On the other hand, a large quasar sample displays correlations of Lyα and Hβ line shapes and widths with the ratio of radio core to radio lobe flux of the sort expected from orientation effects. You see bright core emission when the jet is aimed at you, and since the accretion disk is then face on, the lines are relatively narrow. An additional correlation with radio spectral index also makes sense within unified models (B. Wills).

Netzer noted that the correlation in which Lyα/Hβ is larger for flatter optical spectral indices could have been predicted. So, unfortunately, could have been the opposite correlation (which was in fact his initial guess when first thinking about the problem). The better-known anti-correlation of emission line equivalent widths with continuum flux (Baldwin effect) can be explained in at least five ways, none of which is clearly wrong. These illustrate a theorem due to the late R. O. Redman, "A competent theorist can explain any given set of data using any theory." Longair's corollary (that in most cases he need not even be competent) does not apply in this case.

5.2. *The Big Blue Bump and everything else*

If you fit the continuous spectra of assorted AGNs with the minimum possible number of power laws, you typically end up with a bulge or excess somewhere between the blue and very soft X-ray regimes. This is the Big Blue Bump, and its photon energies mean that it will be a major contributor to photo-ionization (M. Dietrich).

The outstanding issue is where the emission comes from and whether the emitting material is detectable in other ways. G. Kriss pointed out that, if the radiation were mostly free-free, we would see from the same gas O VIII and other emission lines that are not, in fact, detected. Presence of Ne VIII has been proposed as a signature for the BBB being optically thin radiation from a very broad line region. It has probably been seen (R. Cohen). But a detection in another QSO (F. Hamann) was presented as evidence that the BBB gas is the same as the warm X-ray absorber gas. For all I can tell, all three may really be the same stuff.

AGN emission lines are customarily classified as narrow and broad (and why there should be a fairly sharp separation can be explained). M. Brotherton, however, presented evidence for a separate intermediate line region in QSOs. It has (with BLR properties in parentheses), $n = 10^{10}/cm^3$ ($10^{12.5}$), distance from black hole = 1.0 pc (0.1), line width = 2000 km/sec (7000), and velocity off-set from quasar rest frame to the red (blue). The offsets can be different for different lines (Netzer) and none of them are very well understood. M. Gaskell believes that it is possible to account for the blue shifts by electron scattering in expanding atmospheres, without violating the constraint provided

by reverberation mapping of NGC 5548 (Horne, K. Korista), which seems to show that
the gas motion there is not predominantly either outflow or inflow (or rotation). The
presence of dust in the emission line clouds (M. Villar-Martin) is also relevant, because
it may allow us to see clouds on only one side of the central engine, or only one side of
each cloud.

A couple of issues are not in doubt. QSOs interact with their environments in detectable ways (F. Durret). And one category of absorption line (broad ones with redshift
very close to the emission redshift) really is intrinsic to the sources and not the result
of intervening galaxies. This has been further confirmed by direct evidence for radiative
acceleration of the absorbing clouds very close to the quasar centers (N. Arav). The
clouds are dusty (R. Hes).

6. Other interesting objects

The Orion Nebula remains an intensively-investigated HII region. In two dimensions,
its H_2 fluorescence is concentrated in tiny wisps (R. Pogge). Its three dimensional structure can be recovered to a certain extent (Z. Wen), and HST spectroscopy reveals gradients in N^+/O^+ (reflecting some combination of ionization structure and composition
variations, but probably mostly the former, R. Rubin). This could perhaps be described
as a one-dimensional result. N. Cox's work on the Lagoon Nebula can be expected to
reveal a similar richness of structure. This is particularly noteworthy because her observing time was awarded as part of the director's program to share HST with the non-
professional astronomy community.

Another Orion characteristic likely to be shared by other large HII regions is the
presence of protoplanetary disks with perfectly respectable young stellar objects of their
own which are, nevertheless, primarily ionized from outside by nearby OB stars (C. R.
O'Dell). The suggested name of proplyds will need to come with instructions on how to
pronounce, if it is to catch on.

The galactic center also harbors bright, early type stars whose winds manifest themselves as compact He I emission line sources (R. Blum).

Finally, we come to the most interesting objects of all, supernovae and their remnants.
SN 1994D was superficially a fairly typical Ia, but recent spectroscopy (P. Meikle) reveals
the shameful fact that it contains helium in its ejecta, apparently about 0.15 $M_\odot$. Within
popular models, it could have come from rather deep atmospheres on two white dwarfs
or from a helium star donor in a CV-like system with WD accretor.

SN 1987A was, of course, a Type II (that is, with hydrogen lines in its spectrum). The
ring of material expelled when it was a red supergiant is, however, enriched in helium
and nitrogen, and depleted in carbon, relative to the LMC average, just as you would
expect from a 20 $M_\odot$ progenitor or thereabouts (N. Panagia). The ring emission is either
ionization bounded or the gas has already begun to recombine after ionization by the uv
flash at shock outbreak. SN 1987A's best-established contributions to nucleosynthesis
are the 0.075 $M_\odot$ of new iron (from Ni^{56} whose decay was seen) and about 3 $M_\odot$ of fresh
oxygen (Pinto). N 132D in the LMC is also oxygen-rich, and its properties also imply a
progenitor of about 20 $M_\odot$ (W. Blair).

The Crab Nebula is very short of composition anomalies, apart from excess helium
(Woltjer), but its Ni II lines remain anomalously bright. Efforts to blame this on any
particular ionization or excitation peculiarity continue to fail (while succeeding in Orion
and elsewhere). But the fact that the lines are brighter on the side of each emission
feature that faces the pulsar (G. MacAlpine) I feel indicates that the issue is not yet fully
decided in favor of a real composition effect.

A second HST Crab Nebula poster (J. Hester) fully justifies the (rejected) proposal submitted in round 1 by S. van den Bergh and the present author. We asked for time (a) to pin down just how compact the knots are that are unresolved at 0.4″ with CFHT (and some indeed remain unresolved at HST 0.1″ resolution) and (b) to get the "first epoch plates" for an HST proper motion study, on the grounds that ionization structure in the nebula makes it essential that first and second epoch images be taken with the same "plate and filter" combination. The WFPC-2 images (sadly not of the whole nebula) show that indeed no two lines come from the same place in any filament or knot (and that several different sorts of ionization structure exist in different places). Thus any comparison of ground-based images from 1939–66 with HST ones will have motions inextricable from this structure. A chance to get second epoch images is yet another reason to wish HST a long and healthy life. Later ground based images are no longer possible, since both the photographic emulsions and the cameras needed have ceased to exist.

Acknowledgements

It is the traditional privilege of the last speaker at a conference to express the collective thanks of the participants to those who did the work and made the meeting possible. Our gratitude goes first to the Space Telescope Science Institute and its Director, Robert Williams, for their hospitality. Participants who had been at previous ST workshops were immediately aware of a significant difference—the Director was actually present to be thanked. Next, Cheryl Schmidt and the others who took care of local logistic arrangements performed, as always, above and beyond the call of duty. The scientific organization was in the capable hands of Mario Livio. And finally, most important, we are grateful to Michael Seaton and Donald Osterbrock for their world lines, including innovative contributions to the science of astronomy, great generosity to their colleagues, and a very convenient choice of birthdates.

Printed in the United States
By Bookmasters